Performance-Based Seismic Design of Structures

Seismic design of structures is fast turning to performance-based design (PBD) from old codal force-based design (FBD) method. The aim of the book is to expose the readers to the meaning and need of PBD, the evolution of PBD to date, its various forms and applications. Various design philosophies and procedures have been described including modelling aspects and hazard considerations backed by examples. Direct displacement-based design (DDBD) and unified PBD (UPBD) of reinforced concrete (RC) frame buildings, RC dual systems, steel frame buildings and bridge piers have also been explained.

The main features of this book are as follows:

- Illustrates performance-based seismic design to achieve the design target by performance objective-oriented design procedure.
- Covers modern design philosophies, modelling aspects, concepts in nonlinearities and use of supplemental damping devices.
- Contains a chapter on seismic safety of nonstructural components.
- Describes UPBD design procedure and examples of different structural systems.
- Includes application and examples with reference to SAP2000 software.

This book is aimed at graduate students, researchers and professionals in civil engineering, earthquake engineering and structural design.

Performance-Based Seismic Design of Structures

Satyabrata Choudhury

CRC Press
Taylor & Francis Group
Boca Raton London New York

CRC Press is an imprint of the
Taylor & Francis Group, an **informa** business

First edition published 2024
by CRC Press
2385 NW Executive Center Drive, Suite 320, Boca Raton FL 33431

and by CRC Press
4 Park Square, Milton Park, Abingdon, Oxon, OX14 4RN

CRC Press is an imprint of Taylor & Francis Group, LLC

ISBN: 9781032444826 (hbk)
ISBN: 9781032578118 (pbk)
ISBN: 9781003441090 (ebk)

DOI: 10.1201/9781003441090

Typeset in Times
by Newgen Publishing UK

Dedicated to my wife
Smti. Banasree Choudhury
*Who has ever kept on reminding me for the early
completion of the book.*

গাঁথিয়াছি তব তরে প্রথমের মালা
তোমারই দেওয়া ফুলে সুধাগন্ধ ঢালা।
–সত্যব্রত চৌধুরী

**Sweet heart, take this veracious virgin garland of mine
Threaded with the fragrance of flowers that you granted.**
–Satyabrata Choudhury

Contents

Contents

Foreword

Much progress has been made for decades after it was realized by engineers that earthquake generates an inertia force in the structure. This inertia force is the root cause of all damages in human-made structures during strong ground motion. Initially, the quantum of such inertia force was empirically taken as about 10% of the mass of the structure. This led to the development of the code-based seismic design. The central theme of the codes is the response spectrum concept (G.W. Housner, 1941) from which the design spectra evolved (Newmark and Hall, 1982). Initially, the equivalent lateral static force was computed for the structure based on first-mode vibrational characteristics. Later on, the dynamic response spectrum method evolved, which was able to accommodate the effect of many modes. The journey had an upward gradient, but the internal flaw remained that the entire gamut of the code rested on *force* as the primary design criterion.

The failure of several well-built structures designed using prevalent code under strong ground motions led the experts to rethink the efficacy of the force-based design (FBD) philosophy. With time, it was understood that *displacement* is a better choice in design over force. The displacement-based design (DBD) philosophy was conceptualized with this new understanding. In the 1970s, the capacity spectrum method (Freeman) was developed. Then came the series of developments in the form of ATC-40 (graphical capacity spectrum method) and other Federal Emergency Management Agency (FEMA) documents. Today, the DBD philosophy has been absorbed into performance-based seismic design (PBSD) philosophy, where the design is practically under the performance-oriented control of the designers.

I take delight in introducing the book *Performance-Based Seismic Design of Structures* by Prof. S. Choudhury. The author has long years of experience in teaching the subject at the post-graduate and undergraduate levels. His area of research interest is also in this arena. In 20 chapters, the book covers performance-based seismic design philosophies applied to various structures and related topics. The author has also introduced his innovative design philosophy named unified performance-based design (UPBD), which is supposed to be more comprehensive. This method has been applied to various structures, as detailed in the book.

The first six chapters provide the introductory materials. Chapters 7 to 11 deal with the design of various types of buildings including steel buildings. Chapter 12 explains the infill effect on buildings. Chapters 13 to 20 cover other miscellaneous designs and advanced contemporary issues on PBSD.

The structural design world is tending towards PBSD. Against this backdrop, a book like this is welcome. I hope the book will satisfy the needs of students, designers and researchers in the field of PBD.

Sondipon Adhikari (Ph.D.)
Professor of Engineering Mechanics
James Watt School of Engineering
Room No. 801, Rankine Building, Oakfield Avenue
The University of Glasgow

Preface

Structural design, including seismic design, historically and traditionally started with force as the primary focus in design, giving rise to force-based design (FBD). Codal design is mostly force-based. But with experience, two things became clear: (i) structures designed using codal FBD continued to suffer damage and even collapsed under strong ground motions and (ii) the aim of the design is to minimize damage, whereas force is not a direct measure of damage; displacement is the right parameter to quantify the damage. With this understanding, displacement was introduced as the core design parameter. This led to the development of displacement-based design (DBD). With the incorporation of various design criteria, performance indicators and specified hazards in the DBD, a new design named performance-based design (PBD) was conceptualized. PBD, also called performance-based engineering (PBE), can be applied to various fields of engineering. This book will discuss performance-based seismic design (PBSD) for buildings and other structures.

The advantages of PBSD are: (i) target-oriented design, that is, designing with pre-defined target criteria; (ii) design possible for any hazard level; (iii) designs are made possible for various target objectives (multi-objective design); (iv) applicability to various structures; and (vi) design philosophy is clear and sound.

Various codes have, of late, started incorporating PBD; for example, ASCE-SEI-41 incorporated the PBD.

I have taught this subject at the post-graduate and undergraduate levels for more than 15 years. I have observed very keen interest among the students in the subject. This has given me the chance to experiment with the subject, including executing many projects with the students. The net result of these efforts is the present book.

The book is organized into 20 chapters, as detailed below.

Chapter 1 introduces the topic of PBSD, along with the effect of past earthquakes on buildings and nonstructural components, force-based codal design, equal displacement and equal energy principles, and strength-stiffness dependency. Finally, the need for PBD has been discussed.

The historical development of PBSD, various design methods including displacement-based design (DBD), introduction to capacity spectrum method, direct displacement-based design (DDBD), unified performance-based design (UPBD) and capacity design have been taken up in Chapter 2.

Chapter 3 describes hazard considerations. Hazard nomenclature is presented. Hazards as given in SEI-7-16, SEI-41-17, EC-8 (2004) and IS 1893 (2016) have been illustrated. The construction of displacement spectra has been explained.

Modelling aspects have been treated in Chapter 4 at length. Various strength levels, effective stiffness, modelling of beam, column and shear wall have been discussed. Modelling of infill, P-delta effect, ductility considerations, torsional effect, and confined vs. unconfined concrete have been treated. A sufficient number of numerical examples have been provided.

Chapter 5 deals with performance criteria and performance levels. Structural and nonstructural performance levels and their viable combinations have been discussed.

The criteria for member performance levels are explained. Multi-objective design is highlighted.

Wherever necessary, examples have been given to clarify the theoretical concepts. Exercises have been added in all chapters for practice purposes.

The evaluation of structure through analyses is treated in Chapter 6. Linear and nonlinear analyses and P-delta effect, are discussed. The nonlinear static method (pushover analysis) has been discussed in depth. The evaluation of performance point through capacity spectrum method and displacement coefficient method have been discussed in detail.

Chapter 7 discusses the direct displacement-based design (DDBD) for reinforced concrete (RC) frame buildings as per Pettinga and Priestley (2005). Theoretical background, design steps and numerical examples are provided. Chapter 8 similarly deals with the DDBD method for dual system after Sullivan et al. (2006). Frame–wall interaction has been highlighted. Sufficient examples are given to illustrate the applications.

Chapter 9 introduces unified performance-based design (UPBD) for RC frame buildings. UPBD being a relatively new philosophy, the theoretical background, design procedure and implications of the formula for beam depth have been adequately explained. Numerical examples are given to help understand the implementation part of the design. Along a similar line, the UPBD method for dual system is dealt with in Chapter 10. The UPBD method for steel frame buildings, which is quite different from RC buildings, has been explained in Chapter 11.

Along with the exposure to the relevant topics, sufficient hints are available in the book for future research.

The effect of infill in buildings and modelling of infill has been taken up in Chapter 12. The DDBD method for RC buildings incorporating fluid viscous dampers is covered in Chapter 13.

Chapter 14 describes the displacement-based design of bridge piers, while Chapter 15 deals with the UPBD method for bridge piers.

Chapters 16 through Chapter 20 deal with various topics like energy dissipating devices, tuned liquid dampers, tuned mass dampers, non-structural components (operational and functional components) of buildings, preliminary of probability in engineering, fragility analysis and the principle of base isolation.

The effort of the author shall be deemed fruitful if it serves the need of the students, professionals and researchers in a befitting way.

I shall be grateful to the readers including students, professionals and researchers for any suggestion, criticism or correction they forward to me to improve the book in the future.

Satyabrata Choudhury
Email: scnitsilchar@gmail..com

Acknowledgements

In the beginning, I take this opportunity to convey my heartfelt reverence to my Ph.D. supervisors late Prof. D.K. Paul and Prof. Yogendra Singh, both from IIT Roorkee, India, who introduced me to performance-based seismic design. The basic concept of Section 4.2 of this book is derived from class lectures of Prof. Yogendra Singh.

I am thankful to Prof. Sondipon Adhikari, Professor of Engineering Mechanics, The University of Glasgow, for writing a valuable foreword for this book.

I extend my thanks to the research scholars who worked with me and whose research works have influenced the text of this book; to name some of them, Dr. Sunil Singh Mayengbam, Dr. Sourav Das, Dr. Pritam Hait, Dr. Shulanki Pal, Dr. Durga Mibang, Geetopriyo Roy and Tushar Kanti Das. Some of my PG students whose dissertation works have been mentioned in the book are Dr. Partha P. Debnath, Shubhra Das, Suman Banerjee, Vijay K. Kotapaty and P.K. Azeel. All my above-mentioned students encouraged me to write the book and many of them have also provided valuable suggestions.

I am thankful to my son Soham who worked out the derivation of Section 1.5, as well as helped in writing Sections 20.2 and 20.3 of this book. I must mention my daughter Srijita, now in Class X, who helped prepare the Table of Contents, necessary JPG files and kept track of the chapters.

Any such work rests on the shoulder of giants. I acknowledge all the authors of various research papers and books whose thoughts have directly or indirectly influenced my work.

I thank the anonymous reviewers who reviewed the book and provided valuable suggestions. My thanks extend to the starter of the book, Dr. Gagandeep Singh, Senior Publisher (STEM), CRC Press, India and Sub-Saharan Africa, Taylor & Francis Books India Pvt Ltd., who gave valuable suggestions in framing the book. I am thankful to the publication team of CRC Press and other members involved in the production. And many thanks are due to CRC Press (Taylor & Francis) for agreeing to publish this book.

My gratitude goes to my Department of Civil Engineering, National Institute of Technology Silchar, India, and the institute itself for allowing me to float the course of Performance-Based Seismic Design in the curriculum, which enabled me to experiment with the subject and gain experience. In the same tune, I thank all my students whom I have taught this course.

I must mention that my family, especially my wife, constantly encouraged me to write the book.

Last, but not least, I understand that it is the grace of God which is key to the fruition of any project.

Satyabrata Choudhury

Silchar, India, January 2023; email: scnitsilchar@gmail.com

About the Author

Satyabrata Choudhury is a retired professor of the Department of Civil Engineering, National Institute of Technology (NIT) Silchar, India. He had been a faculty of this institute since 1983, first as Lecturer and then as Assistant Professor and finally as Professor and, completed 40 years of service at retirement in January 2023. Prof. Choudhury obtained his bachelor's degree in civil engineering from Regional Engineering College (REC) Silchar (now NIT Silchar), India. He obtained his master's degree in structural engineering from IIT Kharagpur, and Ph.D. in earthquake engineering from IIT Roorkee. He has served in various administrative positions in the institute including as Head of the Department, Dean (P&D) and Coordinator of various activities. He has received various awards including President of India's prize, Dr. Jai Krishna Gold medal (two times), and Institution prize (IEI), among others. He has taught subjects like performance-based seismic design, structural dynamics, concrete and steel design at post-graduate and undergraduate levels. His area of research is performance-based seismic design. Prof. Choudhury is Lead Guest editor of *Practice Periodical on Structural Design and Construction*, ASCE.

He has published 45 papers in international peer-reviewed journals and a number of papers in international and national conferences. He has supervised about 60 M.Tech. dissertations, produced 7 Ph.D. students and is working with another 4 Ph.D. scholars. He was the chairperson of two international conferences and one national conference. He has evolved a new design methodology named as unified performance-based design (UPBD), which accommodates both drift and performance level as the target design objectives. This method also provides member sizes.

Abbreviations

ADRS	Acceleration displacement response spectrum
ANN	Artificial neural network
ATC	Applied Technology Council
CBF	Centrally braced frame
CP	Collapse prevention
CSM	Capacity Spectrum Method
CM	Centre of mass
CR	Centre of resistance
CV	Centre of shear resistance
DL	Dead load
DBD	Displacement-based design
DDBD	Direct displacement-based design
EBF	Eccentrically braced frame
EDD	Energy dissipation devices
ERD	Earthquake resistant design
ESDOF	Equivalent Single degree of freedom
FBD	Force-Based design
FEMA	Federal Emergency Management Agency
GA	Genetic algorithm
HVAC	Heating ventilation air conditioning
IDA	Incremental dynamic analysis
IO	Immediate occupancy
IS	Indian Standard
LL	Live load
LS	Life safety
MDOF	Multiple degree of freedom
NA	Neutral axis
NEHRP	National Earthquake Hazard Reduction Program
OFC	Operational and Functional Components
OVH	Olive View Hospital
PAHO	Pan American Health Organization
PBD	Performance-based design
PBEE	Performance-based Earthquake Engineering
PBSD	Performance-based seismic design
POA	Pushover analysis
POC	Pushover curve
PP	Performance point
RC	Reinforced concrete
SEAOC	Structural Engineer's Association of California
SEAONC	Structural Engineer's Association of Northern California

SDOF (ESDOF)	Single degree of freedom (Equivalent SDOF)
SVR	Support vector regression
UBC	Uniform Building code
UPBD	Unified Performance-Based Design
YPS	Yield point spectra

Symbols

$\ddot{u}_g$	Ground acceleration
$\dot{u}_s$	Velocity of structure
$\ddot{u}_s$	Structural acceleration
$\bar{x}_{CM}$	Distance of centre of mass along x-axis
$\bar{y}_{CM}$	Distance of centre of mass along y-axis
Δ_{c5}	Corner period spectrum displacement corresponding to 5% damping
Δ_E	Displacement of infinitely strong elastic building
Δ_c	Critical storey displacement
$\Delta_{corrected}$	Corrected design displacement
Δ_d	Design displacement
$\Delta_{di,new}$	New design displacement in i-th case
Δ_m	Maximum displacement
Δ_p	Plastic displacement
$\Delta_{roof,PP}$	Roof displacement at performance point
$\Delta_{roof,i}$	Roof displacement at i-th step
Δ_{roof}	Roof displacement
Δ_t	Target displacement
Δ_y	Yield displacement
h_b	Depth of beam
h_c	Height of column
h_i	Height at i-th floor level
h_{inf}	Inflection height of wall
h_{left}	Depth of water on left in tank
h_{right}	Depth of water in right
h_{sc}	Interstorey height for critical storey
h_x	Height of x-th floor from ground
Δ_{di}	Design displacement in i-th case
Δ_i	Displacement at i-th floor
Δ_{iy}	Yield displacement of wall at i-th floor
A_h	Area of cross section of tie bar
A_c	Area of compression concrete
A_{sc}	Area of compression steel
A_{st}	Area of tensile steel
C_0, C_1, C_2, C_3	Constants
C_h	Horizontal seismic coefficient
$C_{c,exp}$	Compression in concrete at expected strength level
$C_{c,extreme}$	Compression in concrete at extreme strength level
$C_{c,k}$	Compression in concrete at characteristic strength level
$C_{c,lim}$	Compression in concrete at limit strength level
C_c	Compression force in concrete
C_i	Damper constant at i-th level
C_m	Effective mass factor to account for higher mode effects
C_v	Seismic coefficient
C_y	Yield strength coefficient

D_c	Core diameter
E_c	Modulus of elasticity of concrete
E_d	Energy dissipated in one hysteretic cycle
E_f	Modulus of elasticity of frame material
E_m	Modulus of elasticity of masonry infill
E_s	Strain energy at peak displacement
F_h	Horizontal seismic force
F_E	Force resisted by infinitely strong elastic building
F_L	Force in liquid
F_b	Base shear
F_{di}	Force in damper at i-th floor
$F_{i,roof}$	Lateral force at roof by distribution of base shear
F_i	Force in the i-th floor
F_o	Peak value of excitation
F_{roof}	Total lateral force at roof
$F_{v,min}$	Minimum vertical seismic force
F_v	Vertical seismic force
F_x	Force in x-direction
F_y	Yield force, Force in y-direction
H_e	Effective height of ESDOF system
I_c	Moment of inertia of column abutting infill
I_{eff}	Moment of inertia of effective (net) section
K_e	Effective stiffness
K_{eff}	Effective stiffness
$K_{elastic}$	Elastic stiffness of the system
K_i	Initial stiffness
L_c	Distance from critical section to point of contraflexure
L_s	strain penetration length
M_{OTF}	Frame over turning moment
$M_{W,i}$	Wall over turning moment at i-th floor level
$M_{W,i+1}$	Wall over turning moment at $(i+1)$-th floor level
M_W	Wall over turning moment
M_{base}	Base moment
M_u	Ultimate moment
M_y	Yield moment
P_i	Axial force in i-th column
Q_i	Lateral design force in i-th floor
R_{OFC}	Response reduction factor for OFC
R_ξ	Spectral reduction factor for damping ξ
S_{D1}	Spectral displacement for 1.0 sec period
S_{DS}	Spectral displacement for short period
$S_{a,5\%}$	Spectral acceleration at 5% damping
$S_{a,\xi\%}$	Spectral acceleration at ξ% damping
S_a	Spectral acceleration
S_{as}	Spectral acceleration corresponding to short period
$S_{d,PP}$	Spectral displacement at performance point
S_d	Spectral displacement
S_p	Structural performance factor
S_v	Spectral velocity

T_1	First mode time period of structure
T_a	Approximate time period
$T_{e,trial}$	Trial effective time period
T_g	A designated time period
T_i	Time period of structure in i-th mode
$T_{initial}$	Initial time period of the structure
T_{target}	Target time period
V_h	Volume of tie
V_w	Wall shear
V_{wi}	Wall shear at i-th floor level
V_b	Base shear
V_c	Volume of core
V_i	Lateral shear at level i, Base shear at step i
V_{ti}	Total shear in wall at i-th floor level
V_y	Yield base shear
W_i	Seismic weight of i-th floor
W_t	Total seismic weight of building
W_x	Seismic weight of x-th floor
a_f	OFC dynamic amplification factor
$c_{critical}$	Critical damping coefficient
c_i	Damping coefficient of i-th damper
d_b	Diameter of bar
d_{bl}	Diameter of longitudinal bar
d_c	Effective depth of compression steel
f_L	Frequency of liquid
$f_{c,exp}$	Expected strength of concrete
$f_{c,exteme}$	Extreme strength for concrete
f_{ce}	Expected material strengths for concrete
f_{ck}	Characteristic strength of concrete
f_{exp}	Expected strength
f_{exteme}	Extreme strength
f_k	Characteristic strength
f_{lim}	Limiting strength
f_r	Radial stress in confined concrete
$f_{s,exp}$	Expected strength of steel
$f_{s,exteme}$	Extreme strength for steel
f_s	Cyclic frequency of structure; stress in steel
f_{sc}	Stress in compression steel
f_{se}	Expected material strengths for steel
f_{st}	Stress in tension steel
f_{sy}	Yield stress of steel at appropriate strength level
f_u	Ultimate strength of rebar
f_y	Yield strength
f_{yh}	Yield strength of tie steel
f_{ye}	Expected yield strength of rebar
k_e	Effective stiffness
k_{eff}	Effective stiffness
k_s	Stiffness of structure
l_b	Length of beam

l_p, L_p	Plastic hinge length
m_L	Mass of liquid
m_{TMD}	Mass of TMD
m_e	Effective mass
m_i	i-th mass
m_s	Mass of structure
p_c	Percentage compression steel, compression steel ratio
p_t	Percentage tension steel, tension steel ratio
r_i	Diagonal length of infill panel
u_{TMD}	Displacement of TMD
u_s	Displacement of structure
v_e	Expected shear strength of masonry
x_i	i-th data
x_{mi}	Distance of i-th mass along x-axis
y_{mi}	Distance of i-th mass along y-axis
β_i	Proportion of shear taken by damper at i-th floor
γ_m	Partial safety factor for material
δ_i	Lateral drift in story i
δ_p	Plastic deformation
δ_u	Ultimate deformation
δ_y	Yield deformation
ε_c	Strain in compression concrete
ε_{cm}	Compressive strain in concrete corresponding to maximum stress
ε_{cm}	Maximum strain in concrete
ε_{st}	Strain in tensile steel
ε_{sc}	Strain in compressive steel
ε_{un}	Compressive strain in unconfined concrete
θ_d	Angular design drift
θ_i	Stability coefficient, Angle of inclination of i-th damper
θ_{max}	Maximum angle of rotation
θ_y	Chord yield rotation of steel beams
$\theta_{yF,i}$	Yield rotation of frame in i-th case
θ_{yF}	Yield rotation of frame
θ_{yW}	Yield rotation of wall
μ_Δ	Displacement ductility
μ_ε	Strain ductility
μ_θ	Rotational ductility
μ_ϕ	Curvature ductility
μ_{ESDOF}	System ductility or ESDOF system ductility
μ_F	Frame ductility
μ_W	Wall ductility
μ_i	Ductility in i-th case
μ_m	Mass ratio
μ_{max}	Maximum coefficient of friction
ξ_h	Hysteretic damping
ξ_D	Damping out of damper
ξ_{ESDOF}	Damping of equivalent SDOF system
ξ_F	Frame hysteretic damping
ξ_W	Wall hysteretic damping

ξ_a	Added damping
ξ_{ei}	Equivalent damping in i-th case
ξ_{eq}	Equivalent damping
ξ_k	Damping ratio for k-th mode
ξ_m	Material damping
ρ_v	Ratio of volume of tie to volume of core
σ_{brick}	Compressive strength of brick
σ_c	Stress in concrete
$\sigma_{c.max}$	Maximum stress in compression concrete
σ_m	Compressive strength of masonry
σ_{mortar}	Compressive strength of mortar
ω_L	Angular frequency of liquid
ω_k	Modal frequency of k-th mode
ω_s	Angular frequency of structure
ϕ_c	Shape coefficient for critical storey
ϕ_i	Shape coefficient for i-th storey
ϕ_{ir}	Relative modal displacement of i-th damper
ϕ_j	Mode shape coefficient of j-th mass
ϕ_l	Diameter of longitudinal bar
ϕ_p	Plastic curvature
ϕ_u	Ultimate curvature
ϕ_y	Yield curvature
ϕ_{yW}	Yield curvature of wall
D	Diameter of pier
E	Modulus of elasticity
e	eccentricity
h	Height of liquid
K	Stiffness, structural construction factor
L	Length of member, Length of tank
L_u	Ultimate limit state factor
M	Magnitude of earthquake, Moment, Bending moment
M_S	Surface magnitude
M_w	Moment magnitude
Q	horizontal shear capacity of bed joint
q	Bed joint shear strength of masonry infill
t	Thickness
X	Direction X
Z	Zone factor
Δ	Linear displacement
λ	Damper parameter
C	Seismic coefficient, structural flexibility factor, total compression force
F	Force
H	Total height of building, height of column, horizontal force
I	Importance factor, moment of inertia of section; OFC performance factor
N	Total number of storeys in a building, number of data
P	Axial force
R	Response Reduction Factor, Strength ratio
T	Time period of structure, tensile force
a	Parameter in infill strut computation

c	Damping coefficient
d	Effective depth
g	Acceleration due to gravity
k	Constant, Stiffness
m	Mass, Slope
n	Number of masses, Total number of storeys in a building
p	Percentage steel
r	Post-yield stiffness ratio
s	Spacing of circular tie
u	Displacement
x	Neutral axis depth
α	Damper parameter
γ	Shearing angle
δ	Angle of inclination, Depth to length ratio of tank water
ε	Strain
η	Tuning ratio
η	Yield force to inertia force ratio
θ	Angle of rotation, angle of inclination
λ	Correction factor, parameter in infill strut
μ	Ductility, mean value
ξ	Critical damping ratio
ρ	Density of liquid
σ	Standard deviation, stress
ω	Natural frequency
ϕ	Curvature twisting angle

1 Force-Based Design and its Limitations

1.1 INTRODUCTION

Ensuring the safety of structures against earthquake events has remained a challenging task for the design engineers. Our understanding of the seismic behaviours of structure evolves with time, so does the seismic design philosophy. Historically, it was a leap forward when engineers understood that the earthquake generates a lateral force in the structure, which is proportional to its mass. Typically, an empirical estimate of such lateral seismic force was taken as 10% of the weight of the structure. This was the beginning of force-based design (FBD) for seismic actions. In 1914, Sano, a Japanese engineer, proposed a seismic design method, which is now called as "seismic coefficient method". Following the 1908 Messina earthquake (Italy), a commission was formed, which gave a recommendation for an earthquake-resistant design of structures, now known as "equivalent static method" (Bozorgnia and Bertero, 2004). The first design code that came into existence was introduced at Santa Barbara following the Santa Barbara earthquake of 1925. Design codes of various countries prescribe design procedures for earthquake-resistant design (ERD). But the structures designed following the prescriptions of such codes are found to be not always safe under severe seismic events, and structures continue to fail or get severely damaged all over the world under major earthquakes. This highlights the limitations of the present codal method of design, which lays emphasis on force alone as the core design parameter. With time, the limitations of codal FBD have become apparent to the engineering community. From 1975 onwards, the concept of performance-based design (PBD) has gradually evolved. In this design philosophy, the structure is designed for a target performance objective under an expected hazard level. Displacement, rather than the force, becomes the guiding parameter in this method of design. Performance objectives may include structural performance level and nonstructural performance level. Drift may be an important design objective for some structures.

The Uniform Building Code (UBC) was adopted in 1927. Maurice Biot in 1932–33 gave the concept of response spectrum, which now is an integral part of all design codes.

The essence of FBD as prescribed in seismic design codes of various countries is to estimate a probable lateral seismic force (called base shear) and then suitably distribute that force over different levels of the structure as design lateral forces. Design is carried out with the prescribed load combinations. All seismic codes give a design spectrum or elastic demand spectrum at 5% damping, which is called 5% damped elastic demand spectrum. Design spectrum is a smoothened average curve of a large number of response spectra of the past earthquakes and probable future artificially

DOI: 10.1201/9781003441090-1

generated earthquakes for a geographical location. Design spectrum is typically an acceleration spectrum. Though fundamentally based on the same principle, the design spectra of countries differ from one another due to geo-tectonic variations and consensus of code makers.

1.2 PERFORMANCE OF BUILDINGS IN PAST EARTHQUAKES

It is worthwhile to scrutinize how the building structures designed as per the codes of different countries performed under earthquakes in the past. In this direction, of particular interest and importance are the hospital buildings, as these had been designed with more stringent design requirements, particularly with higher base shear (namely, with higher importance factor). The damage of hospital buildings leads to losses such as (i) loss of property; (ii) loss of human life (patients, doctors and trained health workers); (iii) downtime loss due to service impairment; (iv) sufferings of patients and injured people who cannot get admitted in damaged or a collapsed hospital; (v) cost of evacuation; (vi) cost of damage of operational and functional components (OFCs), also known as nonstructural components; (vii) cost of repair/retrofitting, if the damage is repairable. Through the damage or collapse of hospital buildings under seismic actions, billions of dollars are wasted almost every year. It may be noted that the cost of OFCs in a hospital is many times more than that of the building structure itself. Nevertheless, the damage to residential buildings and lifeline structures also plays a havoc in major earthquakes.

1.2.1 PERFORMANCE OF BUILDING STRUCTURES IN PAST EARTHQUAKES

In this section, we highlight how well or poorly the building structures performed in past earthquakes. Such a survey is too vast to be covered in the present book. The design of a hospital building needs extra attention due to its high post-earthquake importance. Choudhury (2008) made a thorough review of the performance of hospital buildings in the past earthquakes. Table 1.1 shows the earthquake damage to buildings over a large span of time.

TABLE 1.1
Damage of hospital buildings in past earthquakes

Earthquake	Damage	Source of information
1960 Chile earthquake (M_w 9.6)	Large scale damage due to earthquake and accompanied tsunami	General
1964 Alaska earthquake (M 9.2)	Many buildings and hospitals got damaged.	General
1968 Tokachi-oki earthquake (M 8.3)	Many buildings and hospitals collapsed.	General
1971 San Fernando earthquake (M 6.5)	Olive View Hospital (designed as per UBC-1964 provisions) and Santa Cruz Hospital got severely damaged.	Internet

TABLE 1.1 (Continued)
Damage of hospital buildings in past earthquakes

Earthquake	Damage	Source of information
1976 Guatemala City earthquake (M 7.5)	Several hospitals got destroyed.	PAHO, 2000
1985 Mexico City earthquake (M_w 8.1)	5 hospitals collapsed, 22 hospitals suffered serious damages.	General
1985 Chile earthquake (M 8)	9 hospitals got badly damaged.	General
1988 Armenian earthquake (M 6.8)	All healthcare facilities in Armenia were destroyed and 80% medical staff died or got injured.	Filson *et al.*, 1889
1994 Northridge earthquake (M 6.7)	9 major hospitals got damaged.	
1997 Italy earthquake (M_w 6,7)	11 hospitals got damaged.	Sortis et al., 2000
1999 Central Colombia earthquake (M_L 6.1)	25% of the hospitals in the City of Armenia were out of service.	Sánchez-Silva, 1999
1999 Turkish earthquake (M 7.6)	Many major hospitals got damaged.	General
2001 Bhuj earthquake (M 7.6)	2 District hospitals and many residential buildings got seriously damaged.	Internet
2010 Haiti earthquake (M_w 7)	250,000 residences and 30,000 commercial buildings were severely damaged.	Internet
2011 Tohoku eq. (M_w 9.1)	45,700 buildings were destroyed.	Internet
2015 Hindukush earthquake (M_w 7.5)	109,123 buildings damaged.	Internet
2015 Nepal earthquake (M 7.8)	About half a million buildings damaged.	Internet
2016 Ecuador earthquake (M 7.8)	7000 buildings destroyed.	Internet
2019 Peru earthquake (M_w 8)	883 buildings damaged.	Internet
2020 Oaxaca earthquake (M_w 7.4)	Widespread damage in over 8,000 houses. 213 schools, 15 health centres, three hospitals, seven bridges were damaged.	Internet
2021 Chignik earthquake (Mw 8.2)	Caused minor damages to structures.	General
2022 Qinghai earthquake (Mw 6.8)	Damaged at least 4,914 homes and 25 schools.	Internet
2022 Afghanistan earthquake (Mw 6.2)	6000 buildings either collapsed or severely damaged.	Wikipedia

1.2.2 PERFORMANCE OF NONSTRUCTURAL COMPONENTS IN PAST EARTHQUAKES

Equipment and services also go by the name nonstructural components. These are also known as operational and functional components (OFCs). The functionality of the building comes out of the OFCs. During an earthquake, even if the structure performs well, any poor performance of OFCs will lead to shut down of the facilities housed in the structure. McGavin (1981) discussed the anchorage requirement of different types of electrical and mechanical equipment. The Structural Engineers' Association of California (SEAOC) has discussed the importance, safety and damageability with respect to seismic design of OFCs in the Vision 2000 document. The common nonstructural failure was that of the ceiling system, egress and ingress system and content damage in shelves. In some earthquakes, the nonstructural damage was found to be more than the structural damage (e.g., 1989 Loma Prieta earthquake).

Reports are available on the performances of nonstructural building components in past earthquakes. Extensive damage to glazing, mechanical and electrical equipment, heating ventilation air conditioning (HVAC) system, elevators, pipes, ceilings, sprinkler systems, water supply system, power supply system, building services (building gaps, ducts, ceilings, partitions, connecting pipes) and building content have been reported for both hospitals and general buildings. In the 1994 Northridge earthquake, the Olive View Hospital (OVH), which was rebuilt with enhanced seismic provisions, did not suffer structural damage but the large-scale nonstructural damage led to the evacuation of the hospital. The mechanical and electrical equipment mounted on rubber springs performed poorly. The sprinkler systems were particularly damaged. Lack of adequate bracing in the sprinkler within the false ceiling is attributed as the reason for the failure of the building. In the 1989 Loma Prieta earthquake, many elevators failed to operate due to the failure of guide rail and counterweight dislodging. In the 1999 Chi Chi earthquake of Taiwan, nonstructural damage led to the closure of Shi-Tuwan hospital (Nazer and Elahi, 2004). In the 2001, Nisqually earthquake in Washington State, out of $2 billion loss, a large part was from damage to OFCs (Filiatrault *et al.*, 2001). Dhakal (2010) reported detailed nonstructural damage in the 2010 Darfield earthquake. Biranchi et al. (2019) studied nonstructural safety utilizing traditional methods. Gabbianelli et al. (2020) reported seismic demand profiles of nonstructural components in hospital building.

1.3 CODAL SEISMIC DESIGN PROVISIONS

In the last part of 19th century, the generation of inertia force during earthquake was understood by engineers. It was realized that earthquake caused a horizontal force in structure that led to the damage. Initially, such horizontal force was taken as 10% of total building weight. As mentioned earlier, Sano, a Japanese engineer, in 1914, proposed a method that now goes by the name seismic coefficient method (Bozorgnia and Bertero, 2004). In 1927, the seismic coefficient method was adopted in UBC. In 1948, the Structural Engineer's Association of Northern California (SEAONC) introduced the building period in base shear computation.

In 1957, the committee of the Structural Engineer's Association of California (SEAOC) recommended inclusion of building behaviour factor in the expression for base shear, which highlighted the dissipation of energy.

1.3.1 EC-8 (2004) PROVISIONS

The EC-8 (2004) gives base shear formula as in Eq. (1.3.1):

$$F_b = S_d\left(T_1\right)m\lambda \tag{1.3.1}$$

where
F_b = base shear
$S_d\left(T_1\right)$ = ordinate of design spectrum corresponding to first mode period T_1
m = mass of the building
λ = correction factor which is given as -
$$\lambda = \begin{cases} = 0.85, & \text{if } T_1 \le 2T_C \text{ and the building is has more than 2 stories} \\ = 1.0 & \text{otherwise} \end{cases}$$

The first mode time period is given as

$T_1 = 0.075H^{0.75}$ for moment resistant *concrete* space frame
$T_1 = 0.085H^{0.75}$ for moment resistant *steel* space frame
$T_1 = 0.05H^{0.75}$ for all other structures,

where H is height of building. *Tc* is corner period

1.3.2 INDIAN SEISMIC CODE IS 1893-2016

Indian seismic code, IS 1893-2016, gives seismic design principles for buildings and other structures. In addition, IS 13920-2016 gives ductile detailing requirements for reinforced concrete (RC) structures. For steel structures, the ductile detailing provisions are given in SP 6(6). The highest possible earthquake designated in the code is called maximum considered earthquake (MCE). On the other hand, the design basis earthquake (DBE) is that which can occur during the life span of the structure under which the damage is limited. DBE is taken as half of MCE level. The design spectrum shape is given in the code for three types of soils, namely, hard soil or rock, medium soil and soft soil. Further, design spectrum is given separately for *equivalent static load method* and *response spectrum method*. The whole of India is divided into four seismic zones named as Zone II, Zone III, Zone IV and Zone V. Zone I does not appear here as it was earlier merged with Zone II. The corresponding zone factors (seismicity level) are 0.10, 0.16, 0.24 and 0.36, respectively. The importance factor is 1.0 for ordinary buildings and 1.5 for important buildings such as hospitals. The approximate time period formulae are given for various types of buildings. A base shear correction is applied if the computed base shear corresponding to dynamically calculated time period is less than the base shear obtained by using approximate time period formulae given in the code. The response reduction factor accounts for overstrength and ductility in the system. The value of response reduction factor ranges from 1.5 (for un-reinforced masonry) to 5.0 (for ductile RC structures). Indian code allows (i) equivalent static method, which is allowed only when the time period does not exceed 0.4 s, (ii) dynamic response spectrum method and (iii) dynamic time history analysis method.

The base shear (V_b) is given by Eq. (1.3.2):

$$V_b = \frac{Z}{2} \frac{I}{R} \frac{S_a}{g} W \tag{1.3.2}$$

where Z is the zone factor, I is the importance factor, R is the response reduction factor, $\dfrac{S_a}{g}$ is the spectral acceleration ratio corresponding to time period of the building and W is seismic weight of the building. The approximate time period is given as

$T_a = 0.075H^{0.75}$ for moment resistant *concrete* bare frame buildings
$T_a = 0.085H^{0.75}$ for moment resistant *steel* bare frame buildings
$T_a = 0.08H^{0.75}$ for composite bare frame buildings

$T_a = \dfrac{0.09H}{\sqrt{b}}$ for buildings with infill

where H is height of building in m and b is breadth of building in m in the direction concerned.

1.3.3 OTHER INTERNATIONAL CODES

A comparative analysis of codes of some countries is presented in Table 1.2.

For capacity design, EC-8 stipulates that the sum of moment capacities of columns meeting at joint should be not less than 1.3 times that from beams meeting at the joint in the direction under consideration. The corresponding factor in IS 13920-2016 is 1.4. In many countries, seismic zones are replaced by seismic intensity maps with seismic intensity contours. Modal response spectrum method, equivalent static load method and time history analysis methods are permitted by the New Zealand code. In IBC-2006, seismic zones are specified in terms of seismic intensity maps with intensity contours.

1.4 EQUAL DISPLACEMENT PRINCIPLE

"Equal displacement principle" is an observed phenomenon and not a "principle" as such. This phenomenon has been observed from post-earthquake effects on buildings. The principle may be stated as: *The displacement suffered by an infinitely strong elastic building is approximately equal to the displacement surfed by a real building*

TABLE 1.2

Base shear and its distribution in some international codes (V_b is base shear)

Code	Design base shear	Distribution of shear
1927 UBC	$V_b = CW$, C varies from 0.075 to 0.1.	
1941 NBC	$V_b = CW$, C varies between 0.02 and 0.05 depending on the bearing capacity of the soil, and W is the weight of the building.	The base shear acted laterally at centre of gravity of the structure.
SEAOC, 1980	$V_b = ZIKCSW$, Z = coefficient related to the seismicity of the region K = a quality coefficient for the structural system C = a period-dependent coefficient S = coefficient for site-structure resonance W = the seismic weight.	
1970 NBCC	$V_b = \dfrac{1}{4} RKCIFW$ R is seismic regionalization factor, K depends on type of construction, C is structural flexibility factor, I is importance factor, F is foundation factor and W is weight of structure.	
IS-1893 (Pt. 1)-2016	$V_b = \dfrac{Z}{2} \dfrac{I}{R} \dfrac{S_a}{g}$ V_b is base shear, Z is zone factor, I is importance factor, R is response reduction factor, S_a/g is spectral acceleration coefficient, W is seismic weight.	$Q_i = \dfrac{W_i h_i^2}{\sum_{i=1}^{n} W_i h_i^2}$ W_i, h_i and Q_i are seismic weight, height and lateral force respectively in i-th floor.
UBC 1997	$V_b = \dfrac{C_v I}{RT}$ C_v is seismic coefficient, I is importance factor, factor R accounting for ductility and overstrength, W is seismic weight, T is fundamental period.	$Q_i = \dfrac{(V_b - F_i) W_x h_x}{\sum W_i h_i}$ V_b is base shear, F_x is force at level x, F_i is force at level i, h_i is height of i-th floor.
IBC-2006/ ASCE/SEI 7-05	$V_b = \dfrac{S_{DS}}{R/I} W$ S_{DS} is design elastic response acceleration at short period.	$Q_x = \dfrac{W_x h_x^k}{\sum W_i h_i^k}$ k varies from 1 to 2. Other symbols are as in UBC 1997.

(continued)

TABLE 1.2 (Continued)
Base shear and its distribution in some international codes (V_b is base shear)

Code	Design base shear	Distribution of shear
NZS 4203: 1992	$V_b = CW_t$ $C = C_h(T,\mu)S_p RZL_u$ for ultimate limit state. C_h is basic seismic acceleration coefficient depending on soil type, ductility and period of vibration; S_p is structural performance factor, Z is zone factor, L_u is ultimate limit state factor. W_t is total weight of structure.	$Q_i = Q_t + 0.92 V_b \dfrac{W_i h_i}{\Sigma(W_i h_i)}$ Q_i is design force at level i, W_i and h_i are seismic weight and height from base for floor i. Q_t is force which is equal to $0.08V_b$ at roof level and zero elsewhere. This considers the higher mode effect.
EC-8 (2003) and EC-8 (2004)	$F_b = S_d(T_i)m\lambda$ $S_d(T_1)$ is ordinate of design spectrum, m is total mass of building, λ is correction factor for time period, varying from 0.85 to 1.0.	$Q_i = F_b \dfrac{s_i m_i}{\Sigma s_j m_j}$ F_i is force at level i; s_i, s_j are displacements of masses m_i, m_j.

Note: V_b is base shear.

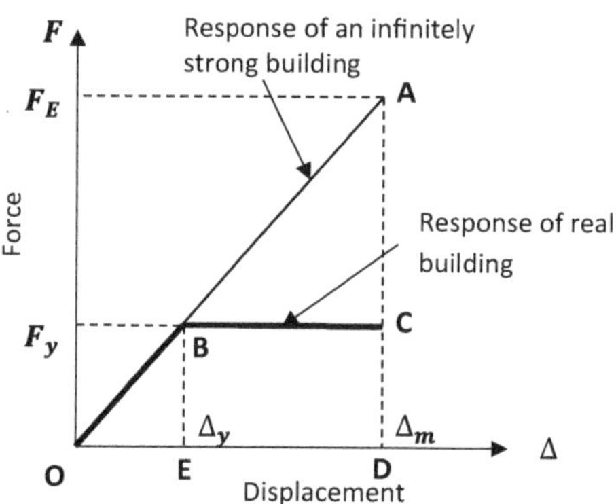

FIGURE 1.1 Equal displacement principle.

with limited strength but with same initial stiffness. An infinitely strong building ever remains elastic without suffering any damage.

On the other hand, a real building has limited strength and it suffers damage after yield force level is reached. We explain this with the help of Figure 1.1. In this figure, OA represents the force–displacement response of an infinitely strong (and hence, always elastic) building. This hypothetical building displays a maximum force of F_E and maximum displacement of OD (Δ_m) under some earthquake. The real building with finite strength yields at force level F_y under the same earthquake. The corresponding displacement at yield is yield displacement Δ_y. But the maximum displacement suffered by this building is also equal to OD.

From similar triangles OAD and OBE, we can write,

$$\frac{AD}{BE} = \frac{OD}{OE}$$

$$\text{Or,} \frac{F_E}{F_y} = \frac{\Delta_m}{\Delta_y}$$

But, by definition of ductility, $\dfrac{\Delta_m}{\Delta_y}$ = displacement ductility (μ). The ratio of peak force to yield force $\dfrac{F_E}{F_y}$ is the response reduction factor, R. So, we can write

$$R = \mu \tag{1.4.1}$$

Eq. (1.4.1) indicates that the response reduction factor is directly proportional to the ductility, which is valid for medium to long period structures only.

Note that we have considered an elasto-plastic behaviour of building (response curve remains flat horizontal after yielding). So, Eq. (1.4.1) is valid for medium to long period buildings with elasto-plastic behaviour only, as shown by dark lines (Figure 1.1).

1.5 EQUAL ENERGY PRINCIPLE

The equal displacement principle is obeyed by medium to long period buildings only. The short period buildings obey what is known as "equal energy principle", which can be stated as: *the strain energy stored by an infinitely strong elastic building under an earthquake is approximately equal to the energy dissipated by a real building with limited strength but with same initial stiffness under the same earthquake.*

To understand this observation, let us consider Figure 1.2. The strain energy stored by the infinitely strong elastic building is given by the area of triangle OAD.

The energy dissipated by the real inelastic building is given by the trapezoidal area OBFG. By equal energy principle, the two areas are same.

$$\text{Area OAD = area OBFG}$$

Deducting common area of both, that is area OBCD, we can write,

$$\text{Area of triangle ABC = area of rectangle DCFG.}$$

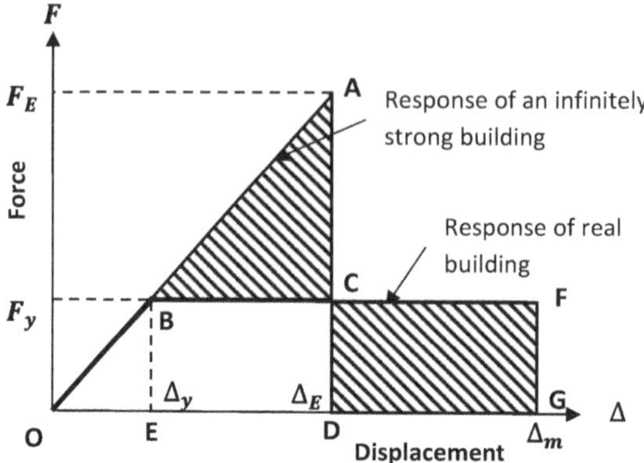

FIGURE 1.2   Equal energy principle.

$$\text{Or,} \quad \frac{1}{2}\,BC \times AC = CD \times DG$$

$$\text{Or,} \quad \frac{1}{2}\,BC \times (AD - CD) = CD \times DG$$

Noting that $BC = ED = \Delta_E - \Delta_y$, $AD = F_E$ and $CD = F_y$, $DG = \Delta_m - \Delta_E$, we can write,

$$\text{Or,} \quad \frac{1}{2}\left(\Delta_E - \Delta_y\right) \times \left(F_E - F_y\right) = F_y \times \left(\Delta_m - \Delta_E\right)$$

$$\text{Or,} \quad \left(\Delta_E - \Delta_y\right) \times \left(F_E - F_y\right) = 2F_y \times \left(\Delta_m - \Delta_E\right)$$

Now dividing both sides by $F_y \Delta_y$,

$$\frac{\Delta_E - \Delta_y}{\Delta_y} \times \frac{F_E - F_y}{F_y} = 2\frac{F_y}{F_y} \times \frac{\Delta_m - \Delta_E}{\Delta_y}$$

$$\text{Or,} \quad \left(\frac{\Delta_E}{\Delta_y} - 1\right) \times \left(\frac{F_E}{F_y} - 1\right) = 2 \times \left(\mu - \frac{\Delta_E}{\Delta_y}\right) \qquad \left[\because \frac{\Delta_m}{\Delta_y} = \mu\right]$$

$$\text{Or,} \quad \left(\frac{F_E}{F_y} - 1\right) \times \left(\frac{F_E}{F_y} - 1\right) = 2 \times \left(\mu - \frac{F_E}{F_y}\right) \qquad \left[\because \frac{F_E}{F_y} = \frac{\Delta_E}{\Delta_y}\right]$$

$$\text{Or,} \quad (R-1) \times (R-1) = 2 \times (\mu - R) \qquad \left[\because \frac{F_E}{F_y} = R\right]$$

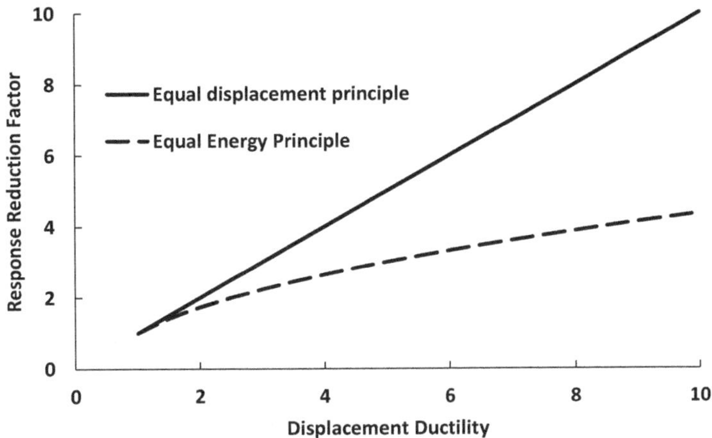

FIGURE 1.3   Response reduction factor based on ductility.

$$\text{Or,}\quad R^2 - 2R + 1 = 2\mu - 2R$$

$$\text{Or,}\quad R = \sqrt{2\mu - 1} \tag{1.5.1}$$

Eq. (1.5.1) indicates that for short period structures, the response reduction factor is not directly proportional to ductility. To understand the two cases, we plot Eq. (1.4.1) and (1.5.1) in Figure 1.3. This figure shows that the response reduction factor is lower for short period structures for similar value of ductility.

1.6   CLASSIFICATION OF STRUCTURES BASED ON TIME PERIOD

The time period of a building indicates whether the building is stiff or flexible. Flexible buildings have a longer time period. An empirical formula for time period of building is given in Eq. (1.6.1).

$$T = \frac{N}{10} \tag{1.6.1}$$

where, T is time period in second and N is the number of stories in the building.

A very general, but not rigid, classification of buildings based on time periods is given below:

Zero period structures: $T < 0.1$ sec.
Such structures are very stiff and vibrate with the same acceleration as that of the ground.

Short period structures: 0.1 s $< T < 0.55$ sec. (these correspond to the horizontal part of spectrum in Indian code):

Long period structures: $T > 0.7$ sec.
Long period structures are flexible in nature.

The zero-period structure vibrates with the ground frequency. As such, there is little chance of energy dissipation through hysteresis. So, the response reduction factor for such structures is taken as unity. In fact, R is a function of time period of structures.

1.7 STRENGTH–STIFFNESS DEPENDENCY

Before entering into the topic, let us refresh ourselves about the moment–curvature relationship. We start with the flexural equation given by Eq. (1.7.1):

$$EI\frac{d^2 y}{dx^2} = M \qquad\qquad (1.7.1a)$$

$$\text{Or,} \quad EI\phi = M \qquad\qquad (1.7.1b)$$

Or, $\quad \phi = \dfrac{M}{EI}$ where, $\phi = \dfrac{d^2 y}{dx^2}$ is the curvature of beam elastic curve.

Here, E is modulus of elasticity, I is moment of inertia of section, M is bending moment and ϕ is curvature.

A plot between M and ϕ (i.e., M vs. $\dfrac{M}{EI}$) is called the moment–curvature diagram. Typically, M–ϕ diagram is nonlinear (Figure 1.4a), but it can be idealized to a bilinear curve as shown in Figure 1.4b.

In Figure 1.4, the variation of yield moment (M_y) is shown with increasing percentage steel in a beam section. The beam sectional size is kept constant. As the steel percentage increases, the yield moment increases, but the yield curvature remains same for the given section. The yield curvature will change if section size is altered. The bilinearized M–ϕ diagram is shown in Figure 1.4b.

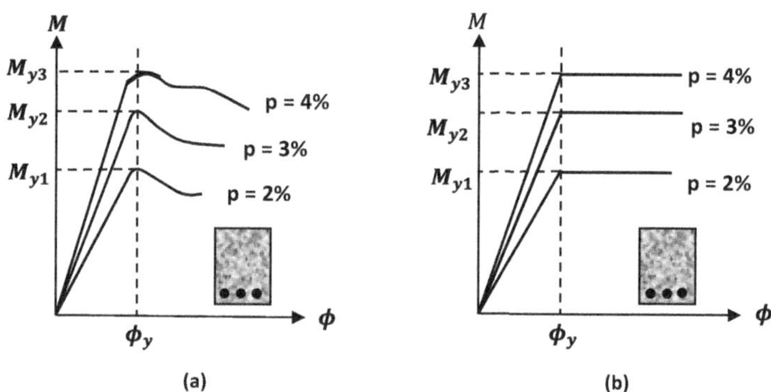

FIGURE 1.4 (a) M–ϕ diagram for a given RC section with varying percentage steel (b) bilinearized M–ϕ diagram (M_y is yield moment, p is percentage steel).

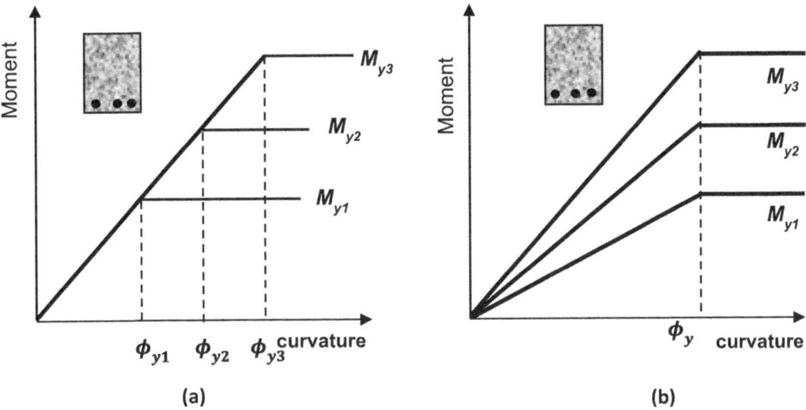

FIGURE 1.5 Moment–curvature relationship: (a) assumed by code (b) in reality.

The code assumes that the stiffness is not affected by steel in the section. So, for a given section, the stiffness does not change with increase of percentage steel. Figure 1.5(a) shows that the slope of M–ϕ diagram is constant for different percentage steel for the given section. As a fall out of such codal assumption, the yield curvature takes different values for the same section, which is against the reality, as has been observed in the preceding paragraph. In reality, we have found that the yield curvature of a section does not change with strength (i.e. percentage steel) for the given section. If the yield curvature has to remain constant at different strengths, we get the diagram 1.5b. In Figure 1.5b, the section assumes different stiffnesses (slopes of M–ϕ lines) for different strengths. This suggests that *the stiffness varies directly as the strength of the section.*

This shows that, in order to get actual stiffness of the section, one needs to calculate the strength of the section. The computation of actual stiffness has been discussed in Section 4.3 of Chapter 4.

1.8 LIMITATIONS OF FORCE-BASED METHOD OF DESIGN

Force-based method of design considers force as the primary design parameter. The traditional codal design is force-based. The following are some of the criticisms against and limitations of this method of design (Priestley et al., 2007).

1. In general, structural design is carried out to prevent damage in the structure or to limit the damage to a desired level. However, force is not a good measure of damage; rather displacement is a direct measure of damage in the structural elements.
2. The design force (base shear) is computed based on initial elastic time period when the system is elastic. However, the actual design force is taken based on some perceived damage and ductility, which state cannot be captured by initial stiffness.

3. The response reduction factor considered in force-based codal method of design is controversial. The assumption that response reduction factor is constant for a particular type of building is not correct. It may vary from building to building of the same class. The assumption that response reduction factor is equal to ductility is based on the model of elastic–perfectly plastic system. The real systems are different from this assumption. The $R = \mu$ proposition is valid for medium period structures. For a short period, structures equal energy principle is valid, when

$$R = \sqrt{2\mu - 1} \qquad (1.5.1)$$

For zero period structures, $R = 1$ (regardless of the value of μ)
$R = \mu$ relationship is generally applicable when $T > 0.7$ sec.
For 0 to 0.7 sec time period, the suggested expression is given by (1.8.1):

$$R = 1 + (\mu - 1)T / 0.7 \qquad (1.8.1)$$

Thus, R is a period dependent parameter, which aspect is generally ignored in the codal method of design.
4. Code assumes that stiffness is independent of strength, but actually it is not so; stiffness varies with strength, as has been discussed in Section 1.7.
5. Design carried out with the initial time period to compute base shear often leads to building unsafe in drift. Hence an iterative solution is required, which takes time.
6. The analysis procedure that considers the effect of first mode may lead to inaccuracy because of ignoring higher mode effects.
7. Design carried out with response reduction factors underestimates the torsional effects.

1.9 CLOSURE

In the traditional codal design method, force is at core criterion of design. The problem with this method is that the damage cannot be quantified from force. Other limitations of the FBD have also been highlighted here. The failure of structures under strong ground motions designed by using codes points to the fact that FBD is not a very reliable design philosophy. A more rational method would be to take displacement as the core parameter in design. It is a reality that stiffness varies with strength, but this is ignored in traditional design codes. Here arises the need for displacement-based design, which has been taken up in the subsequent chapters.

1.10 EXERCISES

Q 1.10.1 Give a brief sketch about the evolution of seismic design of structures.
Q 1.10.2 What is the fall out of "equal displacement principle"?
Q 1.10.3 Some structures obey "equal energy principle". Explain.
Q 1.10.4 Explain how time period can affect the ductility of a structure.

Q 1.10.5 "Response reduction factor is dependent on time period". Explain.

Q 1.10.6 Show that stiffness is dependent on strength.

Q 1.10.7 What are the limitations of force-based method of design?

Q 1.10.8 Why is the safety of nonstructural components important?

Q 1.10.9 Highlight the importance of hospital buildings.

Q 1.10.10 Give an outline of the performance of buildings in past earthquakes.

Q 1.10.11 Give an outline of the performance of nonstructural components in past earthquakes.

Q 1.10.12 Highlight the basic formulae for base shear in international codes.

Q 1.10.13 Explain the constancy of yield curvature for a given section.

FURTHER READINGS

Biranchi, S., Ciurlanti J. and Pampanin S., Seismic vulnerability of nonstructural components: From traditional solutions to innovative low-damage systems, *Conference on Earthquake Risk and Engineering Towards a Resilient World*, 9–10 September 2019, London.

Bozorgnia, Yousef and Bertero V.V. (2004) *Earthquake Engineering from Engineering Seismology to Performance-Based Engineering*, CRC Press, 2004.

Choudhury, S. (2007) Performance-Based Seismic Design of Hospitals, Ph.D. Thesis, Department of Earthquake Engineering, IIT Roorkee, 2008.

Dakhol, R.P. (2020) Seismic Performance of Nonstructural Elements and Learning from Earthquakes: New Zealand Perspective, *Bulletin of the New Zealand Society for Earthquake Engineering*, Vol. 53 No. 3, September.

Dhakal, R.P. (2010) Damage to Non-Structural Components and Contents in 2010 Darfield Earthquake, New Zealand Society for Earthquake Engineering (NZSEE), Vol. 43 No. 4: Special Issue on the 2010 Darfield Earthquake, https://doi.org/10.5459/bnzsee.43.4.404-411

Filiatrault, A., Christopoulos C. and Sterns, C. (2001) Guidelines, Specifications and Seismic Performance Characterization of Nonstructural Building Components and Equipment, PEER Report No. 2002/05, 2002.

Filson, J., Borcherdt, R. D, Langer, C. and Simpson, D. (1989) *Seismology*, EERC, DOI https://doi.org/10.1193/1.1585212

Gabbianelli, G., Perrone D., Brunesi E. and Monterio R. (2020) Seismic Acceleration and Displacement Demand Profiles of Nonstructural Elements in Hospital Buildings, *Buildings*, 10(12): 243. https://doi.org/10.3390/buildings10120243

Gates, W.E., and McGavin, G. (1998) Lessons Learned from the 1994 Northridge Earthquake on Vulnerability of Nonstructural Systems, Proceedings of Seminar on Seismic Design, Retrofit, and Performance of Nonstructural Components, ATC-29-1, CA, USA.

McGavin, G.L. (1981) *Earthquake Protection of Essential Building Equipment Design Engineering Installation*, Wiley, Toronto, 1981.

McGavin, G.L. and Patrucco, H. (1998) Nonstructural Functional Design Considerations for Healthcare Facilities, 6th US Conference on Earthquake Engineering.

Mitchell, D., Paultre P., Tinawi R., Saatcioglu M., Tremblay R., Elwood K., Adams J. and DeVall R. (2010) Evolution of seismic design provisions in the National building code of Canada, *Can. J. Civ. Eng.* 37: 1157–1170.

Nazer, M.H. and Elahi, F.N. (2004) Seismic Vulnerability of Nonstructural Components of Hospitals, 13th World Conference on Earthquake Engineering, Canada, Aug. 1–6, Paper No. 1250, 2004.

Priestley, M.J.N., Calvi, G.M. and Kowalsky, *Displacement-Based Seismic Design of Structures*, IUSS Press, Pavia, Italy, 2007.

Prota, A., Zito M. and Magliulo G. (2022) Preliminary Results of Shake Table Tests of a Typical Museum Display Case Containing an Art Object, *Advances in Civil Engineering*. Volume 2022, Article ID 3975958. https://doi.org/10.1155/2022/3975958

Sánchez-Silva, M., Yamín, L.E., and Caicedo, B. (1999) Lessons of the 25 January 1999 Earthquake in Central Colombia, Earthquake Engineering Research Centre (EERC), Volume 16, Issue 2, https://doi.org/10.1193/1.1586123

SEAOC Vision 2000 Committee (1995) Performance-Based Seismic Engineering, Report prepared by Structural Engineers Association of California, Sacramento, California.

Sortis, A.D., Pasquila, G.D., Orsini, G., Sano, T., Binodi, S., Nuti, C. and Vanzi, L. (2000) Hospital Behaviour During 1997 Earthquake in Umbria and Marche (Italy), 12th World Conference in Earthquake Engineering, Paper No. 2514.

Uniform Building Code (UBC) (1997) Published by International Conference of Building Officials, California.

WCEE, World Conference on Earthquake Engineering (various years).

2 Introduction to Performance-Based Design

2.1 GENERAL

We have seen in Chapter 1 that the FBD cannot always render a safe structure under a severe seismic event. Design codes of various countries prescribe the design procedures for earthquake-resistant design (ERD). But the structures designed following such codes are not always safe under seismic loads. Structures collapse or severely get damaged all over the world due to major earthquakes (see Chapter 1). This points to the limitations of present codal method of design, which lays emphasis on force alone as the design parameter. With time, the limitations of codal method of FBD has become apparent to the engineering community. From 1975 onwards, the concept of performance-based design (PBD) has gradually evolved. In this design philosophy, the structure is designed for some target performance objectives under an expected hazard level. Displacement, rather than the force, is the guiding parameter in this method of design. Performance objectives may include structural performance level, which depends on plastic rotation and, non-structural performance level in terms of damage and loss of functionality. Drift is an important design objective for many structures and non-structural components.

The intuitive application of earthquake resistance in structures may date long back in history. The scientific treatment of the earthquake resistance in structures started with the occurrence of some major earthquakes in modern times like the 1885 Edo earthquake (M_s = 7.0, Japan), 1906 San Francisco earthquake (M_w = 7.9, California, United States), 1908 Messina earthquake (M_w = 7.1, Italy) and 1923 Kanto earthquake (M_w = 8.2, Japan).

In this book, the performance-based seismic design (PBSD; often written as PBD in this book) of various structures has been dealt upon. The other name assigned by some authors is performance-based earthquake engineering (PBEE). The performance-based concept can be applied in other engineering and non-engineering fields as well.

The first tall building designed using PBD approach is the 55-storied high office tower, Torre Mayor. For seismic mitigation, viscous dampers were used in this building. Presently, some countries have partly incorporated PBD in the design codes.

2.2 HISTORICAL DEVELOPMENT OF PBD

The concept of PBD first appeared in the form of capacity spectrum method (CSM) introduced by Freeman et al. (1975) as an evaluation procedure in a pilot project for seismic vulnerability assessment of buildings. However, the root of the technique lies

DOI: 10.1201/9781003441090-2

in the "reserve energy technique" of Blume (Blume et al., 1961). Freeman (1978) improved the technique further. ATC-3-06 (1978) did place the PBD on a quantifiable footing. PBD can be applied to other fields as well like fire engineering and aerospace engineering. It was first applied in the field of earthquake engineering.

CSM was given a formal shape in ATC-40 (1996). This document gave a graphical construction method for getting the performance point. Performance point is the special point where the capacity of the structure is equal to the reduced demand imposed on the structure. The graphical method of CSM became very helpful to engineers in visually understanding the structural dynamic behaviour. ATC-40 prescribes the building performance levels as a combination of structural and non-structural performance levels. The acceptance criteria for component performance levels in terms of plastic rotations are given in FEMA-273 with NEHRP (National Earthquake Hazard Reduction Program) guidelines for rehabilitations of buildings (1997). This document, in parallel with ATC-40, gave the definition of building performance levels and acceptance criteria for member performance levels. In FEMA-356 (2000) document, the coefficient method was proposed for estimating the target displacement, considering the inelasticity, MDOF effect and P-delta effect. The demand capacity ratio was also discussed in this treatise. Kilar and Fajfar (1997) reported pushover analysis for asymmetric building.

A chronological development of PBD is shown in Table 2.1.

2.3 DISPLACEMENT-BASED DESIGN

In 1991, Qi and Moehle reported a displacement-based design (DBD) method for structures. Here, the preliminary design is done by using elastic spectra. The interstory drift is found from an established equation. The design displacement response spectrum is then modified for the drift requirement. The time period is now established. If the time period becomes greater than the characteristic time period, then the design was acceptable, else steps are repeated.

The Blue book (1999), developed by Structural Engineers' Association of Canada (SEAOC), described a DBD methodology. The target displacement was determined from target drift and assumed deflected shape, which method has been used by developers too. The effective period is obtained from ADRS representation of the pushover curve. From the calculated ductility, the effective stiffness was found out. From this, the base shear was computed. Panagiatakos and Fardis (1999) developed a deformation-based design methodology. The structural analysis for earthquake loads was done with uncracked sections and then capacity design was applied. The method focused only on limited safety criteria and could not accommodate the safety of non-structural components.

Browning (2001) reported a design method which was called the "Browning's proportioning method". From a specified relationship between drift and building period, a target period was established. The preliminary member sizes were now decided. If the target time period was not achieved, iterations are carried out. If the base shear turns out to be greater than the yield base shear, the strength of members had to be increased.

TABLE 2.1
Chronological development of PBD

Year	Development in PBD
1961	Blume suggested reserve energy technique.
1975	Freeman *et al.* developed an evaluation procedure in a pilot project for seismic vulnerability of buildings.
1978	ATC-3-06 had introduced physical meaning to parameters in base shear.
1980	ATC-10 gave a procedure to find correlation between earthquake ground motion and building performance.
1993	DDBD was introduced by Priestley.
1995	Vision 2000 document of SEAOC discussed rehabilitation of buildings.
1996	ATC-40 developed capacity spectrum method. It quantified performance levels.
1997	FEMA-273 gave coefficient method for target displacement.
2000	FEMA-368 (NEHRP 2000) provisions for new buildings were given.
2000	FEMA-356 described guidelines for design of new buildings. It quantified performances in terms of plastic deformations.
2004	FEMA-440 improved the coefficient method. Gave equivalent bilinearization method.
2004	FEMA-450 gave seismic regulations for new buildings and other structures.
2005	DDBD for frame buildings was given by Pettinga and Priestley.
2006	DDBD of frame-wall buildings was given by Sullivan *et al.*
2007	UPBD method for RC dual system was developed by Choudhury (2007).
2012	DDBD method for frames with fluid viscous dampers by Sullivan and Lago.
2013	UPBD method for RC frame buildings developed by Choudhury and Singh.
2014	Column size determination in UPBD framework given by Mayengbam and Choudhury
2015	UPBD method for steel buildings by Kotapaty.
2017	Local Building Code based on PBD was introduced in Mexico City.
2019	Influence of effective stiffness on performance buildings in UPBD framework was discussed Das and Choudhury (2019); Das and Choudhury (2019) developed the same in SVR model. Das et al. (2021) gave effective stiffness of columns through GA.
2020	UPBD method for bridge piers was given by Banerjee and Choudhury (2020a, 2020b).
2022	UPBD method applied to Elevated water tank staging by Baruah (2022).

Medhekar and Kennedy (2001) reported a DBD method based on the displacement spectrum where an initial deflected shape was assumed. Vamvastikos and Cornel (2002) reported incremental dynamic analysis (IDA) where the structure is subjected to nonlinear time history analysis under several scaled ground motions. Antoniou and Pinho (2004) reported an adaptive pushover method for displacement-based design. In this method, the applied lateral load was updated at each stage of the change of stiffness keeping in line with the new displacement profile. This is supposed to be more realistic than the normal pushover method with constant pattern load.

2.4   DEFORMATION-CONTROLLED DESIGN BY PANAGIATAKOS AND FARDIS

Panagiatakos and Fardis (1999) proposed a method of seismic design aimed at controlling the deformation. The authors recommended provisions for infill buildings. The steps involved in the design method are generally as follows:

(i)   Elastic analysis using response spectrum at frequent earthquake level is carried out. The sections of elements are considered as uncracked.

(ii)   Capacity design is carried out and proportioning of steel at hinge location is done. Proportioning of steel in other parts of the structure is then carried out.

(iii)   Elastic analysis for maximum earthquake is carried out. Secant stiffness is used at this stage.

(iv)   From time history analysis, the amplifications of chord rotations are obtained.

(v)   Check if chord rotations are within acceptable limits. If needed, modify the steels.

(vi)   Capacity design is to be checked once again.

The method is silent on non-structural damage and safety.

2.5   DISPLACEMENT-BASED DESIGN BY QI AND MOEHLE

Qi and Moehle (1991) developed displacement-based design. The maximum design displacement is given by Eq. (2.5.1).

$$\Delta_{max} = \left(\frac{T}{T_g}\right)^{1.6\eta} \times \Delta_{T_{g,elastic}} \tag{2.5.1}$$

where, Δ_{max} = maximum design displacement

T = elastic time period

η = ratio $F_y / \left(ma_{g,max}\right)$

m = mass of the system

$a_{g,max}$ = maximum ground acceleration

F_y = yield force

T_g = a dividing time period, such that Eq. (2.5.1) is applicable when $T \leq T_g$; and when $T > T_g$ the maximum inelastic displacement may be taken as equal to elastic displacement.

2.6 BROWNING'S PROPORTIONING METHOD OF DESIGN

Browning (1999) proposed the proportioning method as a DBD. This method considers drift as the target criterion, which was supposed to be reflected by a target time period (T_{target}).

The general steps are:

(1) Select the target time period (T_{target}) from target drift limit and displacement spectra corresponding to the design spectrum.
(2) Decide the member sizes. This is called proportioning.
(3) Carry out dynamic analysis to find dynamic initial time period of the structure ($T_{initial}$). In this stage, gross sectional properties are used.
(4) If $T_{initial} \leq T_{target}$, increase the member sizes.
(5) Check if base shear is greater than the lateral capacity of the building. If so, increase the strength by adding steel to those members that suffer damage.
(6) Apply capacity design so that beams yield and not columns.
(7) Apply ductile detailing as per standard specifications.

2.7 YIELD POINT SPECTRA METHOD OF DESIGN

Aschheim and Black (2000) proposed the yield point spectra (YPS) method of design. Here, the yield displacement is the core design parameter. The authors claimed that the yield displacement was more dependable property than the traditional time period of code-based design. The yield point spectra (YPS) is a plot between yield displacement and yield strength coefficient (C_y). The yield strength coefficient is the ratio of the yield strength to total gravity load for a nonlinear oscillator. Whereas normal elastic response spectrum is drawn for a single degree of freedom system (oscillator) having varying time period, the YPS is drawn for an oscillator that yields. The displacement ductility of the oscillator is kept constant. The radial lines in an YPS plot indicate the constant period lines. Several curves are drawn for varying ductilities. Ductility is obtained from the consideration of limiting structural damage and drift. Equivalent single degree of freedom (ESDOF) system properties are obtained using participation factor and mass participation ratio. From the YPS plot, corresponding to the yield displacement and ductility, yield strength coefficient is read out. Yield base shear is given by the product of yield strength coefficient and mass participation ratio. The base shear is distributed over the height of the building as usual and design is carried out.

In fact, YPS is a plot between yield displacement (Δ_y) and inelastic yield strength coefficient (C_y). A typical YPS is shown in Figure 2.1. The steps involved are as follows:

(1) Select design drift and design spectrum.
(2) Construct the YPS corresponding to the perceived seismicity level. This construct will involve plot of Δ_y vs. C_y for various displacement ductilities.
(3) Compute design displacement (Δ_d) corresponding to the design drift.
(4) Arrive at ductility demand and see where it fits in the YPS plot.

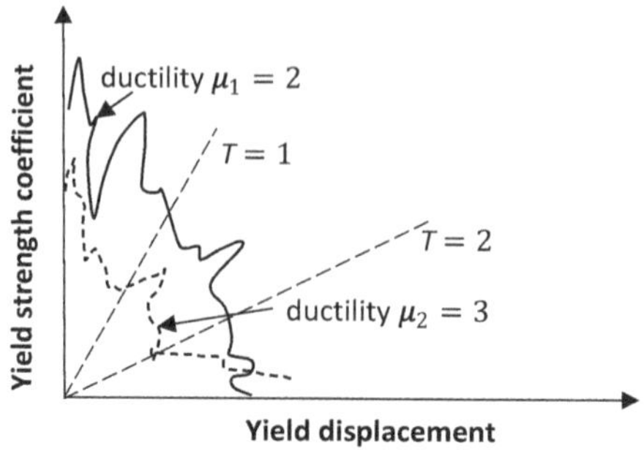

FIGURE 2.1   Typical YPS.

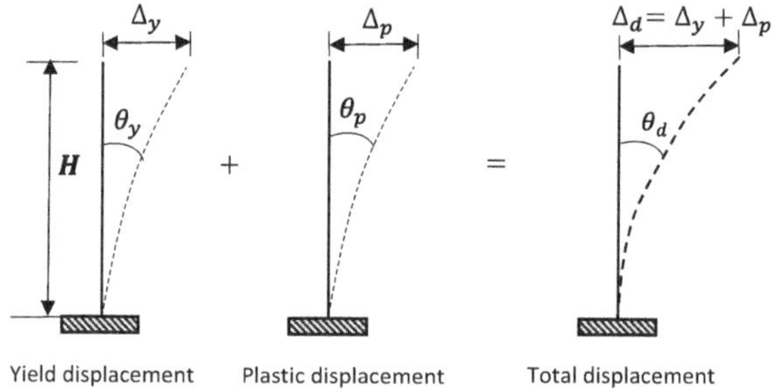

FIGURE 2.2   Design displacement in DBD by Chopra.

(5) Compute yield displacement (Δ_y) and from YPS find C_y and hence F_y.

(6) Design for F_y (as per code of the region).

2.8 DBD BY CHOPRA

Chopra and Goel (2001) proposed a displacement-based design. Referring to Figure 2.2, the system has a height of H, yield displacement Δ_y, design displacement Δ_d, plastic rotation at base is θ_p, Eqs. (2.8.1) to (2.8.3) are readily obtained. The displacement ductility is given by Eq. (2.8.4). The method uses effective stiffness in analysis.

$$\Delta_y = \theta_y H \tag{2.8.1}$$

$$\Delta_p = \theta_p H \tag{2.8.2}$$

$$\Delta_d = \Delta_y + \Delta_p \tag{2.8.3}$$

$$\mu_\Delta = \Delta_d / \Delta_y \tag{2.8.4}$$

$$k = \frac{4\pi^2 m}{T_e^2} \tag{2.8.5}$$

The design steps are as follows:

(1) Find yield displacement. Decide the tolerable value of plastic rotation at the base of the system. Find the displacement ductility as per Eqs. (2.8.1) to (2.8.4). Construct the inelastic spectra for an yielding oscillator.

(2) From the displacement spectra and ductility, read out an effective time period (T_e).

(3) Find stiffness k from Eq. (2.8.5).

(4) Find the design yield force $F_y = k\Delta_y$. Design members for this lateral force (base shear).

(5) For the structure in step (4), find the actual updated stiffness (k_{update}) and actual yield displacement ($\Delta_{y,update}$).

(6) Check if $\Delta_y \approx \Delta_{y,update}$ (i.e., approximately equal). If not, go to step 1. Else, accept the design.

2.9 CAPACITY SPECTRUM METHOD BY FREEMAN

Freeman proposed the capacity spectrum method (CSM) in 1975, which was further refined by Freeman (1998). The CSM is specially oriented for evaluating existing buildings for which the member sizes and strengths are known. As such, there are some gaps as to how the method to be applied to new buildings. The general steps are as follows:

(1) It is assumed that the building member sizes and member strengths are known. The capacity curve is generated between the roof displacement and base shear.

(2) The 5% damped demand spectrum is considered (T vs. S_a). The capacity diagram and demand diagram are converted to an acceleration–displacement format (ADRS format).

(3) Select assumed performance point.

(4) Find damping/ductility associated with the assumed performance point.

(5) Obtain the reduced demand diagram by using reduction factor. Currently, the following formula for reduction factor is in use. Note the point where the reduced demand curve intersects the capacity curve. If the originally assumed performance point differs from the newly obtained one, select another trial performance point between the above two points.

(6) The process is repeated till satisfactory convergence.

(7) The spectral displacement at the final performance point is the design displacement.

(8) Base shear is the product of stiffness and design displacement.

(9) Design is done for this base shear.

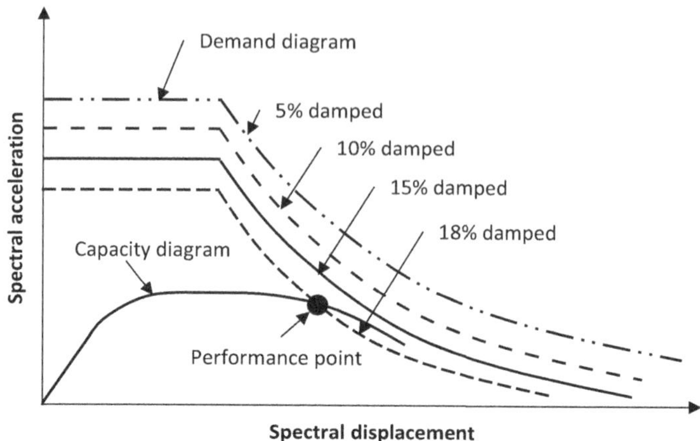

FIGURE 2.3 Capacity spectrum diagram.

A typical CSM scheme is shown in Figure 2.3.
More discussion on CSM is available in Chapter 6.

2.10 DIRECT DISPLACEMENT-BASED DESIGN

Priestley (1993) formulated a direct displacement-based design (DDBD), which was improved over the years (Priestley 1993; Priestley 2000; Priestley 2003a; Priestley 2003b, Priestley et al. 2007). In the context of building, the method involved assumption of a first mode shape and an equivalent SDOF (ESDOF) system for the MDOF building. ESDOF system properties are found from the assumed displacement profile and known masses of the floors. The effective time period is obtained from the secant stiffness connecting the origin and the point of design displacement, using displacement spectra corresponding to the effective damping. Yield displacement is guided by beam depth and yield strain of steel. The design base shear is obtained from design displacement and effective time period. The base shear is distributed suitably over the height of the building and design is carried out with expected strength. The detailed treatment of the method will be done in Chapter 7.

Xue (2001, 2003) reported a DDBD procedure of inelastic structures, which satisfied displacement/drift requirements. Inelastic spectra were utilized with Newmark–Hall reduction factor. An SDOF replacement of structure was necessary. The performance point is determined from displayed ductility. Pettinga and Priestley (2005) developed a DDBD for frame buildings. Sullivan et al. (2005, 2006a, 2006b) developed DDBD method for frame-wall dual system. In this method, the base shear is apportioned between the frame and wall. The yield displacements of walls are found out from suggested equations. Ductility of the frame and the wall are found out. Effective dampings for wall and frame are obtained and effective time period is calculated. The design base shear is now computed. The base shear is now distributed suitably over the height of the building. If necessary, steps are repeated to achieve the design interstory drift. A detailed treatment of DDBD for dual system is done in Chapter 8.

The interstory drift requirement is checked through nonlinear time history analysis (NLTHA) and if necessary, redesign is done. The DDBD method considers interstory drift as the only design criterion.

2.11 UNIFIED PERFORMANCE-BASED DESIGN

Unified performance-based design (UPBD) was introduced by Choudhury (2008) and applied to RC frame-wall buildings of IO performance level. While DBD and DDBD methods considered only drift as the target design objective, UPBD method considered both drift and performance level in the design philosophy. The advantage of the UPBD method over other methods of PBD are: (i) it accommodates two target performance criteria, namely, drift and performance level (in terms of plastic rotation) and (ii) it gives member sizes of the yielding members in the beginning of the design, thus avoiding iteration for member sizes. Choudhury and Singh (2013) developed UPBD method for RC frame buildings. The determination of column size in UPBD framework was treated by Mayengbam and Choudhury (2014). UPBD method has been applied to steel frame buildings (Kotapaty, 2015; Anil, 2021; Azeel, 2022). The UPBD method for RC bridge pier was developed by Banerjee and Choudhury (2020a, 2020b). The UPBD method has been applied to dual system with various performance levels and drifts by Mibang and Choudhury (2019a, 2019b).

The UPBD method for various structures has been discussed in separate chapters in this book.

2.12 CAPACITY DESIGN

The concept of capacity design was developed by Park and Paulay (1975). The need for capacity design arose from the understanding that the failure of different members had different impact on the health of the structure. Failure of some members are more critical to the structure. Also, the hysteretic energy dissipation in different types of failure are different. Capacity design implies designing structural components with such strength allocations that failure of some members is allowed while failure of some other members is prevented. The failure in flexure is gradual in nature and it allows large inelastic deformation without collapse, thus, dissipating energy through hysteresis, cycle after cycle. On the contrary, the failure in shear is sudden and inherently brittle in nature, accompanied with very little energy dissipation through hysteresis. It is judicious that shear failure of the members is avoided and flexural failure is encouraged. The failure of columns is generally through bulging of longitudinal steel following the snapping of confining ties. This type of failure is known as bursting failure, which is essentially sudden and brittle in nature. Columns are also the primary gravity load-carrying members. Failure of columns may easily lead to the collapse of structure, through progressive collapse. If one column fails, the load carried by the column is now shared by the adjacent columns, which in turn become overloaded. This process may continue and the columns may fail progressively. To avoid such situation, the columns are imparted relatively higher strength over beams meeting at a joint so that the beams may fail earlier than the columns. This is known as *weak-beam strong-column* concept.

Unless capacity design is done, the pushover results and time history results *are* not reliable, as undesirable hinge pattern may develop (Choudhury, 2008). In such situation even, the analysis may not progress due to non-convergence. Mathematically, we can express the capacity design requirement by Eq. (2.12.1).

$$\frac{\text{Capacity of non-yielding member}}{\text{Capacity of yielding member}} = \psi > 1.0 \tag{2.12.1}$$

In the context of buildings, in a beam–column joint we need to observe:

$$\frac{\sum M_c}{\sum M_b} = 1.3 \quad \text{as per EC} - 8, 2004 \tag{2.12.2}$$

$$\frac{\sum M_c}{\sum M_b} = 1.4 \quad \text{as per IS} 13920 - 2016 \tag{2.12.3}$$

In Eqs. (2.12.2) and (2.12.3), $\sum M_c$ is sum of limiting moment capacities of columns meeting in the joint in a particular direction, and, $\sum M_b$ is the sum of limiting moment capacities of beams meeting in the joint in the same direction. Limiting moment capacity here means the limit state moment of resistance. As such, the Eqs. (2.12.2) and (2.12.3) give what is called "column-to-beam capacity ratio" (C/B ratio). Obviously, there shall be at least two C/B ratios in a particular joint, along the two major directions of orientation of the column. The requirement of C/B ratio limit need not be applied at the roof level, as there is no column above the roof.

2.13 EQUIVALENT VISCOUS DAMPING

In a simplified way, the equivalent viscous damping (ξ_e) is the sum of material damping (ξ_m) and hysteretic damping (ξ_h), as given by Eq (2.13.1a). It is a simplified statement because there can be added damping (ξ_a) from deliberately introduced energy dissipation devices (Eq. (2.13.1b)). Typically, material damping for concrete is 5% and that of steel is 2% of critical damping.

$$\xi_e = \xi_m + \xi_h \tag{2.13.1a}$$

$$\xi_e = \xi_m + \xi_h + \xi_a \tag{2.13.1b}$$

Damping is physically related to ductility. More ductility will lead to more damping. Hysteretic damping takes place when the structural element undergoes plastic

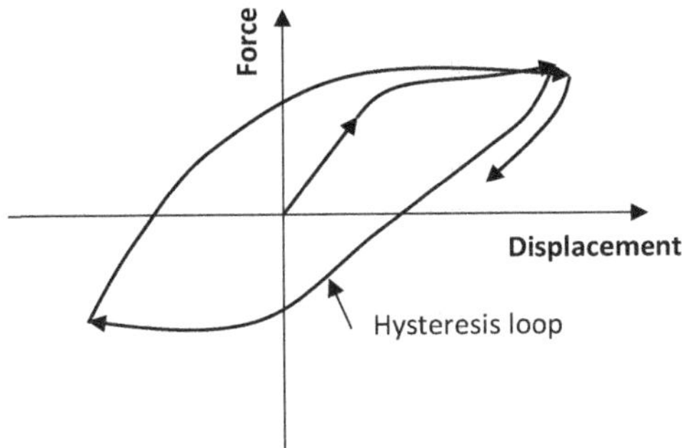

FIGURE 2.4   A typical hysteresis loop.

deformation under reversible actions. The force–deformation diagram is called the hysteresis loop. The area under the loop gives the energy dissipated though hysteresis. In reversible action like earthquake, the hysteresis loops are formed again and again in each cycle, thereby dissipating energy absorbed from earthquake shaking. A typical hysteresis loop is shown in Figure 2.4.

The area of the hysteresis loop gives the energy dissipated through hysteresis. The area of the loop depends on the force applied and deformation that the section undergoes. The force applied cannot exceed the capacity of the section. The deformation that the section can undergo depends on the ductility of the section. The more the ductility more is the energy dissipation. Ductility arises out of detailing in RC structures and joint detailing in steel structures.

The expression for hysteretic damping is given by Eq. (2.13.2):

$$\xi_h = \frac{1}{4\pi} \frac{E_d}{E_{s,max}} \tag{2.13.2}$$

where ξ_h is hysteretic damping, E_d is hysteretic energy dissipated per cycle in hysteresis and $E_{s,max}$ is maximum strain energy through structural displacement.

The shape of the hysteresis loop depends on the material type. The shape is also dictated by section behaviours obtained through experiments. We have Takeda model for RC sections and FEMA model for RC sections and steel sections. The other hysteresis behaviours are Ramberg–Oswgood model, bilinear idealized model and flag-shaped model.

2.14   EXAMPLES

Example 2.14.1 *In a beam–column joint, four beams meet from opposite sides and two columns extend to the upper and lower stories. The beams and column details are given in Figure 2.5. The clear cover in beam is 25 mm and that in column is 40 mm. Concrete grade is M30 and steel grade is Fe500. Find the C/B ratios and comment*

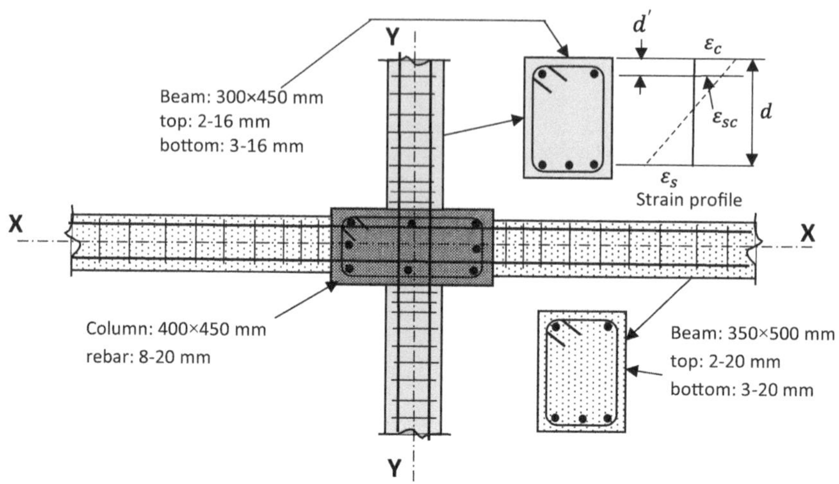

FIGURE 2.5 Beam and column details in Ex2.14.1.

on the values. The axial force in column in the top storey is 300 kN and that in bottom storey is 400 kN. Diameter of stirrups is 8 mm.

Solution: We shall take help of SP.16 of Indian standard for stress–strain relation of rebar and for quick computation of column limiting moments. Here, f_{ck} = 30 MPa, f_y = 415 MPa.

X-directional beams

Top steel = 2-20 mm bars = 628 mm².
Bottom steel = 3-20 mm bars = 942 mm².
b = 350 mm, D = 500 mm, effective depth d = 500 − 25 − 8 − 20/2 = 457 mm, effective depth on compression side d' = 25 + 8 + 20/2 = 43 mm.

Strain in tension steel is $\varepsilon_s = \varepsilon_c \dfrac{d-x}{x}$; strain in compression steel is $\varepsilon_{sc} = \varepsilon_c \dfrac{x-d'}{x}$.
Table Ex2.14.1 is prepared, from which at balanced condition of forces, x = 67 mm. The compression in concrete is given as $C_c = 0.36 f_{ck} bx$.
Moment of resistance = $C_c\left(d - 0.42x\right) + C_s\left(d - d'\right)$ = **173.8 kN-m**.

Y-directional beams

Top steel = 2-16 mm bars = 402 mm².
Bottom steel = 3-16 mm bars = 603 mm².
b = 300 mm, D = 450 m, effective depth d = 450 − 25 − 8 − 20/2 = 407 mm, d' = 25 + 8 + 20/2 = 43 mm.
Table Ex2.14.2 is prepared, from which at balanced condition of forces, x = 58 mm.
Moment of resistance = $C_c\left(d - 0.42x\right) + C_s\left(d - d'\right)$ = **98.4 kN-m**.

TABLE EX2.14.1
Determination of neutral axis depth for Ex2.14.1 (X-beams)

x	ε_{st}	f_{st}	T	ε_{sc}	f_{sc}	C_s	C_c	C	C ≈ T?
mm		$0.87f_s$	kN		MPa	$f_{sc}A_{sc}$	$0.36f_{ck}bx$	$C_s + C_c$	
150	0.007163	435.0	**409.8**	0.002497	385.6	242.2	567	**809.1**	No
100	0.012495	435.0	**409.8**	0.001995	310	194.7	378	**572.7**	No
90	0.014272	435.0	**409.8**	0.001828	296	185.9	340.2	**526.1**	No
70	0.019350	435.0	**409.8**	0.001350	270	169.6	264.6	**434.2**	No
67	0.020373	435.0	**409.8**	0.001254	251	157.5	253.3	**410.7**	Yes

TABLE EX2.14.2
Determination of neutral axis depth for Ex2.14.1 (Y-beams)

x	ε_{st}	f_{st}	T	ε_{sc}	f_{sc}	C_s	C_c	C	C ≈ T?
mm		$0.87f_s$	kN		MPa	$f_{sc}A_{sc}$	$0.36f_{ck}bx$	$C_s + C_c$	
90	0.012328	435.0	**262.3**	0.001828	357	143.5	291.6	**435.1**	No
70	0.01685	435.0	**262.3**	0.00135	270	108.5	226.8	**335.3**	No
60	0.020242	435.0	**262.3**	0.000992	198.3333	79.7	194.4	**274.1**	No
58	0.02106	435.0	**262.3**	0.000905	181	72.8	187.92	**260.7**	Yes

Column at top floor: capacity along X-direction

Column is 400×450 mm in size. Axial force $P = 300$ kN. Factored axial force $P_u = 1.5 \times 300 = 450$ kN. $d' = 40 + 8 + 20/2 = 58$ mm. $d'/D = 58/450 = 0.19$. Area of 8-20 mm bars $= 2513$ mm². Percentage steel in section, $p = A_s/bD = 100 \times 2513/(400 \times 450) = 1.4$. So, $p/f_{ck} = 0.047$.

$$\frac{P_u}{f_{ck}bD} = \frac{450 \times 1000}{30 \times 400 \times 450} = 0.083.$$

From Chart 50 of SP 16, $\dfrac{M_u}{f_{ck}bD^2} = 0.075$.

So, $M_u = 0.09 f_{ck}bD^2 = 0.075 \times 30 \times 400 \times 450^2$ N-mm $= $ **182.2 kN-m**.

Column at top floor: capacity along the Y-direction

Column is 450×400 mm in size. Axial force $P = 300$ kN. Factored axial force $P_u = 1.5 \times 300 = 450$ kN. $d' = 40 + 8 + 20/2 = 58$ mm. $d'/D = 58/400 = 0.15$. Area of 8-20 mm bars $= 2513$ mm². Percentage steel in section, $p = A_s/bD = 100 \times 2513/(400 \times 450) = 1.4\%$. So, $p/f_{ck} = 0.047$.

$$\frac{P_u}{f_{ck}bD} = \frac{450 \times 1000}{30 \times 400 \times 450} = 0.083.$$

From Chart 49 of SP 16, $\dfrac{M_u}{f_{ck}bD^2} = 0.09.$

So, $M_u = 0.09 f_{ck}bD^2 = 0.09 \times 30 \times 450 \times 400^2$ N-mm = **194.0 kN-m**.

Column at bottom floor: capacity along X-direction

Column is 400 × 450 mm in size. Axial force P = 400 kN. Factored axial force $P_u = 1.5 \times 400 = 600$ kN. $d' = 40 + 8 + 20/2 = 58$ mm. $d'/D = 58/450 = 0.19$. Area of 8-20 mm bars = 2513 mm². Percentage steel in section, $p = A_s/bD = 100 \times 2513/(400 \times 450) = 1.4$. So, $p/f_{ck} = 0.047$.

$$\frac{P_u}{f_{ck}bD} = \frac{600 \times 1000}{30 \times 400 \times 450} = 0.11.$$

From Chart 50 of SP 16, $\dfrac{M_u}{f_{ck}bD^2} = 0.08.$

So, $M_u = 0.09 f_{ck}bD^2 = 0.08 \times 30 \times 400 \times 450^2$ N-mm = **194.4 kN-m**.

Column at bottom floor: capacity along the Y-direction

Column is 450 × 400 mm in size. Axial force P = 400 kN. Factored axial force $P_u = 1.5 \times 400 = 600$ kN. $d' = 40 + 8 + 20/2 = 58$ mm. $d'/D = 58/400 = 0.15$. Area of 8-20 mm bars = 2513 mm². Percentage steel in section, $p = A_s/bD = 100 \times 2513/(400 \times 450) = 1.4\%$. So, $p/f_{ck} = 0.047$.

$$\frac{P_u}{f_{ck}bD} = \frac{600 \times 1000}{30 \times 400 \times 450} = 0.11.$$

From Chart 49 of SP 16, $\dfrac{M_u}{f_{ck}bD^2} = 0.1.$

So, $M_u = 0.1 f_{ck}bD^2 = 0.1 \times 30 \times 450 \times 400^2$ N-mm = **216 kN-m**.
The values computed are tabulated in Table Ex2.14.3.

C/B ratio in X-direction = 376.6/347.6 = 1.08.
C/B ratio in Y-direction = 410/196.8 = 2.08.

Comment: The prescribed value of C/B ratio is 1.3 as per EC-8. So, it is found that C/B ratio is not satisfied along the X-direction.

TABLE EX2.14.3
Beam and column moment capacities in Ex2.14.1

	Beam moment (kN-m)			Column moment (kN-m)		
Direction	From left	From right	Sum	From top	From bottom	Sum
X	173.8	173.8	347.6	182.2	194.4	376.6
Y	98.4	98.4	196.8	194.0	216	410.0

[The column needs to be strengthened along the X-direction. This is first done by increasing steel. But steel is generally limited to 3–4% of column gross sectional area. So, if necessary, the column size also may be increased.]

Example 2.14.2 *In a hysteretic curve, the area of the loop is 80 kN-m. The maximum displacement and force are 200 mm and 400 kN, respectively. Find the hysteretic damping.*

Solution: The strain energy $= E_{s,max} = \dfrac{1}{2} F_{max} u_{max} = \dfrac{1}{2} \times 400 \times 0.2 = 40$ kN-m.

Hysteretic damping $\xi_h = \dfrac{1}{4\pi} \dfrac{E_d}{E_{s,max}} = \dfrac{1}{4\pi} \dfrac{80}{40} = 0.159$ or 15.9%.

2.15   CLOSURE

In this chapter, a brief introduction has been given about PBSD. Detailed treatments are available in chapters to follow. The main points of focus in PBD can be expressed as: (i) deciding target design objectives and expected hazard level for the location of the structure and (ii) developing design process in PBD and evaluating the design product through nonlinear analyses. The analyses will reveal whether the targets have been reached or not. The analyses are to be carried out at MCE level. (iii) Implement the procedures in code framing for future world. The structure should be healthy to the extent that the economic loss following any major earthquake should be minimal.

The design should also consider multiple design objectives under various possible hazard levels. The hazard levels may be minor, moderate and major. These are understood in terms of the return periods. The societal, economic and hazardous consequences of failure should be kept in mind.

PBSD is a vast subject. Hence, the literature survey made here is far from complete. Some references are given here; the readers may search for more materials in journals, reports and internet.

2.16   EXERCISES

Q **2.16.1** Give an account of the journey of development of performance-based seismic design.

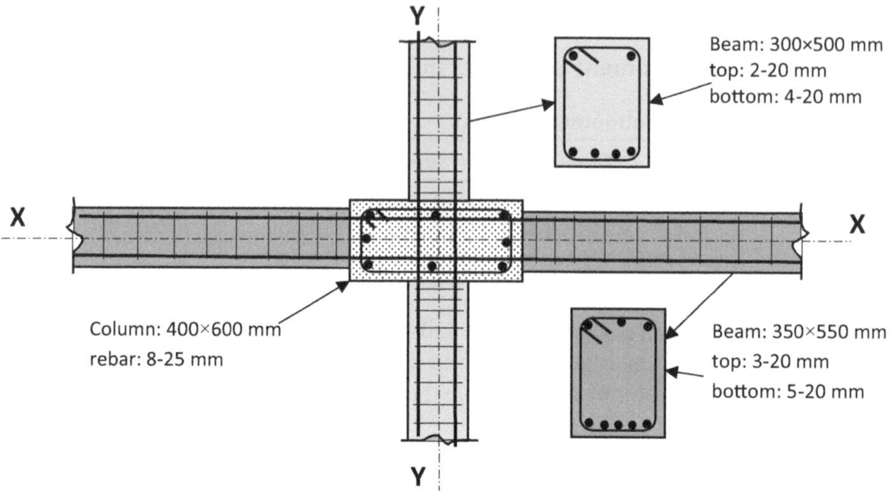

FIGURE 2.6   Beam and column details in Q2.16.4.

Q 2.16.2 Give a detailed account of journey of development of displacement-based seismic design.

Q 2.16.3 Highlight the advantages and applications of Unified performance-based design.

Q 2.16.4 In a beam–column joint, four beams meet from opposite sides and two columns extend to the upper and lower stories. The beams and column details are given in Figure 2.6. The clear cover in beam is 25 mm and that in column is 40 mm. Concrete grade is M30 and steel grade is Fe415. Find the C/B ratios and comment on the values obtained. The axial force in the top storey is 500 kN and that in bottom storey is 600 kN. The diameter of stirrups is 8 mm.

Q 2.16.5 What is yield point spectrum and what are its uses?

Q 2.16.6 Give an outline of capacity spectrum method.

Q 2.16.7 Give an outline of DDBD method.

Q 2.16.8 What is capacity design? Why it is needed?

Q 2.16.9 Explain "Strong-column Weak-beam" principle of design.

FURTHER READINGS

Anil, B. (2021) Unified Performance-Based Design of Steel Frame Buildings, P.G. dissertation supervised by S. Choudhury, Department of Civil Engineering, NIT Silchar, India.

Antoniou, S. and Pinho, R. (2004) Development of a Displacement-Based Adaptive Pushover Procedure, *Journal of Earthquake Engineering*, Volume 8, Issue 5, pp. 643–664.

Aschheim, M. and Black, E. (2000) Yield Point Spectra for Seismic Design and Rehabilitation", *Earthquake Spectra*, Volume 16, Issue 2, pp. 17–336.

ATC-3-06 (1978) Tentative Provisions for the Development of Seismic Regulations for Buildings, Applied Technology Council, California.

ATC-40 (1996) Seismic Evaluation and Retrofit of Existing Concrete Buildings, Applied Technology Council, USA.

Azeel, P.K. (2022) Unified Performance-Based Design of Steel Frame Buildings with Infill, P.G. dissertation, Department of Civil Engineering, NIT Silchar, India.

Banerjee, S. and Choudhury, S. (2020a) An Introduction to Unified Performance-Based Design of Bridge Piers, 17th World Conference on Earthquake Engineering, 17WCEE, Sendai, Japan- September 13 to 18, 2020, Paper ID 9e-0011.

Banerjee, S. and Choudhury, S. (2020b) Design of Bridge Pier using Unified Performance-Based Design method, Challenges of Resilient and Sustainable Infrastructure Development in Emerging Economies (CRSIDE 2020), Organized by ASCE India Section in Association with IIEST, Shibpur and IEI, March 2–4, 2020, p. 354.

Baruah, K.K. (2022) Unified Performance-Based Seismic Design for RC Elevated Water Tank with Frame Staging and Shaft Support, PG dissertation supervised by S. Choudhury, Department of Civil Engineering, NIT Silchar, India.

Blume, J.A., Newmark, N.M. and Coring, L.H. (1961) *Design of Multistory Reinforced Concrete Buildings for Earthquake Motions*, Portland Cement Association, Chicago, IL.

Bozorgnia, Y. and Bertero, V.V. (2004) *Earthquake Engineering from Engineering Seismology to Performance-Based Engineering*, CRC Press, New York.

Browning, J. (1999) Satisfying Performance Criteria for RC Frames Based on Allowable Drift, Report No. *PEER/99-10*, pp. 301–310.

Browning, J.P. (2001) Proportioning of Earthquake-Resistant RC Building Structures, *Journal of Structural Engineering, ASCE*, Volume 127, Issue 2 Feb. pp. 145–191.

Chopra, A.K. and Goel, R.K. (2001) Direct Displacement-Based Design: Use of Inelastic vs. Elastic Design Spectra, *Earthquake Spectra*, Volume 17, Issue 1, pp. 47–64.

Choudhury, S. (2007) Performance-Based Seismic Design of Hospitals, Ph.D. Thesis, Department of Earthquake Engineering, IIT Roorkee, 2008.

Choudhury, S. and Singh, S.M. (2013) A Unified Approach to Performance-Based Design of RC Frame Buildings, *Journal of Institution of Engineers (I)-Series-A*, May 2013, Volume94, Issue 2, pp. 73–82, DOI 10.1007/s40030-013-0037-8

Das, S. and Choudhury, S. (2019) Evaluation of Effective Stiffness of RC Column Sections by Support Vector Regression Approach, *Neural Computing & Applications*, April 2019, DOI 10:1007/s00521-019-04190-0

Das, S., Mansouri, I., Choudhury, S., Gandomi, A.H. and Hu, J.W. (2021) A Prediction Model for the Calculation of Effective Stiffness Ratios of Reinforced Concrete Columns, *Materials Journal* (SCIE). Volume 14, Issue 7, April. https://doi.org/10.3390/ma1 4071792

Deierlein, G.G., Krawinkler, H. and Cornel, C.A. (2003) A Framework for Performance-Based Earthquake *Engineering, 2003 Pacific Conference on Earthquake Engineering*, John A. Blume Earthquake Engineering Center, Stanford University, Stanford, CA.

FEMA-150 (1990) Seismic Considerations – Health Care Facilities, US Federal Emergency Management Agency.

FEMA-273 (1996) NEHRP Guidelines for the Seismic Rehabilitation of Buildings, US Federal Emergency Management Agency, Building Seismic Safety Council, Washington, DC.

FEMA-349 (2000) Action Plan for Performance Based Seismic Design, US Federal Emergency Management Agency, Earthquake Engineering Research Institute.

FEMA-356 (2000) Prestandard and Commentary for the Seismic Rehabilitation of Buildings, US Federal Emergency Management Agency.

FEMA-368 (2001) NEHRP Recommended Provisions for Seismic Regulations for New Buildings and other Structures, US Federal Emergency Management Agency.

FEMA-389 (2004) Premier for Design Professionals, US Federal Emergency Management Agency.

FEMA-440 (2004) Improvement of Nonlinear Static Seismic Analysis Procedures, ATC-55 Project, Applied Technology Council and US Federal Emergency Management Agency.

FEMA-450 (2004) NEHRP Recommended Provisions for Seismic Regulations for New Buildings and other Structures, US Federal Emergency Management Agency.

Freeman, S.A. (1978) Prediction of Response of Concrete Buildings to Severe Earthquake Motion, Publication SP-55, American Concrete Institute, MI 589-606.

Freeman, S.A. (1998) Development and Use of Capacity Spectrum Method, 6th US National Conference on Earthquake Engineering.

Freeman, S.A., Nicoletti, J.P. and Tyrell, J.V. (1975) Evaluation of Existing Buildings for Seismic Risk: A Case Study of Puget Sound Naval Shipyard, Bremerton, Washington, *Proceedings of the US National Conference on Earthquake Engineers*, EERI, Berkeley, CA.

Kamil, Md Waseeuddin (2022) Effect of Infill on Unified Performance-Based Design of Frame-Wall Building, PG dissertation supervised by S. Choudhury, Department of Civil Engineering, NIT Silchar, India.

Kassim, Azeel P. (2022) Unified Performance-Based Design of Steel Frame Building with Masonry Infill Walls, PG dissertation supervised by S. Choudhury, Department of Civil Engineering, NIT Silchar, India.

Kilar, V. and Fajfar, P. (1997) Simple Pushover Analysis of Asymmetric Buildings, *Earthquake Engineering and Structural Dynamics*, Volume 26, pp. 233–249.

Kotapaty, V.K. (2015) Unified Performance-Based Design of Steel Frame Buildings, P.G. dissertation supervised by S. Choudhury, Department of Civil Engineering, NIT Silchar, India.

Mayengbam, S.S. and Choudhury, S. (2014) Determination of Column Size for Displacement-based Design of Reinforced Concrete Frame Buildings, *Journal of Earthquake Engineering & Structural Dynamics*, July 2014, Volume 43, Issue 8, pp. 1149–1172. Article first published online: 21 Nov 2013 | DOI: 10.1002/eqe.2391

Medhekar, M.S. and Kennedy, D.J.L. (2000) Displacement-Based Seismic Design of Buildings – Applications, *Engineering Structures*, Volume 22, pp. 210–221.

Mibang, Durga and Choudhury, Satyabrata (2019a) Performance-Based Design of Dual System, International Conference on Recent Development in Sustainable Infrastructure (Materials and Management) (ICRDSI-2019), Kalinga Institute of Technology, Bhubaneswar, 11–13th July 2019, Paper ID 132.

Mibang, Durga and Choudhury, Satyabrata (2019b) Performance of Dual System Designed Using UPBD Method, 2nd International Conference on Recent Advancements in Interdisciplinary Research (ICRAIR-2019), Asian Institute of Technology Conference Center, Thailand, 1–2 June 2019, Paper ID-69, pp. 327–333.

Panagiatakos, T.B. and Fardis, M.N. (1999) Deformation Controlled Earthquake Resistant Design of RC Buildings, *Journal of Earthquake Engineering*, Volume 3, Issue 4, pp. 498–518.

Park, R. and Paulay, T. (1974) "Reinforced Concrete Structures", John Wiley and Sons.

Pettinga, J.D. and Priestley, M.J.N. (2005) Dynamic Behaviour of Reinforced Concrete Frames Designed with Direct Displacement Pert-Based Design, *Journal of Earthquake Engineering*, Volume 9, Special Issue 2, pp. 309–330.

Priestley, M.J.N. (1993) "Myths and Fallacies in Earthquake Engineering", *Bulletin of NZ National Society for Earthquake Engineering*, Volume 26, Issue 3, pp. 329–341.

Priestley, M.J.N. (2000) Performance Based Seismic Design, *12th World Conference on Earthquake Engineering*, Paper No. 2831.

Priestley, M.J.N. (2003a) Myths and Fallacies in Earthquake Engineering, Revisited, *European School for Advanced Studies in Reduction of Seismic Risk*, 9th Mallet-Milne Lecture.

Priestley, M.J.N. (2003b) Direct Displacement-Based Design, a Rational Seismic Design Approach, *European School for Advanced Studies in Reduction of Seismic Risk*, 9th Mallet-Milne Lecture.

Priestley, M.J.N., Calvi, G.M. and Kowwalsky, M.J. (2007) *Displacement-Based Seismic Design of Structures*, IUSS Press, Pavia, Italy.

Qi, X. and Moehle, J.P. (1991) *Displacement Design Approach for Reinforced Concrete Structures Subjected to Earthquakes*, University of California, Berkeley Earthquake Engineering Research Center.

Sullivan, T.J. and Lago A. (2012) Toward a Simplified Direct DBD Procedure for the Seismic Design of Moment Resisting Frames with Fluid Viscous Dampers, *Engineering Structures*, Volume 35, pp 140–148.

Sullivan, T.J., Priestley, M.J.N. and Calvi, G.M. (2006a) Direct Displacement-Based Design of Frame-Wall Structures", *Journal of Earthquake Engineering,* Volume 10, Special Issue 1, pp. 91–124.

Sullivan, T.J., Priestley, M.J.N. and Calvi, G.M. (2006b) Seismic Design of Frame-Wall Structures", Research Report No. ROSE-2006/02.

Vamvastikos, D. and Cornell, C.A. (2002), Incremental Dynamic Analysis, *Earthquake Engineering and Structural Dynamics*, Volume 31, Issue 3, pp. 491–514.

Xue, Q. (2001) A Direct Displacement-Based Seismic Design Procedure of Inelastic Structures", *Engineering Structures*, Volume 23, pp. 1453–1460.

Xue, Q. and Chen C.C. (2003) Performance-Based Seismic Design of Structures: A Direct Displacement-Based Approach", *Engineering Structures*, Volume 25, pp. 1803–1813.

3   Hazard Considerations

3.1 INTRODUCTION

In seismic design, hazard means the seismicity levels of ground motions. Ground motions are of two categories: (i) weak ground motions, which are of interest to the seismologists, and (ii) strong ground motions, which are of interest to the earthquake engineers and structural engineers. Whether the design is carried out with the codal prescriptive method or using PBD, hazard remains associated in the background. In fact, PBD is the design process accommodating target design criteria to be satisfied at any perceived hazard level. The codal design spectrum indicates some hazard level. The hazard level can be varied by using the seismic zone factor (as in Indian code) or through specifying design seismicity level in terms of g (as in Eurocode).

Seismic hazard has been defined by ASCE-SEI-41-17 as follows:

> Seismic hazard caused by ground shaking shall be defined as acceleration response spectra or ground motion acceleration histories determined on either a probabilistic or deterministic basis.

Here, a refreshing the concept of response spectrum is relevant. *Response spectrum* is the graphical representation of some *peak response quantity* (such as peak displacement, peak velocity or peak acceleration) of a series of *single degree of freedom systems* (also called oscillators) of varying *time periods* with a particular *damping* under a given *earthquake*. According to the peak response quantity chosen, the response spectrum may be a displacement response spectrum, velocity response spectrum or acceleration response spectrum. The response spectra vary with the earthquakes. Again, for a given earthquake, the response spectra will be different for different dampings values associated with the single degree of freedom system. Typically, response spectrum is constructed for 5% damped elastic response. If the oscillator is taken inelastic (that is, it yields), the resultant response spectrum will be called *inelastic response spectrum*. The velocity response spectrum is more zig-zag in nature than the corresponding displacement response spectrum and, acceleration response spectrum is more zig-zag in nature than the velocity response spectrum. A typical acceleration response spectrum is shown in Figure 3.1

It is to be noted that the acceleration response spectrum is of more relevance in design. In the response spectrum, we do not represent actual acceleration response

DOI: 10.1201/9781003441090-3

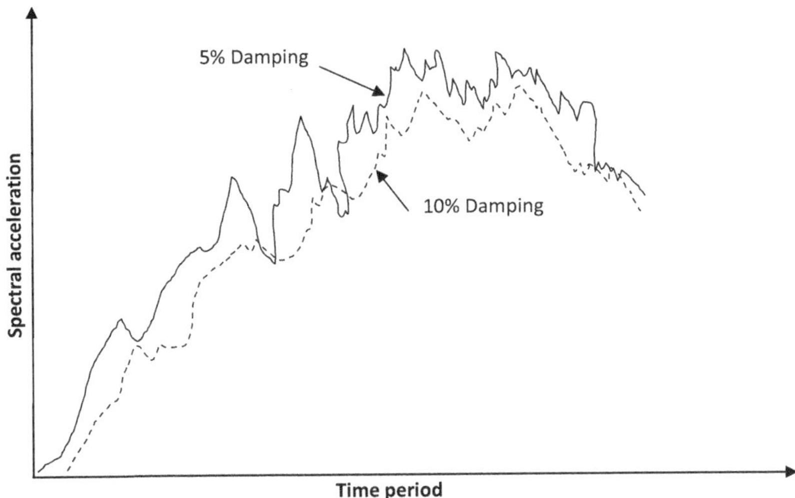

FIGURE 3.1 Typical acceleration response spectra.

but pseudo-acceleration response, or spectral acceleration, denoted by S_a. The spectral velocity is denoted by S_v and spectral displacement is denoted by S_d. The spectral acceleration and spectral displacement are related by Eq. (3.1.1). This equation we use in displacement-based design for the construction of displacement spectra from design spectrum is

$$S_a = \omega^2 S_d \tag{3.1.1}$$

Design spectrum forms the basis for designing structures in a country. Design spectrum is generally a plot between time period and spectral acceleration ratio (S_a/g). Design spectrum is the average smoothened curve for several response spectra (generally at 5% damping) generated from a host of real past earthquakes or artificially generated probable future earthquake for a country. The design spectrum will vary with soil type. In Indian code (IS 1893-2016, part 1), three spectra are given – for hard soil, medium soil and soft soil. Due to soil amplification, the design spectrum for soft soil has higher magnitude of spectral acceleration. The design spectrum is also called as 5% damped elastic demand curve.

Site spectrum is the response spectrum for a particular site. This is used for very important project or very important structures. Actually, the design spectrum given in the code of a country is the average representation of demand on structures. But it does not cover local variations in soil characteristics and seismicity. This may lead to unsafe structure even if designed through codal provisions. So, for important project sites, the geological conditions, soil conditions, seismic hazard source locations (faults) etc. are studied meticulously and based on the data collected, a site-specific response spectrum is drawn. In most cases, the site-specific spectrum gives higher demand than the codal design spectrum. If demand from the site-specific spectrum is lower than that in the codal spectrum, codal demand spectrum is to be used. Site spectra are used for nuclear power projects, other important projects and for the projects the failure of which will lead to hazardous consequences.

3.2 NOMENCLATURE OF HAZARD LEVELS

The hazard levels have been designated in FEMA documents and different national codes based on the mean return period or probability of exceedance. Considering the life span of structure as 50 years, the probability of exceedance of an earthquake as $p\%$ is denoted by $p\%$/50 year. This has an associated mean return period. It may be noted that the return period is an approximate proposition valid when there is one or two sources (fault) of earthquakes at a site. However, this phrase has long been used and may be passed for a qualitative description of the probable return period.

Serviceability earthquake (SE) is that level of hazard which has 50% chance of exceedance in 50 years (50%/50 year). This is frequent earthquake that is likely to occur during the life of the structure. It has mean return period of 75 years.

Design earthquake (DE) is that level of hazard which has a 10% chance of exceedance in 50 years (10%/50 year). This is infrequent earthquake that is likely to occur during the life of the structure. It has a mean return period of 500 years. It may be noted that BSE-1 of FEMA-273 and FEMA-356 corresponds to DE of ATC-40.

Maximum earthquake (ME) is that level of hazard which has 5% chance of exceedance in 50 years (5%/50 year). This is infrequent earthquake that may damage the structure. This definition has no parallel in FEMA-273 and FEMA-356 documents.

FEMA-273 and FEMA-356 have defined Basic Safety Earthquake-1 (BSE-1) as 10%/50 year.

The hazard nomenclature as per various documents are tabulated in Table 3.1.

TABLE 3.1
Hazard nomenclature and probability of exceedance

Earthquake having probability of exceedance	Mean return period (years)	Generally considered rounded off mean return period (years)	Hazard designators		
			ATC-40	FEMA-273	FEMA-356
50%/50 year	72	75	SE	-	-
20%/50 year	225	225	-	-	-
10%/50 year	474	500	DE	BSE-1	BSE-1
5%/50 year	-	-	ME	-	-
2%/50 year	2475	2500	-	BSE-2	BSE-2

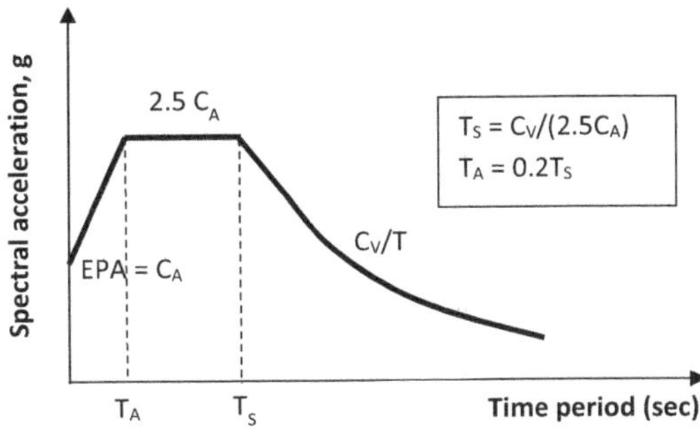

FIGURE 3.2 General elastic 5% damped response spectrum as per ATC-40 (schematic).

3.3 HAZARD CONSIDERATION IN ATC-40 (1996)

ATC-40 classified hazard as serviceability earthquake (SE), design earthquake (DE) and maximum earthquake (ME). ME has been named as maximum capable earthquake (MCE) in some documents. It may be noted that SE = 0.5 × DE and ME = 1.25 × DE to 1.5 × DE. The soil categories are classified as: hard rock, rock, very dense soil and soft rock, for which standard penetration test (SPT) number is greater than 50; stiff soil (SPT between 15 and 50) and soft soil (SPT less than 15). The categories of soil also have corresponding shear wave velocities. Some soil may require site soil investigation.

In ATC-40, the elastic response spectrum is given in terms of two coefficients C_A and C_V, as described in Eqs. (3.3.1) and (3.3.2). The coefficient C_A represents effective peak ground acceleration.

$$C_A = 0.4S_{MS} \qquad (3.3.1)$$

$$C_V = S_{M1} \qquad (3.3.2)$$

where S_{MS} is the spectral acceleration ratio in the short period range for site class B for MCE (BSSC 1996) and S_{M1} is the spectral acceleration at 1.0 second for site class B for MCE (BSSC 1996).

The values of C_A and C_V are given in ATC-40. The value of CA varies from 0.08 to 1.0 depending on the shaking intensity, which is proportional to the shaking intensity. The value of C_V varies from 0.08 to 2.4 depending on the shaking intensity.

The general shape of elastic response spectrum as per ATC-40 is shown in Figure 3.2.

3.4 HAZARD DESCRIPTION IN ASCE-SEI-7-16

3.4.1 SEISMIC RISK CATEGORIES

ASCE-7-16 deals with the design loads including seismic loads. The buildings have been categorized into A, B, C and D and the anticipated risk categories are I to IV.

The *risk categories* are: (i) risk category I – low-risk buildings and other structures, (ii) risk category II – buildings and other structures not belonging to I, III and IV categories, (iii) risk category III – high-risk buildings and other structures, (iv) risk category IV – very high-risk buildings and other structures designated as essential facilities.

3.4.2 Seismic Design Category

The seismic design category of buildings or site classes are A, B, C and D. Seismic design categories are decided based on the spectral acceleration at short period (S_{DS}) and 1 second period (S_{D1}).

The design response spectrum as per ASCE-7-16 is schematically shown in Figure 3.3. The parameters involved in the process are described below in Eqs. (3.4.1) to (3.4.4).

With reference to Figure 3.3,

$$T_0 = 0.2 S_{D1} / S_{DS}$$

where T_0 = fundamental time period

$T_S = S_{D1} / S_{DS}$ = characteristic period or corner period

F_a = short period site coefficient (at 0.2 sec)

F_v = long period site coefficient (at 1.0 sec)

T_L = Long-period transition period

S_{D1} = spectral acceleration at 1.0 sec period in 5% damped design spectrum

S_{DS} = short period spectral acceleration in 5% damped design spectrum

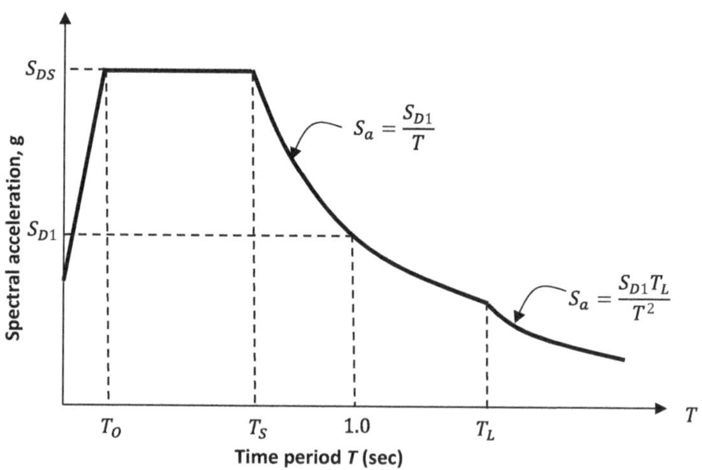

FIGURE 3.3 Design response spectrum as per ASCE-7-16 (schematic).

$$S_{DS} = \frac{2}{3} S_{MS}$$

$$S_{D1} = \frac{2}{3} S_{M1}$$

$$S_{MS} = F_a S_S$$

$$S_{M1} = F_v S_1$$

S_S = mapped maximum considered earthquake spectral acceleration at a short period

S_S = mapped maximum considered earthquake spectral acceleration at 1 sec period

3.5 HAZARD DESCRIPTION IN ASCE-SEI-41-17

The schematic hazard in terms of demand diagram as per ASCE-SEI-41-17 is shown in Figure 3.4. The parameters used in this figure are described below.

S_{xs} = short-period spectral response acceleration parameter

S_{x1} = design spectral response acceleration parameter at a 1-second period; it is design long-period response acceleration parameter

$B_1 = \dfrac{4}{5.6 - \ln(100\xi)}$, where, ξ is effective viscous damping ratio; B_1 is damping

coefficient used to adjust spectral response for the effect of viscous damping

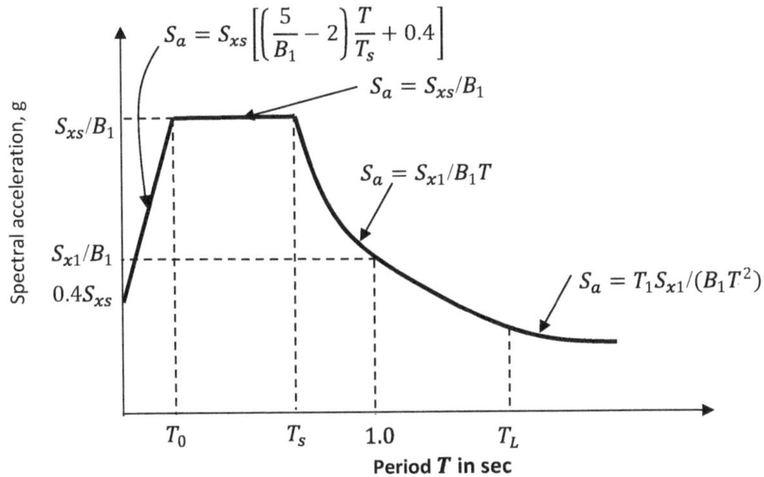

FIGURE 3.4 General horizontal response spectrum as per ASCE-SEI-41-17 (schematic).

T_0 = Period at which the constant acceleration region of the design response spectrum begins,

T = time period

T_s = Characteristic period of the response spectrum, defined as the period associated with the transition from the constant acceleration segment of the spectrum to the constant velocity segment of the spectrum

T_L = long-period transition parameter to be obtained from published maps, site-specific response analysis or any other method approved by the authority having jurisdiction.

3.6 HAZARD DESCRIPTION IN EURO CODE 8 (2004)

3.6.1 Ground Types

Ground types are subdivided as Type A (rock or other rock-like soil), B (very dense sand, gravel or very stiff clay), C (deep dense or medium-dense sand, gravel or stiff clay), D (dense to medium cohesionless soil), E (alluvium layer), S_1 (soft clay/silt) and S_2 (liquefiable soil). There are corresponding shear wave velocity and SPT related to each category.

The design spectrum of EC-8 is shown in Figure 3.5 in a schematic way. The variables involved in this figure are explained below.

S_e = elastic spectral acceleration
S = soil factor
T = time period in seconds.
T_B = period at start of constant acceleration branch

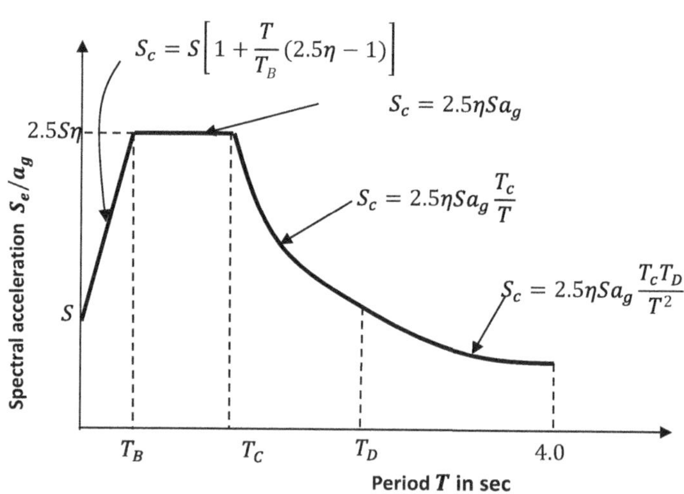

FIGURE 3.5 Design spectrum as per EC-8 (schematic).

T_C= period at end of constant acceleration branch
T_D= period at start of constant displacement range
a_g = design acceleration on type A soil
η = damping correction factor (η = 1.0 for 5% damping)
The numerical values of the parameters are to be found in the code.

3.7   HAZARD DESCRIPTION IN IS 1893 (PT 1) – 2016

Indian Standard IS 1893 (Pt 1) – 2016 has considered four seismic zones in India, namely, Zone II, Zone III, Zone IV and Zone V, in the increasing order of seismicity. Zone I is missing as it was merged with Zone II earlier. The design spectra are given in two categories: (i) for use in equivalent static load method and (ii) response spectrum method. The first one is applicable in limited cases only – namely, building with time period not exceeding 0.4 s. The second method is applicable to all cases. The two categories of design spectrum are given in mathematical expressions for rocky of hard soil, for medium soil and soft soil. The two spectra drawn in excel using expressions of the code are shown in Figures 3.6 and 3.7.

The base shear is given by Eq. (3.7.1):

$$V_b = \frac{Z}{2}\frac{I}{R}\frac{S_a}{g}W \tag{3.7.1}$$

where V_b is the base shear, Z is the zone factor (values are 0.10, 0.16, 0.24 and 0.36 for four zones), I is the important factor, R is response reduction factor, S_a is spectral acceleration, g i s acceleration due to gravity and W is seismic weight.

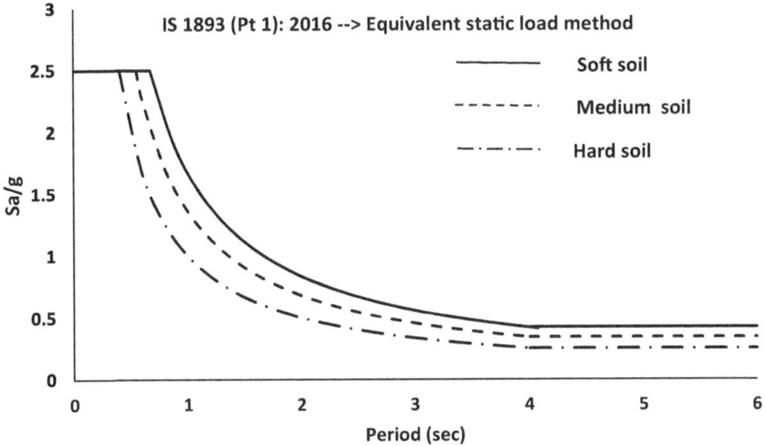

FIGURE 3.6   Schematic design spectrum as per IS 1893 (Pt 1)-2016 for equivalent static method.

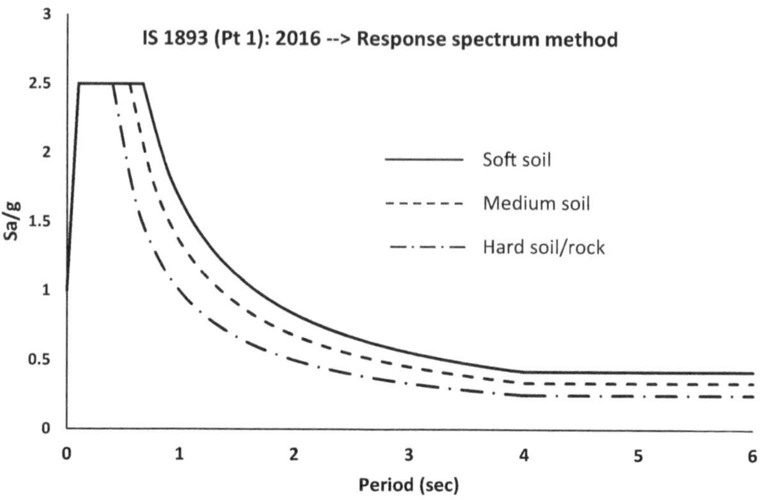

FIGURE 3.7 Schematic design spectrum as per IS 1893 (Pt 1)-2016 for response spectrum method.

3.8 NEW ZEALAND CODE: NZS 1170-5 (2004)

The New Zealand code considers two limit states to be satisfied in the design: the serviceability limit states and ultimate limit states. Ultimate limit state corresponds to probability of fatality as 10^{-6} in the collapse of a building. The code gives seismic design spectra for equivalent static method and response spectrum method (so also for time history analysis) separately. The spectra are given for strong rock (soil class A), rock (soil class B) shallow soil (soil class C), deep or soft soil (soil type D) and very soft soil (soil class E) soil. The design spectra are given in log-log scale. A schematic diagram of the New Zealand design spectra is shown in Figure 3.8.

3.9 CONSTRUCTION OF DISPLACEMENT RESPONSE SPECTRA

In DBD and PBD, displacement response spectra are used instead of acceleration spectra. The acceleration spectra of code are converted to displacement spectra for various dampings.

The acceleration spectra in codes are given for 5% damping in elastic condition. The acceleration spectra at other damping are obtained from Eq. (3.9.1). Here, S_a is spectral acceleration, ξ is critical damping ratio.

$$S_{a,\xi\%} = \sqrt{\frac{10}{5 + \xi\%}} \times S_{a,5\%}$$

(3.9.1)

The conversion of acceleration spectra to displacement spectra is carried out by using Eq. (3.9.2). In this equation, S_d is spectral displacement. Note that in the conversion, the acceleration axis becomes spectral displacement axis. The T-axis does not change.

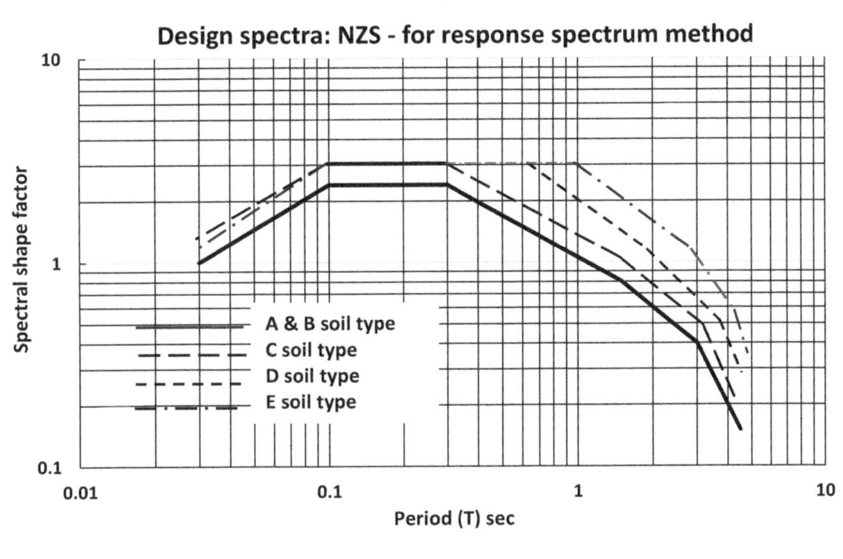

FIGURE 3.8   Design spectra as per the New Zealand code (schematic).

$$S_d = \frac{S_a}{\omega^2} = \frac{T^2}{4\pi^2} S_a \tag{3.9.2}$$

Using Eqs. (3.9.1) and (3.9.2), the displacement spectra for EC-8 for 5% damping for B type soil at various seismicity level is shown in Figure 3.9. For a particular seismicity level (0.45g), the displacement spectra for EC-8 for various dampings are shown in Figure 3.10.

The spectral reduction factor, which is to be applied for getting acceleration spectra or displacement spectra, is be given by Eq. (3.9.3) (EC-8). In Eq. (3.9.3), ξ is is in a ratio form:

$$R_\xi = \sqrt{\frac{0.1}{0.05 + \xi}} \tag{3.9.3}$$

For ground sites with forward directivity velocity pulse, Priestley (2003) suggested Eq. (3.9.4).

$$R_\xi = \left(\frac{0.07}{0.02 + \xi}\right)^{1/4} \tag{3.9.4}$$

The displacement spectra for IS 1893-2016 for medium soil is shown in Figure 3.11.

Following Eq. (3.9.3), the variation of spectral reduction factor is drawn in Figure 3.12.

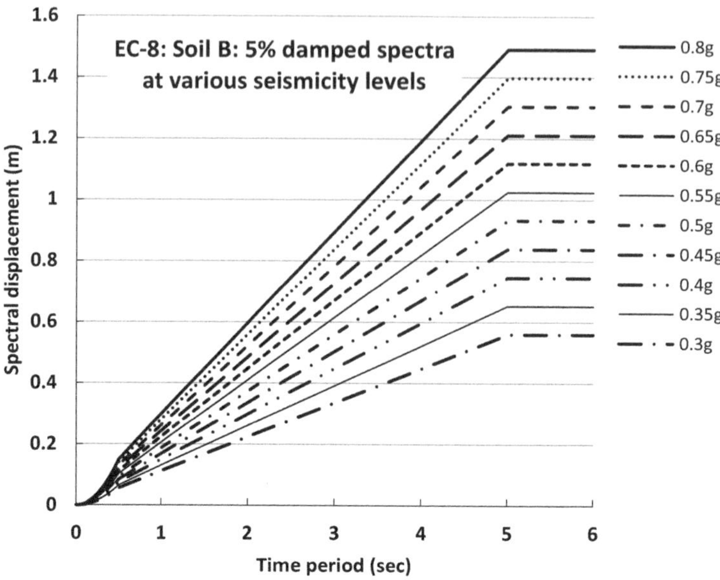

FIGURE 3.9   Displacement spectra for EC-8 at 5% damping and various seismicity levels.

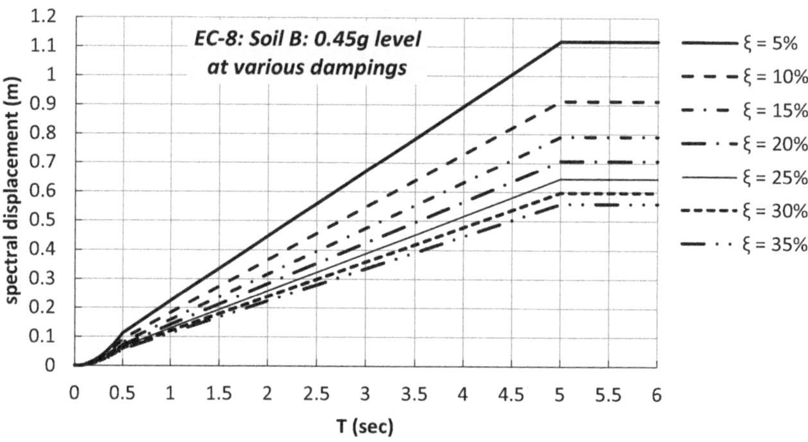

FIGURE 3.10   EC-8 displacement spectra at 0.45g level for various dampings.

3.9.1 STEPS IN THE CONSTRUCTION OF DISPLACEMENT SPECTRA

The steps involved in the generation of displacement spectra are as follows:

(1) Select the design spectrum for which the displacement spectra to be developed. Such design spectrum is developed in terms of time period (T) and spectral acceleration ratio (S_a / g).

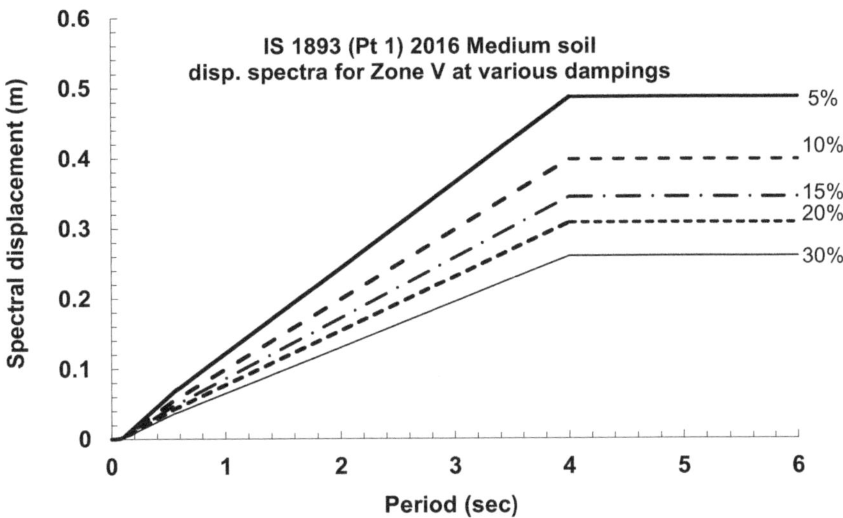

FIGURE 3.11 Displacement spectra for IS 1893-2016 for medium soil.

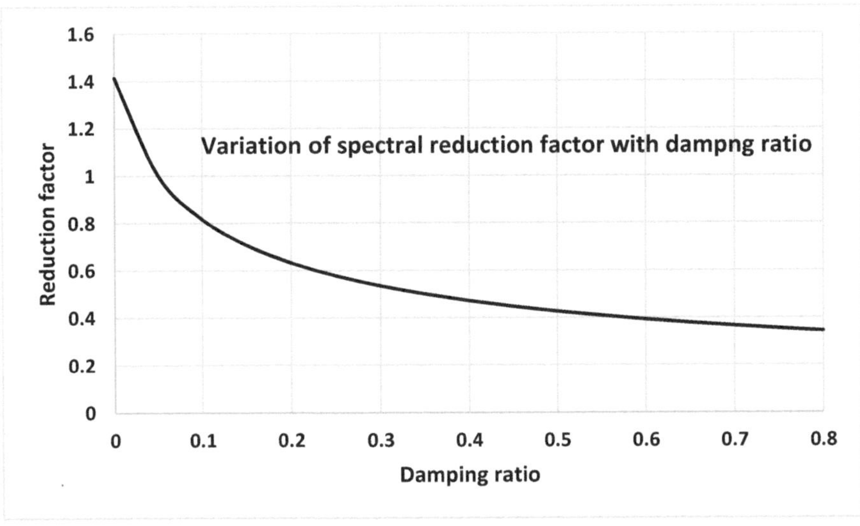

FIGURE 3.12 Variation of spectral reduction factor with damping ratio (Eq. (3.9.3)).

(2) Use the relationship $S_d = \dfrac{T^2}{4\pi^2} S_a$ to get spectral displacement (S_d) for each value of spectral acceleration (S_a) for a given T.

(3) Plot the displacement spectrum with x-axis as T and y-axis as S_d. This completes the displacement spectrum at 5% damping.

(4) For displacement spectra at other damping, use the relationship

$S_{a,\xi\%} = \sqrt{\dfrac{10}{5+\xi\%}} \times S_{a,5\%}$, to get spectral acceleration ratio at other dampings.

Now go to step (2).

3.10 CLOSURE

Hazard analysis is a primary requirement in seismic design. The hazard analysis may be deterministic or probabilistic. For seismic hazard analysis, the readers may refer books on geotechnical earthquake engineering. True hazard analysis involves the study of geology, seismology and geotechnical earthquake engineering. A basic idea about hazard is presented in this chapter. HAZUS (Hazard US) may be consulted for more details.

3.11 EXAMPLES

Ex 3.11.1 *The spectral acceleration ratio corresponding to a design spectrum at 5% damping is 5 m/s². Find the spectral acceleration ratio at 12% damping.*

Solution
Here, $S_{a,5\%} = 5$ m/s².

Now, $S_{a,\xi\%} = \sqrt{\dfrac{10}{5+\xi\%}} \times S_{a,5\%}$

So, $S_{a,12\%} = \sqrt{\dfrac{10}{5+12}} \times 5 = 3.834$ m/s².

Ex 3.11.2 *A ductile RC residential building of seismic weight W with infill is 20 m high. The base dimension is 12 m × 12 m. The building rests on medium soil and falls in seismic Zone V of Indian code. The dynamic analysis shows that the time period is 0.6 sec. Examine the base shear correction requirement.*

Solution
For Indian seismic zone V, zone factor $Z = 0.36$.
For ductile building the response reduction factor $R = 5.0$.
For residential building the importance factor $I = 1.0$.
For medium soil and for response spectrum analysis, the spectral acceleration ratio

$$\frac{S_a}{g} = 1.36/T = 1.36/0.6 = 2.267.$$

Now the base shear $= V_b = \dfrac{Z}{2}\dfrac{I}{R}\dfrac{S_a}{g}W = \dfrac{0.36}{2}\dfrac{1.0}{5.0} \times 2.267W = 0.082\ W.$

The approximate time period formula for buildings with infill is

$$T_a = \frac{0.09H}{\sqrt{B}}$$ where H is building height and B is building width, both in metres.

$$T_a = \frac{0.09 \times 20}{\sqrt{12}} = 0.520 \, \text{sec.}$$

The spectral acceleration ratio corresponding to this time period is

$$\frac{S_a}{g} = 1.36 / T_a = 1.36 / 0.52 = 2.615.$$

The base shear $= \bar{V}_b = \dfrac{Z}{2} \dfrac{I}{R} \dfrac{S_a}{g} W = \dfrac{0.36}{2} \dfrac{1.0}{5.0} \times 2.615 W = 0.094 \, W.$

As $\bar{V}_b > V_b$, base shear correction is necessary.

The base shear correction factor $= \bar{V}_b / V_b = 0.094W/0.082W = 1.146.$

All the response quantities like bending moment and shear has to be now multiplied by 1.146 for design purposes.

Ex 3.11.3 *A ductile RC residential building has initial elastic time period as 0.7 s. At damage state for which it is designed, it shows an effective damping of 12% at a displacement of 0.35 m. Find the design base shear as per displacement-based design. Take EC-8 spectrum for B type soil at 0.45g seismicity level.*

Solution

Consider the displacement spectra in Figure 3.10.

Corresponding to displacement (spectral) of 0.35 m and damping of 12%, the effective time period is 2.2 sec. Take effective mass as 500 kg.

The base shear is given by

$$V_b = k_e \Delta_d = \frac{4\pi^2 m_e}{T_e^2} \Delta_d \quad \left[\text{see Chapter 7}\right]$$

$$= \frac{4\pi^2 \times 500}{2.2^3} \times 0.35 = 4078 \, \text{kN}.$$

3.12 EXERCISES

Q3.1 Explain the terms: hazards, response spectrum, design spectrum, site spectrum, design earthquake, basic safety earthquake, maximum earthquake.

Q3.2 Briefly discuss the hazard as per (i) ATC-40, (ii) ASCE-SEI-7, (iii) ASCE-SEI-41-17, (iv) EC-8 and (vi) IS 1893.

Q3.3 Write down the steps involved in the construction of displacement spectra.

Q3.4 What is the advantage of using displacement spectra? [see Chapters 7–10].

FURTHER READINGS

ASCE-SEI-7 (2010) Minimum Design Loads and Associated Criteria for Buildings and Other Structures.

ASCE-SEI-41-17 (2017) Seismic Evaluation and Retrofit of Existing Buildings.

ATC-40 (1996) Seismic Evaluation and Retrofit of Existing Concrete Buildings, Applied Geochronology Council.

Euro Code 8 (EC-8) (2003) Design of Structures for Earthquake Resistance, Part 1: General Rules, Seismic Actions and Rules for Buildings, prEN 1998-1. December.

FEMA-273 (1996) NEHRP Guidelines for the Seismic Rehabilitation of Buildings, US Federal Emergency Management Agency, Building Seismic Safety Council, Washington DC.

FEMA-349 (2000) Action Plan for Performance Based Seismic Design, US Federal Emergency Management Agency, Earthquake Engineering Research Institute.

FEMA-356 (2000) Prestandard and Commentary for the Seismic Rehabilitation of Buildings, US Federal Emergency Management Agency.

FEMA-368 (2001) NEHRP Recommended Provisions for Seismic Regulations for New Buildings and other Structures, US Federal Emergency Management Agency.

FEMA-389 (2004) Premier for Design Professionals, US Federal Emergency Management Agency.

FEMA-396 (2003) Incremental Seismic Rehabilitation of Hospital Buildings, December, US Federal Emergency Management Agency.

FEMA-450 (2004) NEHRP Recommended Provisions for Seismic Regulations for New Buildings and other Structures, US Federal Emergency Management Agency.

IBC 2003 (2006) International Building Code, International Code Council.

IS 1893 (Part 1) – 2016: Criteria for Earthquake Resistant Design of Structures, BIS.

Kramer S.L. (1996) *Geotechnical Earthquake Engineering*, Prentice Hall International Series.

NZS 1170-5 (2004) Structural Design Actions: Earthquake Actions.

Priestley, M.J.N. (2003) Myths and Fallacies in Earthquake Engineering, Revisited, *European School for Advanced Studies in Reduction of Seismic Risk*, 9th Mallet-Milne Lecture.

4 Modelling Aspects

4.1 INTRODUCTION

It is difficult to exactly replicate a real structure in a computer model. Some simplifying assumptions are made to make a viable model of the prototype structure. The model should be able to capture the behaviour of real structure as realistically as possible. Modelling is necessary for (i) analysis and design and (ii) evaluation of the designed structure. Evaluation may involve (a) elastic response and (b) inelastic response. Inelastic model is also called the nonlinear model. Nonlinearity may arise out of (a) material nonlinearity (response beyond elastic limit of the material), (b) geometric nonlinearity (mostly P-delta effect and large deflection) and (c) nonlinearity due to contact and boundary. Material nonlinearity occurs due to yielding of material and subsequent formation of plastic condition in sections of elements of the structure, either partly or fully. In this chapter, we discuss various aspects of nonlinear modelling.

An analysis is said to be nonlinear when the relationship between the applied action and response (displacement) is nonlinear. In solid mechanics, nonlinearity means a nonlinear relationship between stress and strain. Any material will undergo yielding when sufficient force is applied in the section. This will lead to material nonlinearity. Material nonlinearity, however, is not restricted to yielding only. Creep and visco-plasticity also will lead to material nonlinearity. Various nonlinear effects may occur in combination also. Material nonlinearity will make change in stiffness. With increase of nonlinearity, the stiffness reduces. Contact nonlinearity is demonstrated by friction devices.

Geometric nonlinearity involves large deformation or large strains. If the magnitude of the deformation is such that it alters the response of the structure, it is a case of geometric nonlinearity. P-delta effect is a typical example of geometric nonlinearity. Nonlinear analysis needs to go step by step as the stiffness or deformation will change every time.

Nonlinear analysis may be dynamic when the applied force is time-dependent; or it may be static nonlinear analysis when the applied force is static, but changes from step to step. Nonlinear time history analysis (NLTHA) can be performed by various methods like the Newmark method, Wilson's method, Hilber–Hughes method,

DOI: 10.1201/9781003441090-4

collocation method and Chung-Hulbert method. Nonlinear static analysis is also popularly known as "pushover analysis" (POA).

4.2 STRENGTH LEVELS

Construction materials like concrete, rebar and structural steel are designated in terms of their strengths. Strength used in such definition is called *nominal strength*. Generally, materials are designated in terms of their *characteristic strength*. Inherently, material strengths are probabilistic in nature. Material strength is determined by laboratory testing. It is a natural phenomenon that in every test result, the strength is different with some amount of variation. One may take the average value of the strengths, but then one also needs to know the dispersion of the result reflected by the standard deviation of the results. The standard deviation σ is given by Eq. (4.2.1) when sample size is limited and by Eq. (4.2.2) when the sample size is large. In these two formulae, μ is mean value of the data and x_i is i-th observed data and N is the number of data. The mean value (or expected value) is given by Eq. (4.2.3). The square of standard deviation σ^2 is called *variance*.

When sample size is limited,

$$\sigma = \sqrt{\frac{1}{N}\sum_{i=1}^{N}\left(\mu - x_i\right)^2} \tag{4.2.1}$$

When sample size is large,

$$\sigma = \sqrt{\frac{1}{N-1}\sum_{i=1}^{N}\left(\mu - x_i\right)^2} \tag{4.2.2}$$

The mean value is given as

$$\mu = \frac{x_1 + x_2 + x_3 + \ldots + x_n}{N} \tag{4.2.3}$$

4.2.1 CHARACTERISTIC STRENGTH

The nominal strength by which the material is designated is called the characteristic strength. A formal definition is: *characteristic strength is the strength below which not more than 5% of the test results are likely to fall.* Characteristic strength, in general, may be designated by the symbol f_k .

To understand the meaning of the term, we take the help of the probability density function (PDF) for normal distribution given by Eq. (4.2.1.1).

Probability density function of a continuous random variable is a function whose value at any given sample in the sample space gives likelihood that the random variable would equal that sample.

$$f(x) = \frac{1}{\sqrt{2\pi}} e^{-x^2/2} \tag{4.2.1.1}$$

The PDF for normally distributed data in parametrized form is given by Eq. (4.2.1.2):

$$\text{PDF} = \frac{1}{\sigma\sqrt{2\pi}} e^{-\frac{1}{2}\left(\frac{x-\mu}{\sigma}\right)^2} \tag{4.2.1.2}$$

where x is any data value, μ is mean value and σ is the standard deviation of data set.

The plot of a PDF is parabolic as shown in Figure 4.1. The area under the parabolic curve is unity or 100%. As shown in Figure 4.1(a), 50% of the area corresponds to the highest value of PDF at the centre. This point corresponds to mean value of the data. It may be noted that in probability curves in Figure 4.1, there is no strength axis. The probability curve gives the probability of occurrence of a value. It is our convenience as to how we define strengths based on the probability of its occurrence (or degree of confidence).

In Figure 4.1(b) the 5% probability point is the 5% area (shaded) on the left of the point. Corresponding to this point, the strength level is the called characteristic strength. The characteristic strength of concrete is given by f_{ck} (notation in India) and characteristic strength of rebar steel is given by f_y (yield strength).

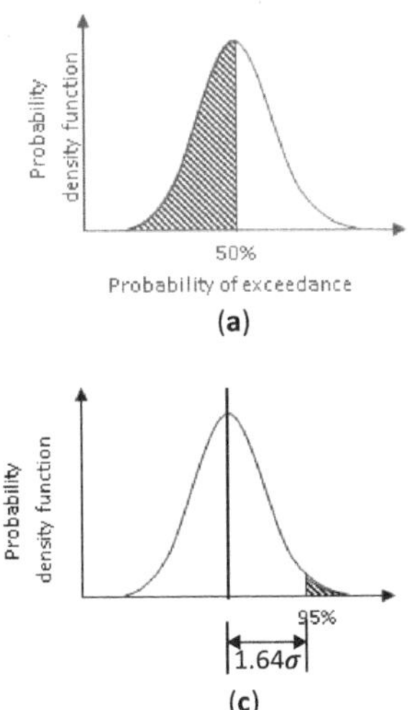

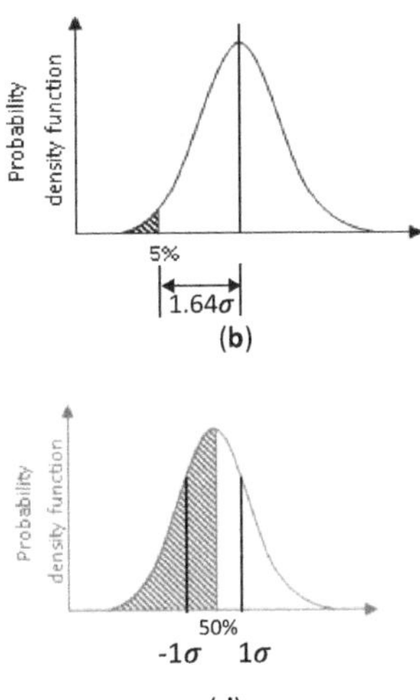

FIGURE 4.1   Probability density function plot.

As mentioned, the characteristic strength is used for nominal designation of materials. So, the basic strength is characteristic strength. Other strengths are obtained from this strength by adding the statistical distances from characteristic strength point.

4.2.2 Limit Strength

Limit strength or limiting strength is the design level strength described by the design code. Limit strength (f_{lim}) is obtained by dividing the characteristic strength by partial safety factor of material (γ_m) and is given by Eq. (4.2.2.1). In Indian code (IS 456-2000), partial safety factor for concrete is 1.5 and that for rebar it is 1.15.

$$f_{lim} = f_k / \gamma_m \qquad (4.2.2.1)$$

Using the limiting strength of concrete and rebar, we can compute the limiting moment of resistance of a section.

4.2.3 Expected Strength

Expected strength is also known as *mean strength*. It is the mean value of test data. *It may be defined as that strength below which not more than 50% of test results are expected to fall.* In Figure 4.1(a), the middle point of PDF corresponds to this strength. Needless to say, 50% of test results are also expected to lie above the mean value. The horizontal axis of PDF plot is designated also in terms of standard deviation as shown in Figure 4.1(d). The 5% area point is at a distance of 1.64 σ from the central line. So, to go to mean level point from characteristic level point, we have to travel a statistical distance of 1.64 σ. Thus, the expected strength (f_{exp}) is related to the characteristic strength by Eq. (4.2.3.1). It may be noted that the 1 σ distance corresponds to *84 percentile* value.

$$f_{exp} = f_k + 1.64\sigma \qquad (4.2.3.1)$$

The expected strength of concrete $(f_{c,exp})$ and that for rebar $(f_{s,exp})$ can now be expressed by Eqs. (4.2.3.2) and (4.2.3.3), respectively.

$$f_{c,exp} = f_{ck} + 1.64\sigma_c \qquad (4.2.3.2)$$

$$f_{s,exp} = f_y + 1.64\sigma_s \qquad (4.2.3.3)$$

where, σ_c is the standard deviation of concrete and σ_s is is standard deviation of steel. The standard deviation for particular concrete mix is available in IS 456 for

India, and concrete design codes of other countries. The standard deviation of rebar for all grades may be taken as 50 MPa.

The expected strength has great significance in PBSD as the design is damage-based and nonlinear evaluation is also damage-based. So, in PBSD expected strength is used in design and nonlinear analysis.

4.2.4 Extreme Strength

Figure 4.1(c) gives a point at a statistical distance of 1.64(sigma) towards right from the mean position. This corresponds to the strength beyond which not more than 5% results are expected to lie. In other words, 95% of samples will have strength less that value which corresponds to this level. This is called extreme strength (f_{exteme}). It is clear that extreme strength level is at a statistical distance of $2 \times 1.64\sigma$ from the characteristic strength level point. Thus, extreme level strength is given by Eq. (4.2.4.1). For concrete and rebar, the corresponding formulae are Eqs. (4.2.4.2) and (4.2.4.3), respectively.

$$f_{exteme} = f_k + 2 \times 1.64\sigma \qquad (4.2.4.1)$$

$$f_{c,extreme} = f_{ck} + 2 \times 1.64\sigma_c \qquad (4.2.4.2)$$

$$f_{s,extreme} = f_y + 2 \times 1.64\sigma_s \qquad (4.2.4.3)$$

The question of what is the utility of extreme strength may arise. In capacity design, this strength becomes useful. In capacity design, there is an yielding element and an elastic element. We wish that the yielding element only yields under high load but not the elastic element. This can be ensured if we set the condition as given in Eq. (4.2.4.4). When such condition is set, there is all likelihood that the yielding element will yield first.

Extreme strength of yielding element ≤ *limiting strength* of elastic element
$$(4.2.4.4)$$

4.2.5 Strength of Concrete

Strength of concrete is expressed by 28-day characteristic cube (15 cm cube) strength or 28-day characteristic cylinder strength. Cylinder strength is typically 80% of the cube strength due to buckling effect of length. European codes and US codes use cylinder strength whereas Indian code uses cube strength. But we have to remember that all strengths are obtained by laboratory tests. So, laboratory test conditions affect the test strength. Remember that when a concrete is used in a structure, it is in the site condition and laboratory conditions are no more available. A cube or a cylinder is pressed between two steel plattens driven hydraulically in the compression machine. The two steel plattens confine the test specimen. Such confinement effect increases the test strength. In the site, there is no such confining effect and concrete is stressed by external actions in a *free* ambience. As such, the strength exhibited in the site is lower than that in the laboratory. Further, the size of the specimen also affects the compressive strength. This is called size effect. Together with platten effect and size effect, the laboratory test specimen shows higher strength, which not available in the site condition. Such discrepancy is adjusted by taking the site strength as two-third of that in the laboratory. This is given by Eq. (4.2.5.1).

Strength to be taken in design = 0.67 × cube strength (or cylinder strength)

$$(4.2.5.1)$$

Over and above, the partial safety factor for materials will apply.

4.2.6 Strength of Reinforcing Steel

Reinforcing steel bar (rebar) for reinforced concrete (RC) can be mild steel or high-strength steel. The stress–strain characteristics of both are different. Mild steel shows specific yield point in the laboratory tensile test. Though there is a lower yield point and higher yield point with minor variation for mild steel, on a bilinearized stress–strain curve, the yield point can be easily located (see Figure 4.2a). From the yield point, we get the yield strength.

High-strength steel, on the other hand, does not show any specific yield point. It follows some curved path. Through consensus, it is considered that the yield point is at the intersection of stress–strain curve and a straight line drawn at strain 0.002, parallel to the initial tangent curve (see Figure 4.2b). Since the stress–strain curve is nonlinear, the points in the curve must be supplied by the code. For Indian condition, SP.16 gives the stress–strain tables for various grades of steel.

Figure 4.3 shows the stress–strain curves for rebar at different strength levels.

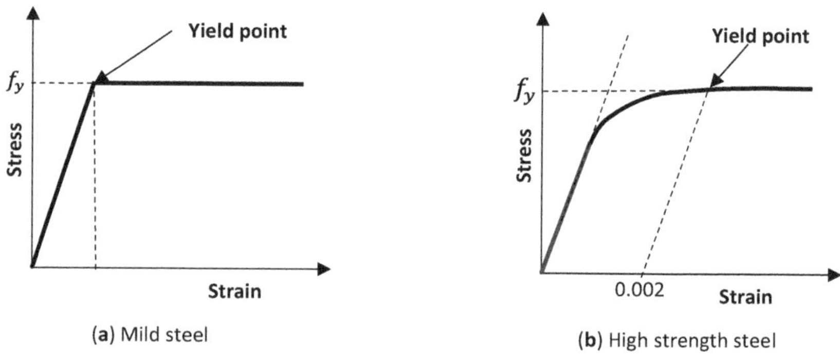

FIGURE 4.2 Yield point of rebar: (a) mild steel and (b) high-strength steel.

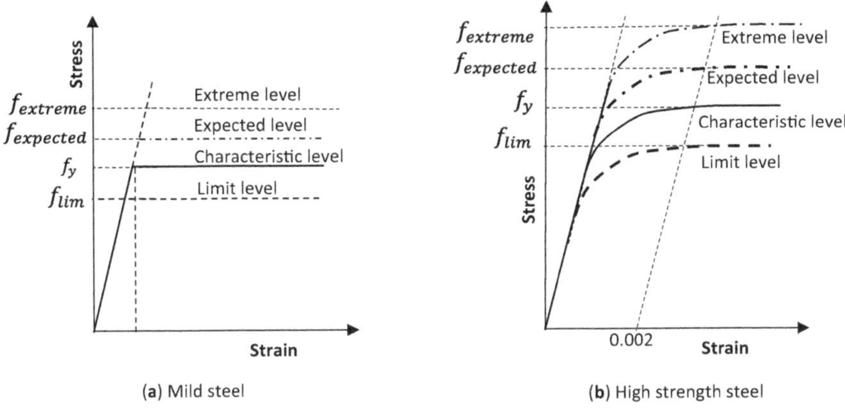

FIGURE 4.3 Stress–strain curves at different levels: (a) mild steel and (b) high-strength steel.

4.2.7 MOMENT OF RESISTANCE COMPUTATION FOR CONCRETE SECTION

The moment of resistance (MR) for concrete section can be computed at different strength levels. As per IS 456-2000, the compressive strength offered by compression concrete at limiting level is given by Eq. (4.2.7.1). This expression has embedded partial safety factor for concrete as 1.5. So, the compression offered in characteristic level is given by Eq. (4.2.7.2), with a multiplication factor 1.5. Similarly, for expected level Eq. (4.2.7.3) holds good and, for extreme strength level Eq. (4.2.7.4) is to be used. Here, x is neutral axis depth. The factor 0.36 has a component 0.67 and a parabolic-rectangular stress block in the background (as per IS 456). The factor 0.67 should work with only f_{ck} and not its variational increment to other strength levels.

$$C_{c,lim} = 0.36 f_{ck} bx \tag{4.2.7.1}$$

$$C_{c,k} = 1.5 \times 0.36 f_{ck} bx \tag{4.2.7.2}$$

$$C_{c,exp} = 1.5 \times 0.36 \left(0.67 f_{ck} + 1.64 \sigma_c\right) bx \tag{4.2.7.3}$$

$$C_{c,extreme} = 1.5 \times 0.36 \left(0.67 f_{ck} + 2 \times 1.64 \sigma_c\right) bx \tag{4.2.7.4}$$

The compression in compression steel (C_s) is given by Eq. (4.2.7.5). Total compression C is given by Eq. (4.2.7.6). The tension in tensile steel (T) is given by Eq. (4.2.7.7). For equilibrium of forces in the section, Eq. (4.2.7.8) works, which gives the unknown value of x. The value of x is to be assumed in the beginning and one needs to test if Eq. (4.2.7.8) is satisfied or not. Iteration may be required. Finally, moment of resistance is given by Eq. (4.2.7.9). Here, f_{sc} is stress in compression steel; f_{st} is stress in tension steel; d_c is depth of centre of compression from top of the section; d is effective depth of section.

$$C_s = f_{sc} A_{sc}$$
(4.2.7.5)

$$C = C_s + C_c$$
(4.2.7.6)

$$T = f_{st} A_{st}$$
(4.2.7.7)

$$C = T$$
(4.2.7.8)

$$MR = C_s \left(d - d_c\right) + C_c \left(d - 0.42x\right)$$
(4.2.7.9)

Calculation of f_{sc} and f_{st} needs some effort. These are to be obtained by noting the strain level in the steel. If the strain level exceeds yield strain, the stresses are obtained as given in SP.16 for Indian case (values of stresses in nonlinear strain range for Fe415 (yield strength 415 MPa) and Fe500 (yield strength 500 MPa) steel). The yield strain of rebar (ε_{sy}) is given by Eq. (4.2.7.10). Here, f_{sy} is yield stress of steel at various strength levels as given below:

For limit strength, $f_{sy} = 0.87 f_y$

For characteristic strength, $f_{sy} = f_y$

For expected strength, $f_{sy} = f_y + 1.64\sigma_s$

For extreme strength, $f_{sy} = f_y + 2 \times 1.64\sigma_s$.

$$\varepsilon_{sy} = \frac{f_{sy}}{E_s} + 0.002 \quad \left[\text{as per Indian code}\right]$$
(4.2.7.10)

The extreme compression fibre stress at top is taken as failure strain of concrete, which is equal to 0.0035 (Indian code). The strain profile gives relation as shown in Eqs. (4.2.7.11) and (4.2.7.12).

$$\varepsilon_{st} = 0.0035 \frac{d - x}{x}$$
(4.2.7.11)

$$\varepsilon_{sc} = 0.0035 \frac{x - d_c}{x}$$
(4.2.7.12)

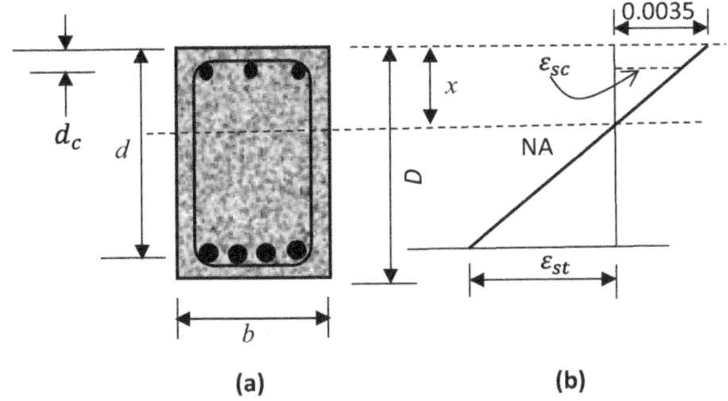

FIGURE 4.4 Beam section and strain profile.

If ε_{st} or ε_{sc} is greater than ε_y, then the corresponding steel has yielded, else not yielded. Once yielded, the stress is equal to yield strength of given level, else it is to be found from the tabular values of Table A of SP.16 (for Indian case). If the strain is less than elastic strain, the stress can be found from Hooke's law of proportionality of stress–strain. Beam section and strain profile are shown in Figure 4.4.

4.3 EFFECTIVE STIFFNESS

Gross sectional stiffness refers to the condition when the full sizes (i.e. longer and shorter dimensions) of the cross-section of the member is used in computing stiffness. In RC members, due to tension developed below the neutral axis, full section is not available to render stiffness. This is for the reason that concrete is weak in tension and hence, it is neglected below the neutral axis. Thus, only a part of the gross section will be available in offering stiffness against some action. Such sections are called "cracked" section. In steel section, this problem does not arise, as steel is equally strong in tension and compression. We define effective stiffness factor (or effective stiffness ratio, or, stiffness reduction factor) by Eq. (4.3.1).

$$\text{Effective stiffness factor, } r = \frac{\textit{Actual available stiffness of member}}{\textit{Gross stiffness of member}}$$

$$= \frac{I_{net}}{I_g} \tag{4.3.1}$$

where, I_{net} and I_g are moments of inertia of net section and gross section respectively.

The development of tension in beam and column section is explained in Figure 4.5. Beam is generally under transverse bending; but a column can have uniaxial or biaxial bending. Further, column is under axial force, which helps the concrete to be in compression. So, the stiffness reduction factor of column is higher (i.e. reduction is lower) than that of beam.

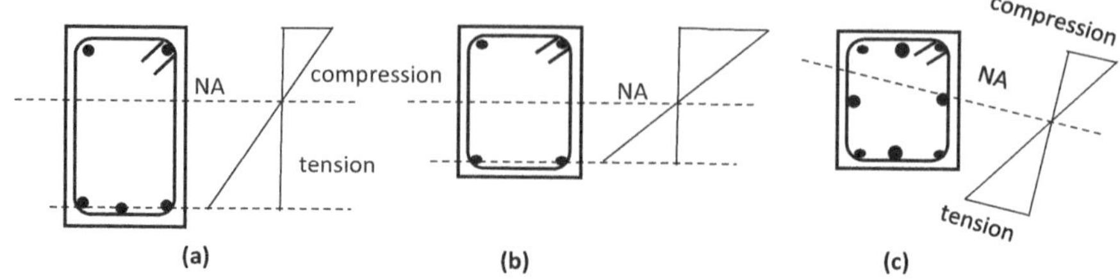

FIGURE 4.5 Tension in sections: (a) beam under transverse bending, (b) column under uniaxial bending and (c) column under biaxial bending (NA is neutral axis).

The expressions of stiffness for various common actions are given in Eqs. (4.3.2). It is found from Eqs. (4.3.2) that in most of the cases EI (called flexural rigidity) term is common. Here, A is area of section, E is modulus of elasticity, L is length of member, I is moment of inertia of gross section of member.

$$\text{Axial stiffness} = \frac{AE}{L} \tag{4.3.2a}$$

$$\text{Rotational stiffness for beam} = \frac{4EI}{L} \tag{4.3.2b}$$

$$\text{Stiffness under deflection of beam end} = \frac{6EI}{L^2} \tag{4.3.2c}$$

$$\text{Lateral stiffness of column for side sway} = \frac{12EI}{L^3} \tag{4.3.2d}$$

$$\text{Cantilever stiffness with point load at end} = \frac{3EI}{L^3} \tag{4.3.2e}$$

Denoting EI_{eff} as effective flexural rigidity and EI_g as gross flexural rigidity, we can write the expression for stiffness reduction factor as in Eq. (4.3.3).

$$r = \frac{I_{\text{eff}}}{I_g} = \frac{EI_{\text{eff}}}{EI_g} \tag{4.3.3a}$$

From Eq. (1.7.1b) of Chapter 1, at yield condition,

$$EI_{\text{eff}} = \frac{M_y}{\phi_y} \tag{4.3.3b}$$

Combining Eqs. (4.3.3a) and (4.3.3b), we get Eq. (4.3.4).

$$r = \frac{M_y / \phi_y}{EI_g} \tag{4.3.4}$$

The yield moment of a section (M_y) cannot be found from basic mechanics of RC. It is obtained from bilinearization of moment–curvature diagram (see Section 1.7). SAP2000 software gives the yield moment in hinge results. The yield curvature for various sections are available in the literature and are summarized in Table 4.1 (see Priestley et al., 2007). The yield rotations of frame are also given in this table. It may be mentioned that for nonlinear analyses, one has to use the expected strength of materials.

The stiffness that can be taken in nonlinear analyses are one of the followings types:

1. Gross stiffness based on gross sectional property
2. Effective stiffness as given in FEMA-356 and in ASCE-SEI-41-17
3. Effective stiffness based on strength.

The effective stiffness values given by FEMA-356 and ASCE-SEI-41-17 are furnished in Table 4.2.

Das and Choudhury (2020) have highlighted the effect of using these three types of stiffnesses. It may be noted that the stiffness based on strength is the "exact" value of stiffness. So, in all nonlinear analyses, stiffness based on strength is to be used. In fact, one has to use the stiffness reduction factor in the computer model for each element.

4.3.1 Effective Stiffness for Columns

For columns, it is not possible to get any yield moment, as the failure is through an interaction between axial force and bending moment(s). It actually fails on a yield surface, which is constituted by P–M (axial force–moment) pair of values at each point. To overcome this problem for columns, Pettinga and Priestley (2005) suggested to take the yield moment of column as moment corresponding to the axial load in column arising out of gravity load (dead and live load) in the interaction curve. We need to have the column interaction curves. Further, as the nonlinear analyses are done on expected strength level, the expected interaction diagram for column needs to be drawn. The process is explained in Figure 4.6.

With reference to Figure 4.6, let the axial force under dead load and live load for a column is P', which is shown by ordinate OA. The corresponding moment as per failure surface is M' given by abscissa OC. This may be considered as the yield moment for the column.

$$r = \frac{1}{EI_g}\frac{M_y}{\phi_y} \text{ where, } M_y = M', \phi_y = 2.1\frac{\varepsilon_y}{h_c} \text{ for column.} \qquad \text{[Table 4.1]}$$

TABLE 4.1
Yield curvature/yield rotation for standard sections

Section type	Drawing	Yield curvature/ yield rotation
Circular concrete column	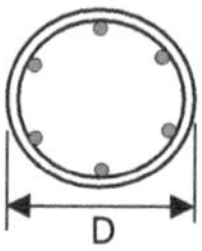	$\phi_y = 2.25\dfrac{\varepsilon_y}{D}$ D = diameter
Rectangular concrete column	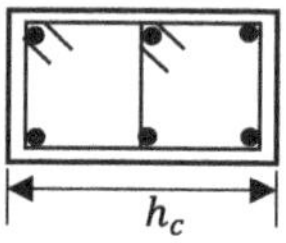	$\phi_y = 2.1\dfrac{\varepsilon_y}{h_c}$
Rectangular concrete shear wall	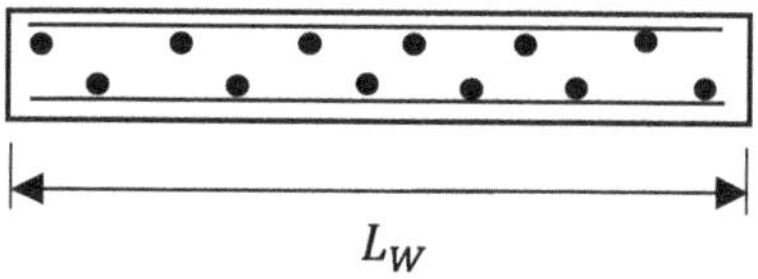	$\phi_y = 2.0\dfrac{\varepsilon_y}{L_w}$
Concrete flanged shear wall	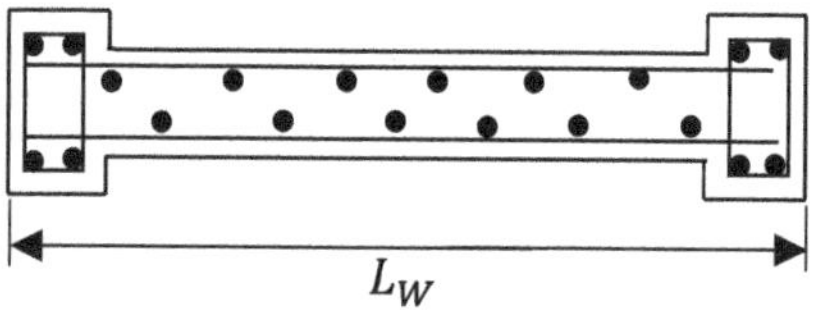	$\phi_y = 1.5\dfrac{\varepsilon_y}{L_w}$
Symmetrical steel section	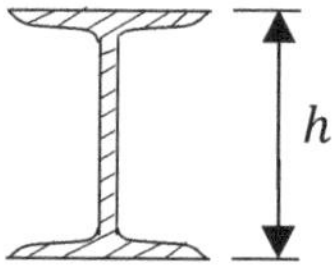	$\phi_y = 2.1\dfrac{\varepsilon_y}{h}$
Flanged concrete beam	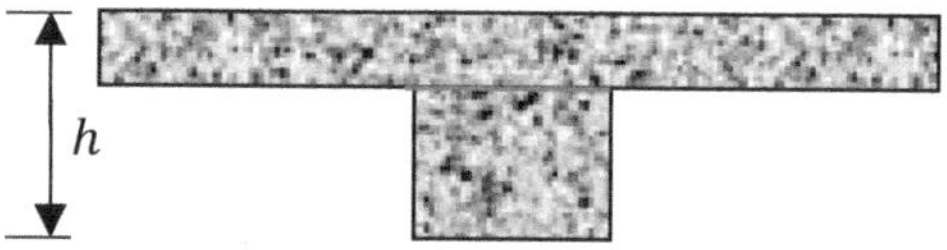	$\phi_y = 1.7\dfrac{\varepsilon_y}{h}$
Rectangular RC beam	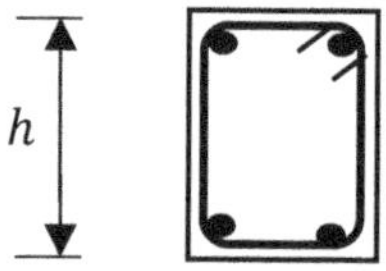	$\phi_y = 1.87\dfrac{\varepsilon_y}{h}$
Rectangular masonry wall	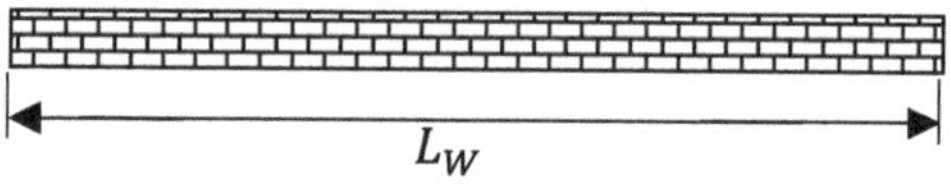	$\phi_y = 2.1\dfrac{\varepsilon_y}{L_w}$
RC frame	Yield rotation	$\theta_{yF} = 0.5\dfrac{\varepsilon_y l_b}{h_b}$
Steel frame	Yield rotation	$\theta_{yF} = 0.65\dfrac{\varepsilon_y l_b}{h_b}$

TABLE 4.2
Effective stiffness for RC members (flexural)

Item	FEMA-356	ASCE-SEI-41-17
Beam – non-prestressed	$0.5\ E_c I_g$	$0.3\ E_c I_g$
Beam – prestressed	$E_c I_g$	$E_c I_g$
Column with compressive gravity load $\geq 0.5 A_g f_c$	$0.7\ E_c I_g$	$0.7\ E_c I_g$
Column with compressive gravity load $\leq 0.3 A_g f_c$, or in tension	$0.5\ E_c I_g$	$0.3\ E_c I_g$ (gravity load $\leq 0.1 A_g f_c$)
Walls – uncracked (on inspection)	$0.8\ E_c I_g$	-
Walls – cracked (on inspection)	$0.5\ E_c I_g$	-

N.B. E_c is modulus of elasticity of concrete, I_g is moment of inertia of gross section, f_c is compressive strength of concrete, A_g is gross section area.

Shear rigidity is $0.4\ E_c A_w$. Axial rigidity in column is $E_c A_g$; A_w is area of web of beam.

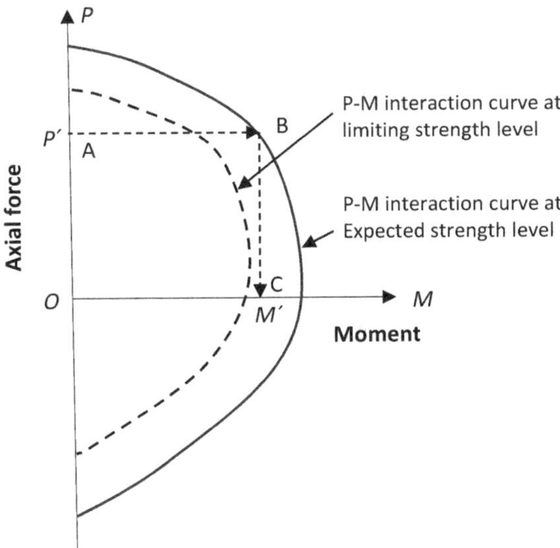

FIGURE 4.6   P–M Interaction curves at different strength levels.

Since the interaction diagram at expected strength is involved, the computation of stiffness reduction factor for column is tedious. Das and Choudhury (2019) and Das et al. (2021) have developed artificial neural network (ANN), support vector regression (SVR) and genetic algorithm (GA) models for computation of stiffness reduction factor for columns, which eases out the process.

Example 4.3.1.1 *The yield moment for a beam section of size 300 mm × 500 mm is 210 kN-m. The yield strength of rebar is 415 MPa. Find the stiffness reduction factor for transverse bending about major axis. Concrete characteristic strength is 25 MPa.*

As per given data: f_y = 415 MPa, f_{ck} = 25 MPa; My = 210 kN-m; width of beam b = 300 mm, depth of beam h_b = 500 mm, modulus of elasticity for steel (E_s) = 2 × 10⁵ MPa.

$$\varepsilon_y = \frac{f_y}{E_s} = 415/2 \times 10^5 = 0.002075.$$

For RC beam section, yield curvature $\phi_y = \dfrac{1.87\varepsilon_y}{h_b}$ [Table 4.1]

$$= 1.87 \times 0.002075/500 \text{ per mm} = 7.8 \times 10^{-6} \text{ per mm}.$$

$$I_g = 300 \times 500^3/12 = 3125 \times 10^6 \text{ mm}^4.$$

$$r = \frac{1}{EI_g}\frac{M_y}{\phi_y} = \frac{1}{25000 \times 3125 \times 10^6}\frac{210 \times 10^6}{7.8 \times 10^{-6}} = 0.345.$$

It is found that the FEMA-356 specified value is much higher (0.5). The ASCE-SEI-41-17 value has a better match (0.3).

Example 4.3.1.2 *For a column section of size 400 mm × 500 mm, the gravity load is 400 kN and corresponding to this load in expected interaction diagram for column the moment about major axis is 420 kN-m. The rebar yield strength is 500 MPa. Find the stiffness reduction factor for bending about the major axis.*

As per given data: f_y = 500 MPa, depth of column section h_c = 500 mm; modulus of elasticity for steel (E_s) = 2 × 10⁵ MPa.

$$\varepsilon_y = \frac{f_y}{E_s} = 500/2 \times 10^5 = 0.0025.$$

For RC column section, yield curvature $\phi_y = 2.1\varepsilon_y/h_c$ = 2.1×0.0025/500 per mm = 1.05×10⁻⁵ per mm.

$$I_g = 400 \times 500^3/12 = 4167 \times 10^6 \text{ mm}^4.$$

$$r = \frac{1}{EI_g}\frac{M_y}{\phi_y} = \frac{1}{27386 \times 4176 \times 10^6}\frac{420 \times 10^6}{1.05 \times 10^{-5}} = 0.350.$$

Here the column compressive gravity load $\leq 0.3A_g f_c$. The value of r is almost matching with ASCE-SEI-41-17 value.

4.4 PLASTIC HINGE CONCEPT

A physical hinge in a structure is characterized by zero bending moment. In contrast to this, the plastic hinge is a section where the materials are stressed to their maximum capacity and the moment of resistance offered by the section is maximum. This is called the plastic moment of resistance (M_p). A section in an element remains under stress due to external loading. As the external load increases, the stress also increases and the stress may reach the yield stress at the extreme fibre. This condition is called as the yield level. If the external load is further increased, the inner fibres gradually reaches the yield stress. The steel section can thus attain a plastic condition where all the fibres are under yield stress. In concrete section, the nature of the stress block varies with codes. In Indian code (IS 456), the stress block is parabolic-rectangular. In the American code, it is rectangular. Also, in steel, the yield stress is clearly defined (95% chance of exceedance) and is denoted by f_y. But in concrete section, the corresponding stress in concrete is taken as $0.446 f_{ck}$, embedding the partial safety factor. When the parabolic-rectangular stress block formation is complete, the section is said to have reached the fully plastic condition (Figure 4.7).

As the plastic condition is reached, the section undergoes plastic deformation (non-regainable deformation). This will be visible as damage in the section. When the damage is more or less is confined in small length of the member, it is said to be *lumped plasticity*. In the computer model, lumped plasticity is a point plasticity. In contrast, when the damage is spread over a finite length of the member, it is said to be *distributed plasticity*. Modelling of distributed plasticity in computer is much more complicated than modelling of lumped plasticity. Localized or lumped plasticity and distributed plasticity are explained schematically in Figure 4.8. The length of distributed plasticity is denoted by l_p.

Length of plasticity is available in literature in empirical forms (e.g., Priestley et al.,2007):

$$l_p = kL_c + L_s \geq 2L_s \tag{4.4.1}$$

$$\text{where } k = 0.2\left(\frac{f_u}{f_y} - 1 \right) \leq 0.08$$

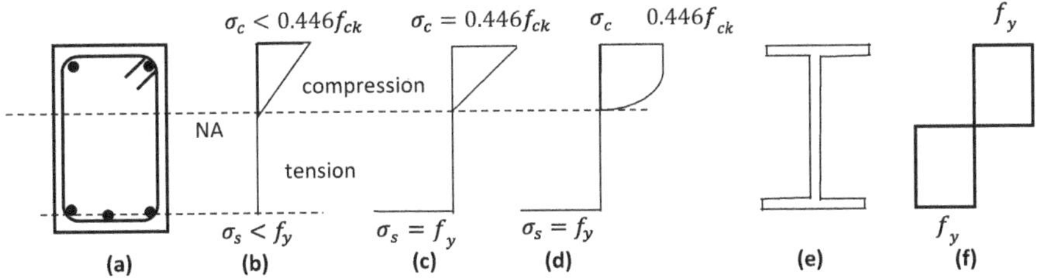

FIGURE 4.7 (a) RC section, (b) elastic stress, (c) extreme stress in concrete, (d) plastic stress block, (e) steel section and (f) plastic stress block in steel section.

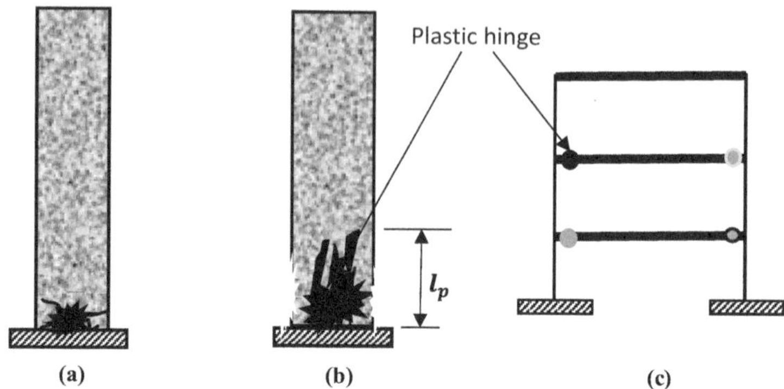

FIGURE 4.8 (a) RC column with localized plasticity, (b) RC column with distributed plasticity, (c) lumped plastic hinges in beams (colours may indicate different degree of damage).

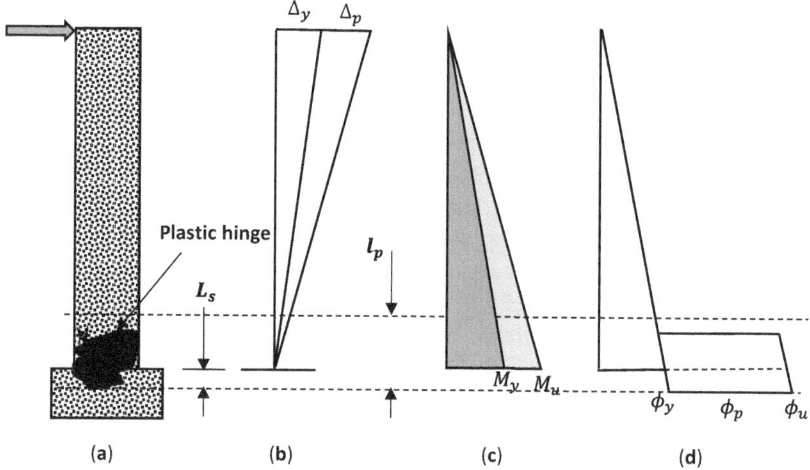

FIGURE 4.9 (a) Plastic hinge in column, (b) displacement at top, (c) BMD and (d) curvature diagram.

$$L_s = 0.022 f_{ye} \phi_l$$

In Eq. (4.4.1), l_p is plastic hinge length, L_s is strain penetration length, L_c is distance from critical section to point of contraflexure in the member, f_u is ultimate strength of rebar and f_y is yield strength of rebar, f_{ye} is expected yield strength of rebar, ϕ_l is diameter of the longitudinal bar in mm. Strain penetration means the penetration of plasticity inside the foundation below the footing level of column. The length of plastic zone to be taken is at least equal to Ls.

The plastic hinge is further discussed below by taking a column as shown in Figure 4.9. The height of the column is H. Up to the yield condition, the treatment is linear. Beyond yielding, the system is nonlinear.

The ultimate curvature (ϕ_u) consists of yield curvature (ϕ_y) and plastic curvature (ϕ_p). This is given by Eq. (4.4.2).

$$\phi_u = \phi_y + \phi_p \qquad (4.4.2)$$

Noting that rotation = curvature × length and displacement = rotation × length, we can write,

$$\Delta_p = \phi_p l_p H \qquad (4.4.3)$$

$$\Delta_y = \phi_y \left(H + L_s \right)^2 / 3 \qquad (4.4.4)$$

Eq. (4.4.4) is valid for cantilever column of height H.

Example 4.4 *A cantilever RC column 400 mm × 400 mm in size, is 3.5 m in height. The rebar is of diameter 20 mm and of yield strength 500 MPa and ultimate strength 580 MPa. Find plastic hinge length. Also find the plastic displacement and yield displacement.*

Solution: Here, f_y = 500 MPa, f_u = 580 MPa, H = 3.5 m; expected yield stress f_{ye} = 1.25 f_y = 625 MPa, ϕ_u = 0.004, ϕ_l = 20 mm.
   Strain penetration length = L_s = $0.022 f_{ye} \phi_l$ = 0.022 × 625 × 20 = 275 mm.

$$k = 0.2 \left(\frac{f_u}{f_y} - 1 \right) = 0.2 \left(\frac{625}{500} - 1 \right) = 0.05 \leq 0.08.$$

Length of plastic hinge = $l_p = kL_c + L_s$ = 0.05×3.5×1000+275 = 450 < $2L_s$.

So, l_p = $2L_s$ = 2×275 = 550 mm.

For RC column section, yield curvature $\phi_y = 2.1\varepsilon_y / h_c$ = 2.1(500/(2 × 10^5))/ 400 = 1.3 × 10^{-5} per mm.

Plastic displacement = $\Delta_p = \phi_p l_p H$ = (3 × 1.3 × 10^{-5}) × 550 × 3500 = 75 mm.

Yield displacement = $\Delta_y = \phi_y \left(H + L_s \right)^2 / 3$ = 1.3 × 10^{-5} × (3500+275)²/3 = 61.8 mm.

Total displacement = 75 + 61.8 = 136.8 mm; this is at the top of the column.

Shear wall
For shear wall, the plastic hinge length is given by Eq. (4.4.5).

$$L_p = 0.022 f_y d_b + 0.054 h_{inf} \qquad (4.4.5a)$$

$$L_p = 0.2L_W + 0.03h_{inf} \qquad\qquad (4.4.5b)$$

In Eq. (4.4.5), h_{inf} is height of inflection of shear wall. The lower value of L_p out of Eq. (4.4.5a) and (4.4.5b) is to be adopted.

4.5 MODELLING OF BEAM, COLUMN AND SHEAR WALL

4.5.1 MODELLING BEAMS AND COLUMNS

RC beams are generally modelled as rectangular sections. This is valid for end section of the beam. At the mid-section, the RC beam is a T-beam. It is possible to model T-beam using section designer of SAP2000 software. A steel beam has a uniform cross-section all through its length. An RC column section may be square or rectangular. When it is rectangular, its orientation is to be decided. A column section can have longitudinal steel at opposite two faces, or equally spread over all faces. For steel column section also, the orientation is important. It is possible to model the stirrups in beams and ties in columns with appropriate software.

4.5.2 MODELLING OF SHEAR WALL

Shear wall is a thin (thickness 100 mm to about 350 mm) structural wall, which carries the lateral load in its own plane. It is weak in out-of-plane action. The shear wall can be modelled in computer software as: (i) planar shell element, (ii) wide column or (iii) layered shell element. Depending on the modelling, not all characteristics of shear wall can be captured. When it is modelled as a wide column, sectional dimension of the wide column is same as that of the shear wall (that is why the name wide column). The problem with this model is that the non-linear behaviour of the column and shear wall are not the same. Even the range of plastic rotations for IO, LS and CP are also different for the two cases. One way to partly tackle this problem is to edit the hinge result of the wide column and match it manually with the data of shear wall. Such a nonlinear hinge is called "user-defined hinge". However, the behaviours of shear wall and column are different. Layered shell element consists of layers of concrete and reinforcing steel. It is said that this model is better than other models and is applicable to large shear wall too. But the problem with this model in SAP2000 is this that it does not show lumped plasticity hinge. Instead, it gives a distributed plastic state. In software like Perform-3D, shear wall can be modelled as a planar element.

Shear walls are to be provided symmetrically in opposite faces of the plan of any building. Also, there should be shear wall(s) in both the mutually perpendicular directions of the plan. Shear walls put at the periphery are able to resist more torsion. Shear walls are also provided centrally as lift case. Steel in shear wall runs horizontally to take shear and, vertically to take the cantilever moment. Shear wall is actually a cantilever deep beam. Figure 4.10 gives details for a shear wall.

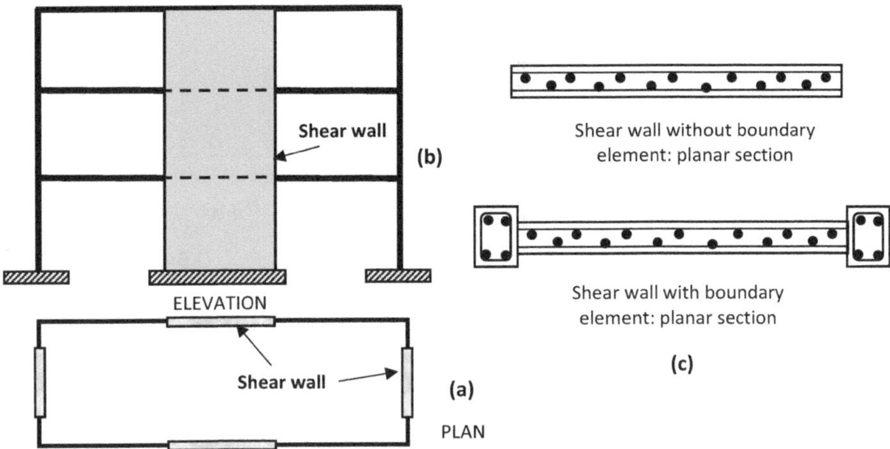

FIGURE 4.10   Shear wall: (a) plan, (b) elevation and (c) cross-section.

4.5.2.1   Computation of Plastic Rotation in Layered Shell Element Model

As mentioned already, layered shell element model does not give plastic rotation in the lumped plasticity form. Here, we tackle this in a different way. From section cut of wall at base, the moment–rotation diagram can be obtained, as shown in Figure 4.11. From moment–rotation diagram, maximum rotation angle can be read out. From maximum rotation angle, the yield rotation is subtracted to get the plastic rotation (Eq. (4.5.2.1)).

$$\theta_{pW} = \theta_{max} - \theta_{yW} \tag{4.5.2.1}$$

$$\theta_{yW} = \phi_{yW}\,\frac{h_{inf}}{2} \tag{4.5.2.2}$$

$$\phi_{yW} = \frac{2\varepsilon_y}{L_w} \quad \left(\text{from Table 4.1}\right) \tag{4.5.2.3}$$

Here, θ_{max} is the maximum rotation of wall shown in section cut at base (Figure 4.11), θ_{pW} is plastic rotation of wall, h_{inf} is inflection height of wall, θ_{yW} is yield rotation of wall, ε_y is yield strain of rebar, ϕ_{yW} is yield curvature of wall, L_w is horizontal length of wall.

4.5.3   BRIDGE PIER

Figure 4.12 shows a bridge pier, which has undergone plastic hinge formation at the bottom. A part of plasticity enters the base and its length is called *plastic strain penetration length* (L_s).

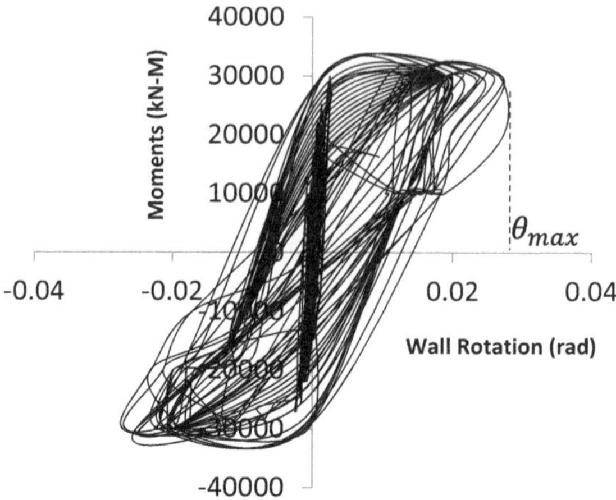

FIGURE 4.11 Moment–rotation hysteresis curves of section cut at base of shear wall.

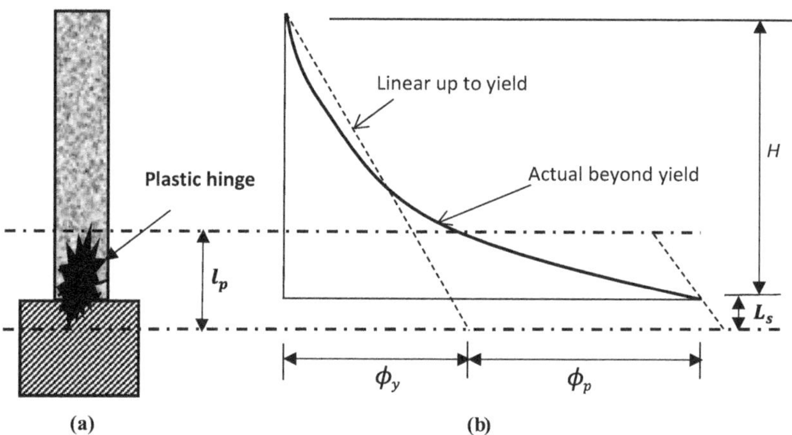

FIGURE 4.12 (a) RC bridge pier with plasticity and (b) curvature diagram.

Total curvature ϕ consists of yield curvature ϕ_y and plastic curvature ϕ_p.

$$\text{Plastic rotation, } \theta_p = \phi_p l_p \qquad (4.5.3.1)$$

$$\text{Plastic displacement at top of pier, } \Delta_p = \theta_p H = \phi_p l_p H \qquad (4.5.3.2)$$

$$\text{Yield displacement at top of pier (Priestley et al., 2007), } \Delta_y = \phi_y H^2/3 \qquad (4.5.3.3)$$

$$\text{Total displacement at top of pier} = \Delta_{max} = \Delta_y + \Delta_p$$

$$\text{Displacement ductility} = \mu_\Delta = \frac{\Delta_{max}}{\Delta_y} = \frac{\Delta_y + \Delta_p}{\Delta_y} = 1 + \frac{\Delta_p}{\Delta_y} = 1 + \frac{\phi_p l_p H}{\phi_y H^2 /3}$$

$$\text{Or,} \quad \mu_\Delta = 1 + 3\frac{\phi_p l_p}{\phi_y H} \tag{4.5.3.4}$$

4.6 MODELLING INFILL WALL

Infill walls are generally made up of brick masonry. These are non-load-bearing and nonstructural members. Infills serves functions like protection from sun, wind, theft and sound. These also provide privacy to occupants. Infill being made of bricks and mortar joints, it cannot take tension, and develops cracks under the tensile load. But infill is good in compression and takes a large force in compression. It is difficult to incorporate the infill as an area element, requiring finite element meshing. So, infill is replaced by equivalent compression strut. The thickness of the equivalent diagonal strut is equal to the thickness of the infill and the effective width of the strut is available in literature. During the to-and-fro shaking in earthquakes, the diagonal in the infill panel alternatively undergoes tension and compression as shown in Figure 4.13.

Now we shall discuss the computation of width of equivalent diagonal strut from literature.

4.6.1 FEMA-356 Provisions for Infill Strut

The width of infill strut (w) as per FEMA-356 in *inches* is given by Eq. (4.6.1).

$$a = 0.175\left(\lambda h_c\right)^{-0.4} r_i \tag{4.6.1}$$

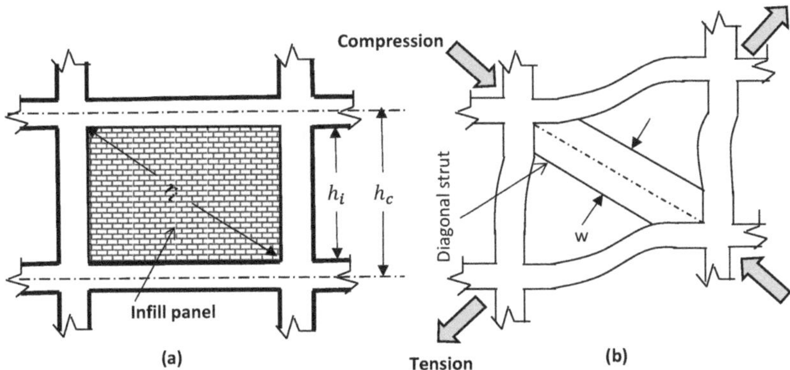

FIGURE 4.13 (a) Infill panel and (ii) diagonal strut concept when under dynamic loading.

where

$$\lambda = \left[\frac{E_m t \sin 2\theta}{4 E_f I_c h_i}\right]^{0.25}$$

h_c = height of column between centres lines of beams (in inches)
r_i = diagonal length of infill panel (in inches)
E_m = modulus of elasticity of masonry infill (in psi)
E_f = modulus of elasticity of frame material (in psi)
t = thickness of infill panel (in inches)
θ = angle of inclination of diagonal strut of infill panel with horizontal
I_c = moment of inertia of column abutting infill (in inches^4)
h_i = height of infill (in inches)

4.6.2 PROVISIONS OF IS 1893-2016 FOR INFILL STRUT

IS 1893 (Part 1) -2016 has given similar expression as FEMA-356, with parameters in SI units (Eq. 4.6.2).

$$a = 0.175\left[h\left(\frac{E_m t \sin 2\theta}{4 E_f I_c h}\right)^{0.25}\right]^{-0.4} \tag{4.6.2}$$

where h is height of the column.

In a computer model, for nonlinear analyses, the infill strut is to be provided along both the diagonals; the compression diagonal will be active at any particular instant. Very often, we provide weight in beams manually to avoid meshing in the slab. Now, over and above, infill strut is provided. This may increase the weight over what is actual. To avoid this, the infill material may be defined with zero density and modulus of elasticity same as that of infill material.

4.6.3 STRENGTH OF INFILL

FEMA-356 has given the compressive strength of masonry as 900 psi (good condition), 600 psi (fair condition) and 300 psi (poor condition) depending on the quality of masonry. The corresponding shear strength in psi are 27, 20 and 13 psi, respectively.

IS 1893 (Part 1) – 2016 has given the compressive strength of masonry by Eq. (4.6.3.1).

$$f_m = 0.433 f_b^{0.64} f_{mo}^{0.36} \tag{4.6.3.1}$$

where, f_m is masonry compressive strength in MPa, f_b is compressive strength of brick in MPa, and f_{mo} is compressive strength of mortar in MPa.

The code also gives modulus of elasticity (E_m) as Eq. (4.6.3.2).

$$E_m = 550 f_m \left(\text{units in MPa}\right) \tag{4.6.3.2}$$

The expected strength of masonry is 1.3 times its designated strength. More discussions on infill is available in Chapter 12.

4.6.4 AXIAL CAPACITY OF INFILL STRUT

In computer modelling of infill strut, it is necessary to provide its axial strength. In fact, the axial-carrying capacity of infill compression strut arises out the horizontal shear strength of the infill masonry. The horizontal shear failure occurs along the bed joint. Consider Figure 4.14.

With reference to Figure 4.14,

Q = horizontal shear capacity of bed joint
q = bed joint shear strength of masonry infill
θ = angle of inclination of diagonal with horizontal
A = net area of cross-section of infill
t = thickness of infill
L = horizontal length of an infill panel

Here, $A = tL$, $Q = qA$.

Resolving the forces horizontally,

$$F \cos \theta = Q$$

$$F = \frac{Q}{\cos \theta}$$

$$\text{Or,} \quad F = \frac{qtL}{\cos \theta} \tag{4.6.4}$$

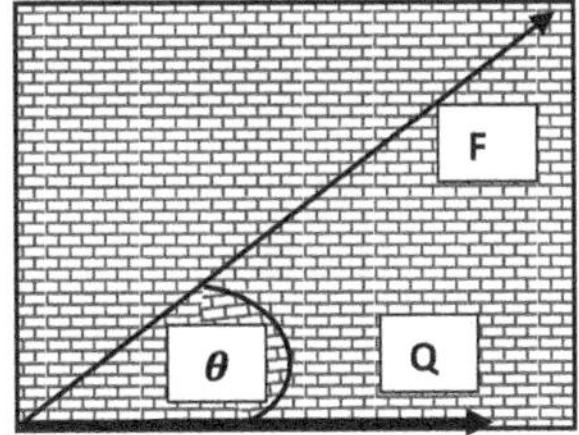

FIGURE 4.14   Infill strength.

The axial capacity of the equivalent infill strut is given by Eq. (4.6.4). Such axial force will be required in modelling of axial force–deformation behaviour of the infill strut.

4.7 P-DELTA EFFECT

P-delta ($P - \Delta$) effect is said to be a geometric nonlinearity. The column in normal state may be under some axial load and bending moment. When there is sway of the column, the axial load changes position by an amount Δ with respect the lower end, as a result of which additional moment equal to PΔ comes into play (Figure 4.15). The total moment now has more potential to cause damage in the column. The damage at the bottom of the column will increase its sway and hence, will create more P-delta effect.

P-delta effects are classified as: (i) static P-delta effect and (ii) dynamic P-delta effect.

4.7.1 Static P-Delta Effect

This involves static procedure as per FEMA-356. It is to be included in both linear static and nonlinear static cases.

For linear procedure, the stability coefficient (θ_i) as given by Eq. (4.7.1) is to be calculated:

$$\theta_i = \frac{P_i \delta_i}{V_i h_i} \tag{4.7.1}$$

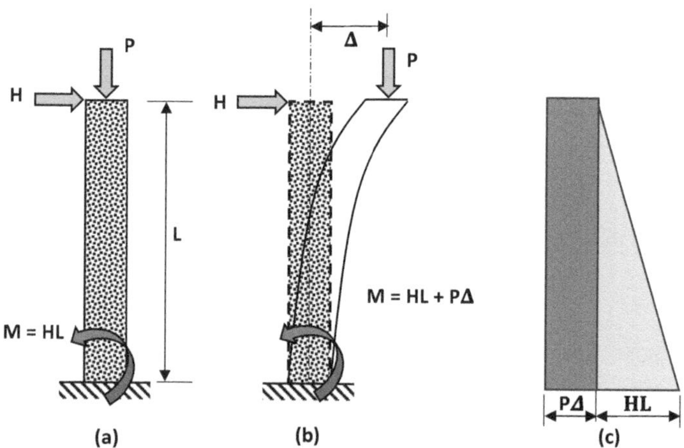

FIGURE 4.15 P-delta effect: (a) column under load, (b) column sway and moments and (c) BMD.

where P_i = is portion of the total weight of the structure arising out of DL, permanent LL and 25% of transient LL acting on columns at floor level i.

V_i = the total calculated lateral shear force in the direction under consideration at storey i due to earthquake response to the selected ground shaking level, as indicated by the selected linear analysis procedure

h_i = height of storey i between centre line of slabs

δ_i = Lateral drift in story i, in the direction under consideration, at its centre of rigidity, using the same units as for measuring h_i

The stability coefficient is to be calculated for all storeys. If θ_i is less than 0.1 in all storeys, the P-delta effect need not be considered. When θ_i is between 0.1 and 0.33, all seismic force in the storey concerned to be increased by the factor $1/(1 - \theta_i)$. If it is greater than 0.33, the structure is unstable and needs redesign/retrofit.

For nonlinear procedures, static P-delta effect shall be incorporated in the analysis by including within the mathematical model the nonlinear force–deformation relationship of all elements and components subjected to axial forces.

4.7.2 Dynamic P-Delta Effect

This considers hysteresis and stiffness degradation during dynamic loading. This is partially accommodated by introducing pseudo lateral load. In nonlinear dynamic analysis, the P-delta effect is automatically considered.

P-delta effect can cause strength degradation and stiffness reduction of the system as shown in Figure 4.16. In fact, the stiffness may become negative (i.e. slope is negative). Strength degradation is synonymous with reduction in capacity. Stiffness degradation will lead to more displacement.

4.8 TOTAL DAMPING

Total damping in the system is also called as equivalent damping (ξ_{eq}) and it arises out of material damping (ξ_m), hysteretic damping (ξ_h) and added damping (ξ_a). This

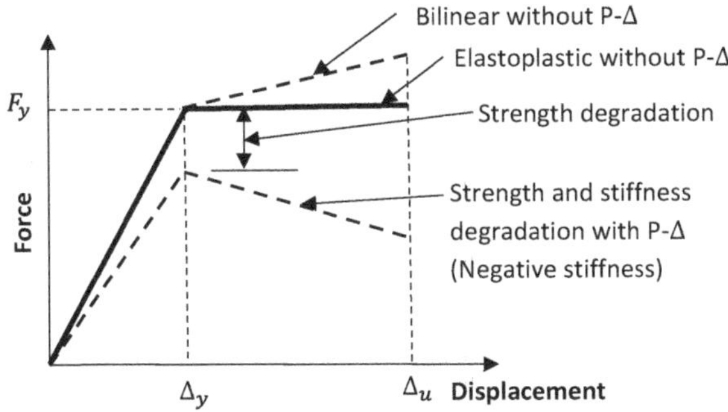

FIGURE 4.16 P-delta effect: strength and stiffness degradation.

is given in Eq. (4.8.1). Material damping is inherent property of the material. For concrete $\xi_m = 5\%$, for steel $\xi_m = 2\%$.

$$\xi_{eq} = \xi_m + \xi_h + \xi_a \tag{4.8.1}$$

The hysteretic damping (also see sec. 2.13) arises out of damage and hysteresis loop arising out of to-and-fro or cyclic motion. Hysteretic damping is given by Eq. (4.8.2).

$$\xi_h = \frac{1}{4\pi}\frac{E_d}{E_s} \tag{4.8.2}$$

where
$\quad E_d$ = energy dissipated in one hysteretic cycle (area of hysteresis loop)
$\quad E_s$ = strain energy at peak displacement.

Equivalent damping can be approximated in terms of ductility (μ) as shown in Eq. (4.8.3) (Priestley *et al.* (2007):

$$\xi_{eq} = 0.05 + a\frac{\mu - 1}{\mu\pi} \tag{4.8.3}$$

The values of a are available in Priestley *et al.* (2007).

4.9 DUCTILITY CONSIDERATIONS

4.9.1 BASIC CONCEPTS OF DUCTILITY

Ductility is the ability of a structure or structural element to undergo large deformation without significantly losing its strength. Ductility helps the structure survive from a collapse under strong ground motions. Ductility also enables us to apply response reduction factor in base shear formula, thereby reducing the design force and leading to economy. Ductility arises out of proper detailing in RC structures and from appropriate connection details in steel structures. As a mathematical expression, ductility (μ) is the ratio of ultimate deformation (δ_u) to the yield deformation (δ_y), given by Eq. (4.9.1.1) and explained in Figure 4.17.

$$\mu = \frac{\delta_u}{\delta_y} \tag{4.9.1.1}$$

Here the deformation means a generalized deformation involving linear displacement (Δ), rotation (θ), shearing (γ), twisting (ϕ), strain (ε), curvature (ϕ) etc. Based

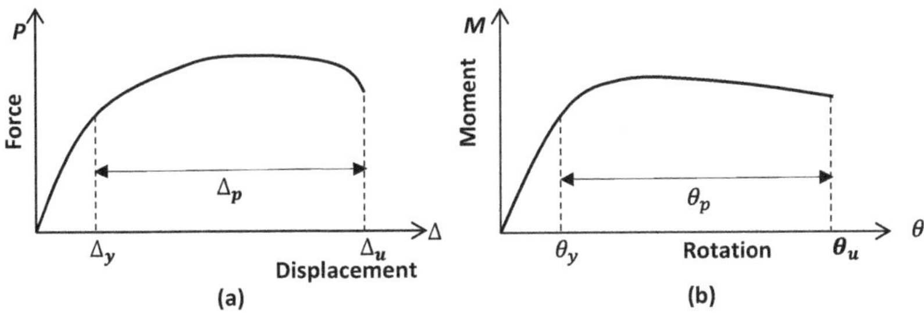

FIGURE 4.17 Deformation response: (a) translational deformation and (b) rotational deformation.

on these deformation parameters, the ductility also can be described differently. Corresponding to the yield point we have yield deformation. Beyond yield deformation is the plastic deformation zone (δ_p). It is evident that the ductility will be more when plastic deformation is more. This is also evident from Eq. (4.9.2).

$$\mu = \frac{\delta_u}{\delta_y} = \frac{\delta_y + \delta_p}{\delta_y} = 1 + \frac{\delta_p}{\delta_y} \tag{4.9.1.2}$$

Eq. (4.9.1.2) also suggests that ductility is more than unity. A ductility value of unity implies an elastic system (as is δ_p zero). Eq. (4.9.1.2) also shows that ductility can be increased by enhancing the plastic deformation zone length. This is achievable through ductile detailing. We shall now describe ductility based on the deformation types.

Referring to Figure 4.17, we can now give expressions for displacement ductility (μ_Δ) and rotational ductility (μ_θ).

$$\mu_\Delta = \frac{\Delta_u}{\Delta_y} = 1 + \frac{\Delta_u}{\Delta_y} \tag{4.9.1.3}$$

$$\mu_\theta = \frac{\theta_u}{\theta_y} = 1 + \frac{\theta_p}{\theta_y} \tag{4.9.1.4}$$

Referring to Figure 4.18, we can give expressions for strain ductility (μ_ε) and curvature ductility (μ_ϕ).

$$\mu_\varepsilon = \frac{\varepsilon_u}{\varepsilon_y} = 1 + \frac{\varepsilon_u}{\varepsilon_y} \tag{4.9.1.5}$$

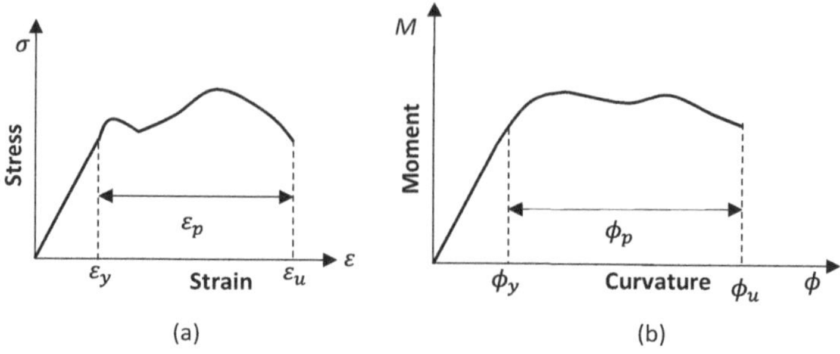

FIGURE 4.18   Deformation response: (a) strain deformation and (b) curvature deformation.

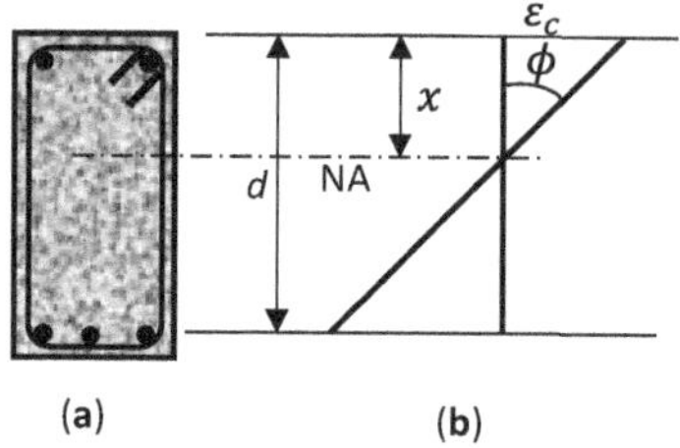

FIGURE 4.19   Concrete member: (a) section and (b) strain profile.

$$\mu_\phi = \frac{\phi_u}{\phi_y} = 1 + \frac{\phi_p}{\phi_y} \tag{4.9.1.6}$$

Similarly, the shearing ductility and twisting ductility can be defied.

4.9.2   CURVATURE DUCTILITY FROM STRAIN PROFILE

The curvature ductility can be expressed in terms of strain profile parameters as discussed below. Let x denote the depth of neutral axis for a section (Figure 4.19).

Flexural equation is

$$EI\frac{d^2 y}{dx^2} = M$$

$$\Rightarrow EI\phi = M \quad \left[\text{as} \frac{d^2 y}{dx^2} = \phi, \text{the curvature} \right]$$

$$\Rightarrow \phi = \frac{M}{EI}$$

$$\Rightarrow \phi = \frac{M/x}{E(I/x)} \quad \left[\text{dividing numerator and denominator or by } x \right]$$

$$\Rightarrow \phi = \frac{M}{(I/x)}\frac{1}{Ex} \quad [\text{re-arranging}]$$

$$\Rightarrow \phi = \frac{M}{Z}\frac{1}{Ex} \quad \left[\text{as } \frac{I}{x} = Z, \text{ the section modulus}\right]$$

$$\Rightarrow \phi = \sigma\frac{1}{Ex} \quad \left[\text{as } \frac{M}{Z} = \sigma, \text{ the bending stress}\right]$$

$$\Rightarrow \phi = \frac{\sigma}{E}\frac{1}{x} \quad [\text{re-arranging}]$$

$$\Rightarrow \phi = \frac{\varepsilon_c}{x} \quad \left[\text{as } \frac{\sigma}{E} = \varepsilon_c\right] \tag{4.9.2.1}$$

Eq. (4.9.2.1) gives curvature in terms of strain in section and neutral axis depth. It may be noted that the magnitudes of the individual ductilities are different, but they represent the same ductility of a structure or element. Magnitude wise we can write: $\mu_\phi > \mu_\varepsilon > \mu_\theta > \mu_\Delta$.

4.10   BASE OF THE BUILDING

In modelling, confusion often arises as to where the base of the building to be considered. This is clarified by Figure 4.20. The base of a building is the junction of the column and the footing.

The physical base of a building is bottom of the foundation. But as the foundation is infinitely rigid, so the base is considered at the bottom of the column for computer model of building.

4.11   A NOTE ON PLASTIC ROTATIONS AS PER ASCE-SEI-17

The FEMA-356 gives plastic rotation limit states for structural and nonstructural components separately. But in the ASCE-SEI-41-17 document, the plastic rotation limit states are given for nonstructural components only. It is found that SAP2000 software implemented rotation limits as per first row of table (instead of average value of compliant buildings). The provisions of ASCE-SEI-17 are LS is 0.75 times of deformation at CP; IO is not greater than 0.67 times deformation at LS. Considering plastic deformation at first row of table in ASCE-41-17 (as is also taken in SAP2000 software), for primary members: $a = 0.025$ radian. So, for CP the plastic rotation for primary members is 0.025 radian. For LS it is $0.75 \times 0.025 = 0.01875$ radian. $IO = 0.67 \times 0.01875 = 0.0125$ radian. These are explained in Figure 4.21. Thus, while

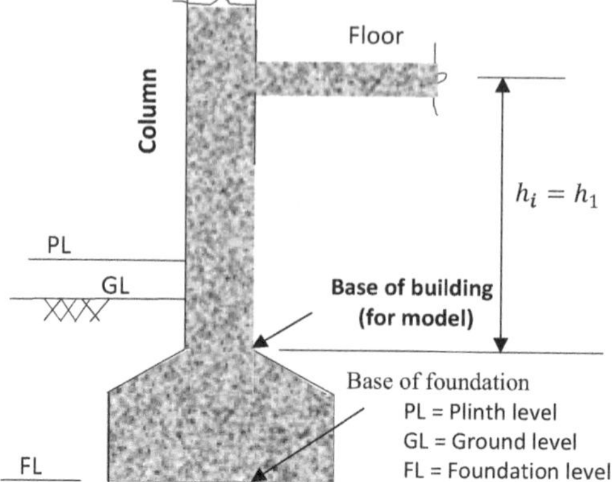

FIGURE 4.20   Height of floor from base.

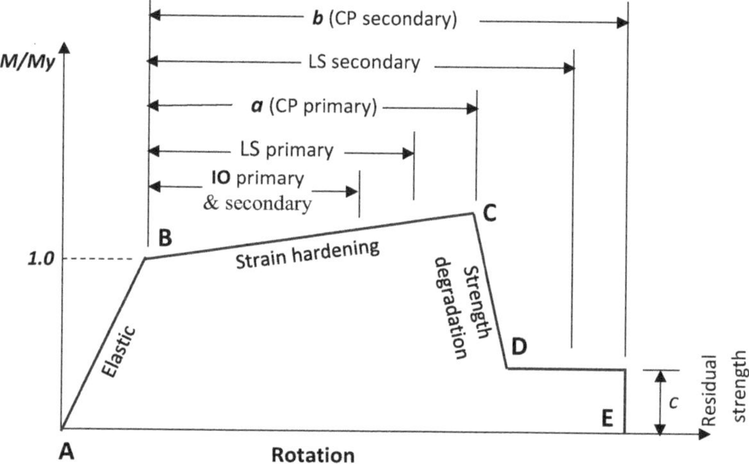

FIGURE 4.21   IO, LS, CP locations are per ASCE-SEI-41-17 (schematic).

using new versions of SAP2000, one has to supply user-defined hinge for structural members.

4.12 TORSIONAL EFFECT IN STRUCTURES

The eccentricity of centre mass and centre of rigidity cause torsion in the structure. Similarly, the eccentricity of centre of shear strength and centre of mass also leads to torsion. The eccentricity may be along both the major directions of plan of structures. Consider Figure 4.22.

Let, CM sand for centre of mass, CR stand for centre of rigidity and CV denote centre of shear resistance.

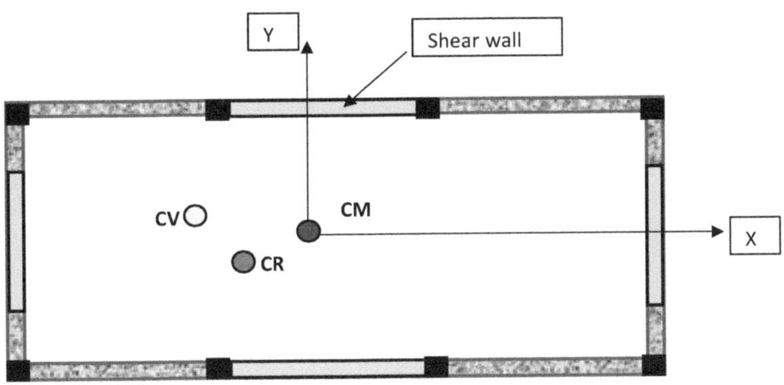

FIGURE 4.22 Centre of mass, centre of rigidity and centre of shear resistance.

The centre of mass ($\overline{x}_{CM}$, $\overline{y}_{CM}$) can be obtained from locations of masses (m_i) as (x_{mi}, y_{mi}) with respect to some reference axis by Eq. (4.12.1).

$$\overline{x}_{CM} = \frac{\sum m_i x_{mi}}{\sum m_i} \tag{4.12.1a}$$

$$\overline{y}_{CM} = \frac{\sum m_i y_{mi}}{\sum m_i} \tag{4.12.1b}$$

Similarly, the centre of rigidity ($\overline{x}_{CR}$, $\overline{y}_{CR}$) can be obtained from locations of stiffnesses (k_i) as (x_{ki}, y_{ki}) with respect to same reference axis, by Eq. (4.12.2).

$$\overline{x}_{CR} = \frac{\sum k_i x_{xi}}{\sum k_i} \tag{4.12.2a}$$

$$\overline{y}_{CR} = \frac{\sum k_i y_{ki}}{\sum k_i} \tag{4.12.2b}$$

The centre of shear resistance ($\overline{x}_{CV}$, $\overline{y}_{CV}$) can be obtained from locations of design shear (V_i) as (x_{Vi}, y_{Vi}) with respect to the same reference axis by Eq. (4.12.3).

$$\overline{x}_{CV} = \frac{\sum V_i x_{Vi}}{\sum V_i} \tag{4.12.3a}$$

$$\overline{y}_{CV} = \frac{\sum V_i y_{Vi}}{\sum V_i} \tag{4.12.3b}$$

Once the locations of CM, CR and CV are determined, the eccentricities with respect to CM can be calculated. These eccentricities matter in computing torsional effects.

The best solution for avoiding torsion in structure is to pan and design members so that torsional effect is avoided or put to a minimum. However, a change in displacement may be applied for mitigating torsional effects (Priestley et al., 2007) as shown in Eq. (4.12.4).

$$\Delta_{corrected} = \Delta_c - \theta\left(x_c - e\right) \tag{4.12.4}$$

where

$\Delta_{corrected}$ = corrected design displacement
Δ_c = drift-controlled displacement at critical element
x_c = distance of critical element from CM
θ = design drift
e = eccentricity between CM and CV in the direction concerned.

4.13   CONFINED VERSUS UNCONFINED CONCRETE

In all practical situations, the concrete in beams or columns is confined by the stirrups and ties. This confinement increases the compressive strength and ductility to a large extent. Normal design ignores this effect, leading to conservative design. It is possible to model the confined concrete in Section Designer of SAP2000 software. The behaviour of unconfined and confined concrete is demonstrated in Figure 4.23.

Unconfined concrete is not confined by any reinforcement. This is just like plain concrete. So far as structural use is concerned, unconfined concrete is a hypothetical case. Unconfined concrete exhibits the bare compressive strength of concrete.

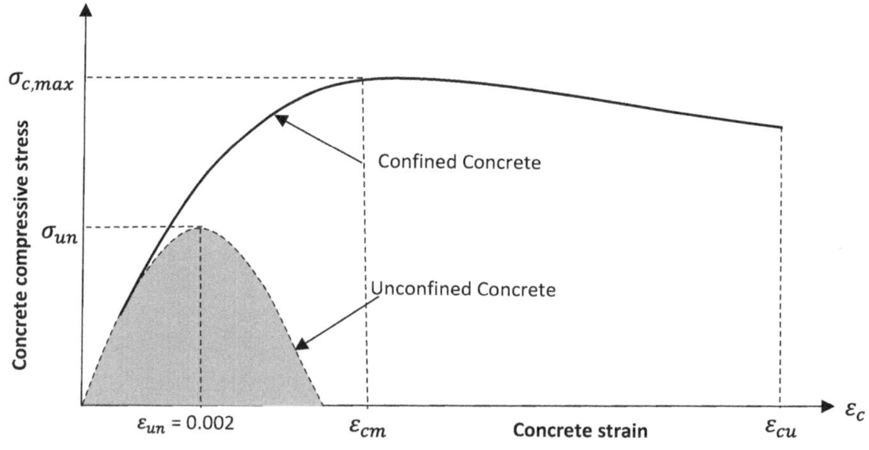

FIGURE 4.23   Stress–strain curve of confined and unconfined concrete.

The rising arm of unconfined concrete is given by Eq. (4.13.1) (Park and Paulay, 1974):

$$\sigma_c = f_{ck}\left[\frac{\varepsilon_c}{0.002} - \left(\frac{\varepsilon_c}{0.002}\right)^2\right] \tag{4.13.1}$$

where ε_c is the concrete strain ≤ 0.002. σ_c is the compressive stress in concrete.

Mander et al. (1988a, b) proposed a model for confined concrete as follows:

$$\sigma_{c,logitudinal} = \frac{\sigma_{c.max}\, xr}{r - 1 + x^r}$$

where

$\sigma_{c,logitudinal}$ = longitudinal compressive stress in concrete

$\sigma_{c.max}$ = maximum concrete compressive stress as shown in Figure 4.23

x = strain ratio = $\varepsilon_c / \varepsilon_{cu}$

ε_c = concrete compressive strain

ε_{cu} = ultimate compressive strain in concrete

$$r = \frac{E_c}{E_c + E_m}$$

$$E_c = 5000\sqrt{\sigma_c}$$

$$E_m = \frac{\sigma_{c.max}}{\varepsilon_{cm}}$$

ε_{cm} = compressive strain in concrete corresponding to maximum stress

$$\varepsilon_{cm} = \varepsilon_{un}\left[1 + 5\left(\frac{\sigma_{c.max}}{\sigma_{un}} - 1\right)\right]$$

ε_{un} = compressive strain in unconfined concrete corresponding to the maximum unconfined compressive stress, σ_{un}.

As seen from Figure 4.23, the confined concrete exhibits much higher strength and ductility (as ultimate strain is much higher). A confined concrete section has the inner core confined by steel and outside the confining steel remains the unconfined concrete. With Section Designer of SAP2000 software, it is possible to model such section.

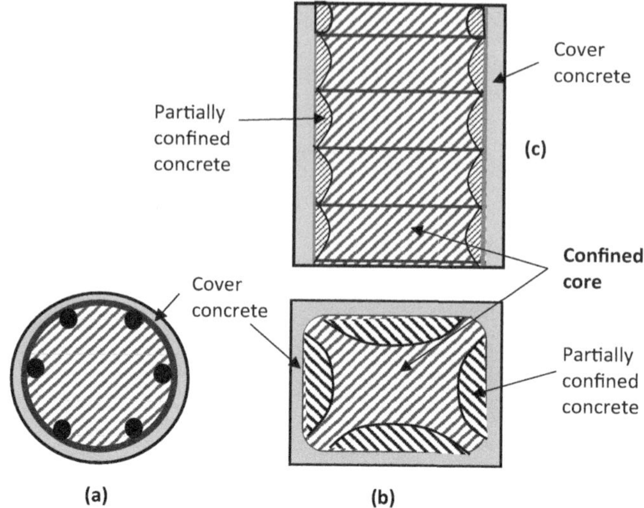

FIGURE 4.24 Confined columns: circular section, (b) square/rectangular section and (c) longitudinal section.

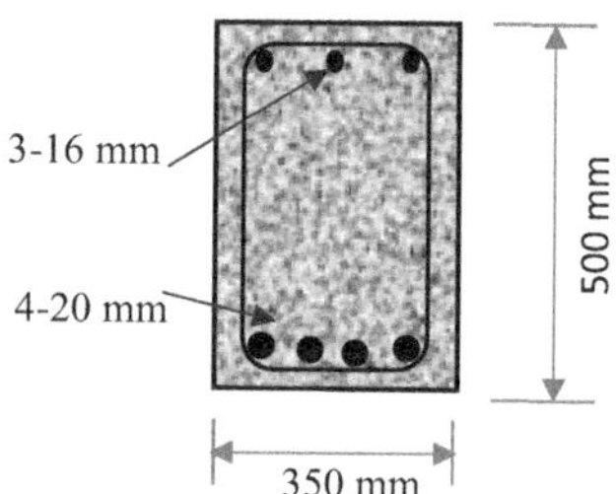

FIGURE 4.25 Beam section in Example 4.14.1.

Figure 4.24 shows the confinement effect in column. Due to the confining effect of tie, three types of concrete are developed, namely, confined core concrete, unconfined outer cover concrete and partially confined concrete. The unconfined concrete spall easily during earthquake shaking.

4.14 EXAMPLES

4.14.1 EXAMPLES RELATED TO SECTION 4.2

Example 4.14.1 *For the beam section shown in Figure 4.25, find the moment of resistance at (1) limiting level, (2) characteristic level, (3) expected level and (4) extreme level. The concrete characteristic strength is 30 MPa and yield strength of steel is 500 MPa. The bottom steel comprises of 4–20 mm bars and top steel comprises 3–16 mm bars. The clear cover is 30 mm. Diameter of stirrup is 8 mm.*

Solution: $A_{st} = 4 \times \dfrac{\pi}{4} 20^2 = 1257$ mm²; $A_{sc} = 3 \times \dfrac{\pi}{4} 16^2 = 603$ mm². $f_{ck} = 30$ MPa,

$f_y = 500$ MPa. $d = 500 - 30 - 8 - 20/2 = 452$ mm; $d_c = 30 + 8 + 16/2 = 46$ mm.

(1) *Limit moment capacity*

$$f_{sy} = 0.87 f_y$$

$$\varepsilon_{sy} = \frac{f_{sy}}{E_s} + 0.002 = \frac{0.87 f_y}{E_s} + 0.002 = \frac{0.87 \times 500}{2 \times 10^5} + 0.002 = 0.004175.$$

$$\varepsilon_{st} = 0.0035 \frac{d - x}{x} = 0.0035 \frac{452 - x}{x}$$

$$\varepsilon_{sc} = 0.0035 \frac{x - d_c}{x} = 0.0035 \frac{x - 46}{x}$$

We assume values of x and tabulate the values in Table 4.3.

Explanation for Table 4.3: Here, ε_{st} is always greater than ε_y; so, tension steel always yields; $f_{st} = 0.87\ f_y = 435$ MPa. From Table A of SP 16, the lowest strain after which nonlinearity starts in Fe500 steel is 0.00174. In step 1 to 3, $\varepsilon_{sc} > 0.00174$. So, f_{sc} is computed by interpolation from Table A of SP 16. In step 4, $\varepsilon_{sc} < 0.00174$. So, $f_{sc} = \varepsilon_{sc} E_s = 342.2$ MPa. $C_{c,\,lim} = 0.36 f_{ck} bx$.

At $x = 80$ mm we get $C = T$. So, Moment capacity is given by -

$$MR = C_s \left(d - d_c \right) + C_c \left(d - 0.42x \right) = 206.4(452 - 46) + 340.2(452 - 0.42 \times 90)$$

kN-mm = **224.7** kN-m.

TABLE 4.3
Computation of *limiting* moment capacity in Example 4.14.1

Step	x	ε_{st}	f_{st}	T	ε_{sc}	f_{sc}	C_s	C_c	C	C = T?
	(mm)		(MPa)	(kN)		(MPa)	(kN)	(kN)	(kN)	
1	100	0.012320	435	546.8	0.001890	363.4	219.1	378.0	597.1	C>T
2	95	0.013153	435	546.8	0.001805	354.6	213.8	359.1	572.9	C>T
3	92	0.013696	435	546.8	0.001750	348.8	210.3	347.8	558.1	C>T
4	**90**	0.014078	435	546.8	0.001711	342.2	206.4	340.2	546.6	**C~T**

TABLE 4.4

Computation of *characteristic* moment capacity in Example 4.14.1

Step	x	ε_{st}	f_{st}	T	ε_{sc}	f_{sc}	C_s	C_c	C	$C = T?$
	(mm)		(MPa)	(kN)		(MPa)	(kN)	(kN)	(kN)	
1	150	0.007047	500	628.5	0.002427	457.9	276.1	850.5	1127	C>T
2	100	0.01232	500	628.5	0.001890	363.4	219.1	567.0	786.1	C>T
3	90	0.014078	500	628.5	0.001711	342.2	206.4	510.3	716.7	C>T
4	80	0.016275	500	628.5	0.001488	297.5	179.4	453.6	633.0	C~T

(2) *Characteristic moment capacity*

$$f_{sy} = f_y$$

$$\varepsilon_y = \frac{f_{sy}}{E_s} + 0.002 = \frac{f_y}{E_s} + 0.002 = \frac{500}{2 \times 10^5} + 0.002 = 0.0045.$$

$$\varepsilon_{st} = 0.0035 \frac{d - x}{x} = 0.0035 \frac{452 - x}{x}$$

$$\varepsilon_{sc} = 0.0035 \frac{x - d_c}{x} = 0.0035 \frac{x - 46}{x}$$

We assume the values of x and tabulate the values in Table 4.4.

Explanation for Table 4.4: Here, ε_{st} is always greater than ε_y; so, tension steel always yields; $f_{st} = f_y = 500$ MPa. From Table A of SP 16, the lowest strain after which nonlinearity starts in Fe500 steel is 0.00174. In steps 1 and 2, $\varepsilon_{sc} > 0.00174$. So, f_{sc} is computed by interpolation from Table of A of SP 16. Further, the table values in SP 16 are for limit strength. So, the obtained stresses are divided by 0.87 to raise them to the characteristic level. In steps 3 and 4, $\varepsilon_{sc} < 0.00174$. So, $f_{sc} = \varepsilon_{sc} E_s$. $C_{c,k} = 1.5 \times 0.36 f_{ck} bx$.

At $x = 80$ mm we get C ~ T. So, Moment capacity is given by -

$$MR = C_s \left(d - d_c \right) + C_c \left(d - 0.42x \right) = 179.4(452 - 46) + 453.6(452 - 0.42 \times 80) \text{ kN-mm} = \mathbf{262.6} \text{ kN-m.}$$

(3) *Expected moment capacity*

$$f_{sy} = f_y + 1.64 \sigma_s$$

TABLE 4.5
Computation of *expected* moment capacity in Example 4.14.1

Step	x	ε_{st}	f_{st}	T	ε_{sc}	f_{sc}	C_s	C_c	C	$C = T?$
	(mm)		(MPa)	(kN)		(MPa)	(kN)	(kN)	(kN)	
1	100	0.01232	582	731.6	0.001890	426	256.9	798.3	1055	C>T
2	90	0.014078	582	731.6	0.001711	342.22	206.4	718.5	924.8	C>T
3	80	0.016275	582	731.6	0.001488	297.50	179.4	638.7	818.0	C>T
4	72	0.018472	582	**731.6**	0.001264	252.78	152.4	574.8	**727.2**	C~T

$$\varepsilon_y = \frac{f_{sy}}{E_s} + 0.002 = \frac{f_y + 1.64\sigma_s}{E_s} + 0.002 = \frac{500 + 1.64 \times 50}{2 \times 10^5} + 0.002 = 0.00491.$$

$$\varepsilon_{st} = 0.0035\frac{d - x}{x} = 0.0035\frac{452 - x}{x}$$

$$\varepsilon_{sc} = 0.0035\frac{x - d_c}{x} = 0.0035\frac{x - 46}{x}$$

We assume values of x and tabulate the values in Table 4.5.

Explanation for Table 4.5: Here, ε_{st} is always greater than ε_y; so, tension steel always yields; $f_{st} = f_y + 1.64\,\sigma_s = 500 + 1.64 \times 50 = 582$ MPa. From Table A of SP 16, the lowest strain after which nonlinearity starts in Fe500 steel is 0.00174. In step 1, $\varepsilon_{sc} > 0.00174$. So, f_{sc} is computed by interpolation from Table of A of SP 16. Further, the table values in SP 16 are for limit strength. So, the obtained stresses are divided by 0.87 to raise them to characteristic level. And then 1.64×5 is added. In step 2 to 4, $\varepsilon_{sc} < 0.00174$. So, $f_{sc} = \varepsilon_{sc}E_s$. $C_{c,\,exp} = 1.5 \times \frac{0.36}{0.67}\left(0.67f_{ck} + 1.64\sigma_c\right)bx$.

At $x = 78$ mm we get C ~ T. So, moment capacity is given by

$MR = C_s\left(d - d_c\right) + C_c\left(d - 0.42x\right) = 152.4(452 - 46) + 574.8(452 - 0.42 \times 78)$ kN-mm = **304.3** kN-m.

(4) *Extreme moment capacity*

$$f_{sy} = f_y + 2 \times 1.64\sigma_s$$

$$\varepsilon_y = \frac{f_{sy}}{E_s} + 0.002 = \frac{f_y + 2 \times 1.64\sigma_s}{E_s} + 0.002 = \frac{500 + 2 \times 1.64 \times 50}{2 \times 10^5} + 0.002 = 0.00532.$$

TABLE 4.6
Computation of *extreme* moment capacity in Example 4.14.1

Step	x	ε_{st}	f_{st}	T	ε_{sc}	f_{sc}	C_s	C_c	C	C = T?
	(mm)		(MPa)	(kN)		(MPa)	(kN)	(kN)	(kN)	
1	100	0.01232	664	834.6	0.001890	434	261.7	1029.6	1291	C>T
2	90	0.014078	664	834.6	0.001711	342.2	206.4	926.7	1133	C>T
3	80	0.016275	664	834.6	0.001488	297.5	179.4	823.7	1003	C>T
4	67.5	0.019937	664	**834.6**	0.001115	222.9	134.4	695.0	**829.4**	C~T

TABLE 4.7
Summary of results of Example 4.14.1

Sl	Strength level	Moment of resistance (kN-m)
1	Limit strength	**224.7**
2	Characteristic strength	**262.6**
3	Expected strength	**304.3**
4	Extreme strength	**349.0**

$$\varepsilon_{st} = 0.0035\frac{d-x}{x} = 0.0035\frac{452-x}{x}$$

$$\varepsilon_{sc} = 0.0035\frac{x-d_c}{x} = 0.0035\frac{x-46}{x}.$$

We assume values of x and tabulate the values in Table 4.6.

Explanation for Table 4.6: Here, ε_{st} is always greater than ε_y; so, tension steel always yields; $f_{st} = f_y + 2 \times 1.64\,\sigma_s = 500 + 2 \times 1.64 \times 50 = 664$ MPa. From Table A of SP 16, the lowest strain after which nonlinearity starts in Fe500 steel is 0.00174. In step 1, $\varepsilon_{sc} > 0.00174$. So, f_{sc} is computed by interpolation from Table of A of SP 16. Further, the table values in SP 16 are for limit strength. So, the obtained stresses are divided by 0.87 to raise them to characteristic level. And then $2 \times 1.64 \times 5$ is added. In step 2 to 4, $\varepsilon_{sc} < 0.00174$. So, $f_{sc} = \varepsilon_{sc} E_s$. $C_{c,\,extreme} = 1.5 \times 0.36(f_{ck} + 2 \times 1.64\sigma_c)bx$.

At $x = 67.5$, mm we get $C \sim T$. So, moment capacity is given by –

$$MR = C_s\left(d - d_c\right) + C_c\left(d - 0.42x\right) = 134.4(452 - 46) + 695(452 - 0.42 \times 78)$$

kN-mm = **349** kN-m. Table 4.7 gives summary of results.

4.14.2   EXAMPLES RELATED TO SECTION 4.5

Example 4.14.2 *A shear wall of horizontal length 5 m has inflection height as 20 m. In a section cut result at the base of the wall, maximum rotation was found as 0.004 radian. If IO, LS and CP for shear wall correspond to 0.005, 0.01 and 0.02 radian respectively, determine the plastic state of the wall. The rebar is of grade Fe415.*

Solution: Here, L_w = 5 m, h_{inf} = 20 m, θ_{max} = 0.004 radian, f_y = 415 MPa.

Now, $\phi_{yW} = \dfrac{2\varepsilon_y}{L_w}$ = 2(415/200000)/5 = 0.00083 per m.

Yield rotation = $\theta_{yW} = \phi_{yw}\dfrac{h_{inf}}{2}$ = 0.00083 × 20/2 = 0.0083 radian.

$\theta_{pW} = \theta_{max} - \theta_{yW}$ = 0.004 – 0.0083 = 0.0043 radian.

So, the plastic rotation being less than 0.005, it is in IO range.

Example 4.14.3 *An RC bridge pier is 5 m high and is 0.8 m in diameter. The longitudinal rebar of diameter 20 mm is of grade Fe415 with ultimate strength 480 MPa. Find the displacement ductility of the pier. Given that total curvature is limited to 0.08 per m.*

Solution: $k = 0.2\left(f_u\Big/f_y - 1\right) = 0.2\left(\dfrac{480}{415} - 1\right) = 0.031 \le 0.08$.

$L_s = 0.022 f_{ye}\phi_l$ = 0.022 × (1.25 × 415) × 20 mm = 228 mm.

$l_p = kL_c + L_s$ = 0.031 × 5000 + 228 = 383 mm $< 2L_s\,(= 456\,mm)$

So, l_p = 456 mm.

From Table 4.1, $\phi_y = 2.25\dfrac{\varepsilon_y}{D}$ = 2.25 × (415/200000)/800 = 5.836 × 10⁻⁶ per mm.

ϕ_{total} = 0.08 per m = 8 × 10⁻⁵ per mm.

$$\mu_\Delta = 1 + 3\frac{\phi_p l_p}{\phi_y H}.$$

Here, $\dfrac{\phi_p}{\phi_y} = \dfrac{\phi_p}{\phi_y} + 1 - 1 = \dfrac{\phi_p + \phi_y}{\phi_y} - 1 = \dfrac{\phi_{total}}{\phi_y} - 1 = \dfrac{8\times10^{-5}}{5.836\times10^{-6}} - 1 = 12.7$.

$$\mu_\Delta = 1 + 3\frac{\phi_p l_p}{\phi_y H} = 1 + 3 \times 12.7 \times \frac{456}{5000} = 4.48.$$

N.B. We could also use the relation: $\phi_p = \phi_{total} - \phi_y$.

Example 4.14.4: *An RC bridge pier is 1500 mm in diameter. The longitudinal bars have diameter 20 mm with yield strength 415 MPa. The design drift is 3.0%. Effective pier height is 9.5 m. Find the ductility and equivalent damping.*

$$\Delta_y = \phi_y \left(H + L_s\right)^2 / 3 \qquad\qquad \text{[Eq. (4.4.4)]}$$

Given data:

D = diameter of pier = 1500 mm

H = effective height of pier = 9500 mm

f_y = 415 = yield strength of longitudinal bar = 415 MPa

f_{ye} = expected strength of rebar = 1.25 f_y = 1.25 × 415 = 518.75 MPa

ε_y = yield strain of longitudinal rebar = f_y / E_s = 415/ 2 × 10^5 = 0.002075

ϕ_l = diameter of longitudinal bar = 20 mm

ϕ_y = yield curvature of pier

$$\phi_y = 2.25\frac{\varepsilon_y}{D} = 2.25 \times \frac{0.002075}{1500}$$

$$= 3.1125 \times 10^{-6}/\text{mm}.$$

From Eq. (4.4.4), yield displacement $\Delta_y = \phi_y \left(H + L_s\right)^2 / 3$.

L_s = strain penetration length = $0.022 f_{ye}\phi_l$ = 0.022 × 518.75 × 20 = 228.3 mm.

$$\Delta_y = \frac{\phi_y \left(H + L_s\right)^2}{3} = 3.1125 \times 10^{-6} \left(9500 + 228.3\right)^2 / 3 = 98.2\text{mm}.$$

Design displacement at top of pier based on drift = $\Delta_d = \theta_d h = 0.03 \times 9500 = 285$ mm.

$$\text{Ductility } \mu = \frac{\Delta_d}{\Delta_y} = \frac{285}{98.2} = 2.9.$$

Equivalent viscous damping at peak response

$$\xi_{eq} = 0.05 + a\frac{\mu - 1}{\mu\pi} =$$

$$= 0.05 + 0.444\frac{2.9 - 1}{2.9\pi} \text{ [value of } a \text{ from Table in Priestley et al., 2007]}$$

$$= 0.142 \ (14.2\%).$$

4.14.3 Example Related to Section 4.6 (Infill)

Example 4.14.5: *An infill panel is shown in Figure 4.26. The centre-to-centre of floor slabs is 3.2 m. The horizontal distance between columns is 4 m. The thickness of infill is 200 mm. Beam size is 350 mm × 450 mm and column size is 400 mm × 400 mm. Modulus of elasticity of concrete is 362,000 psi. Shear strength of masonry is 20 psi. Find (i) the width of equivalent infill strut (ii) axial capacity of the infill strut.*

Solution: Horizontal length of infill panel = 4 − 0.4 = 3.6 m = 141.73 inches.
Vertical height of infill panel = 3.2 − 0.45 = 2.75 m = 108 inches.
t = 200 mm = 7.87 inches.

Angle of inclination of infill strut with the horizontal = $\tan^{-1}\dfrac{2.75}{3.6}$ = 37.4 degree.

Modulus of elasticity of masonry $E_m = 550 f_m = 550 \times 600 = 330000$ psi, for fair condition.

Width of equivalent infill strut, (see Figure Ex4.14.5).

$$a = 0.175\left(\lambda h_c\right)^{-0.4} r_i$$

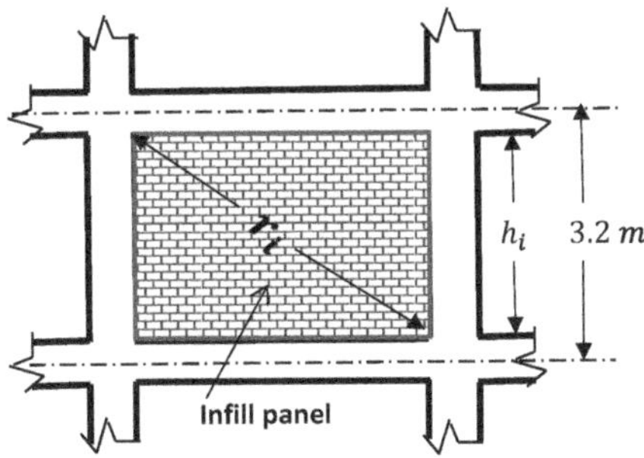

FIGURE 4.26   Infill panel in Ex4.14.5.

$h_c = 3.2 \text{ m} = 125.98 \text{ inches}$

$r_i = \sqrt{2.75^2 + 3.6^2} \text{ m} = 4.53 \text{ m} = 178.4 \text{ inches}.$

Column size = 400 mm = 15.75 inches.

Moment of inertia of column $I_c = \dfrac{15.75 \times 15.75^3}{12} = 5127 \text{ inch}^4.$

where $\lambda = \left[\dfrac{E_m t \sin 2\theta}{4 E_f I_c h_i}\right]^{0.25} = \left[\dfrac{330000 \times 7.87 \times \sin(2 \times 37.4)}{4 \times 362000 \times 5127 \times 108}\right]^{0.25} = 0.042$

$$a = 0.175(\lambda h_c)^{-0.4} r_i = 0.175 \times (0.042 \times 125.98)^{-0.4} \times 178.4$$

$$= 16.6 \text{ inches}.$$

Axial capacity of infill strut = $F = \dfrac{qtL}{\cos\theta} = \dfrac{20 \times 7.87 \times 141.73}{\cos 37.4^\circ} = 28081 \text{ pounds}.$

4.15 CLOSURE

General modelling issues for civil engineering structures have been discussed in this chapter. Accuracy in getting true structural behaviour depends vastly on correct modelling. It may be noted that the modelling takes more time than analysis and design do. Various strength levels have been discussed on statistical terms. In nonlinear analyses and damage-based design, expected strength is to be used.

4.16 EXERCISES

Q4.16.1 Explain different types of strength levels of materials.

Q4.16.2 Explain practical use of different types of strength levels of materials.

Q4.16.3 Explain how the expected moment of resistance of a concrete section is determined.

Q4.16.4 Find the moment of resistance at: (1) limiting level, (2) characteristic level, (3) expected level and (4) extreme level for the beam section shown in

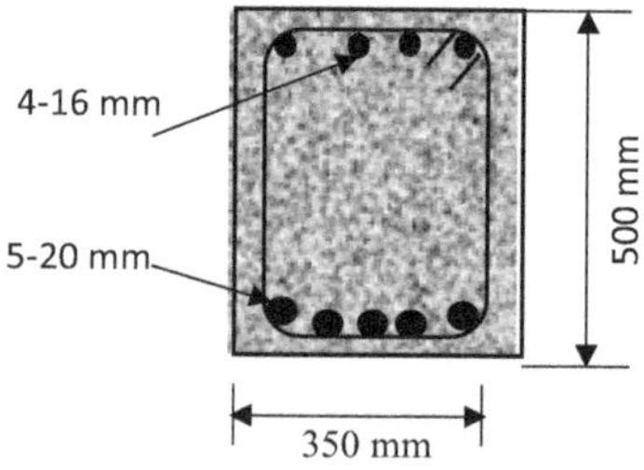

FIGURE 4.27 Beam section in Q4.16.4.

Figure 4.27. The grade of concrete is M35 and grade of steel is Fe415. The bottom steel comprises of 5–20 mm bars and top steel comprises of 4-16 mm bars. The clear cover is 30 mm. Diameter of stirrup is 8 mm.

Q4.16.5 What do you mean by effective stiffness? How it can be determined? Give numerical example.

Q4.16.6 A cantilever RC column 600 mm × 600 mm in size, is 4.5 m in height. The rebar is of diameter 25 mm and of grade Fe500 with ultimate strength 580 MPa. Find plastic hinge length. Also find the plastic displacement and yield displacement. The plastic curvature may be taken as three and half times the yield curvature.

Q4.16.7 Explain how shear walls are modelled in computer.

Q4.16.8 Explain different ductility indices.

Q4.16.9 Explain the effect of confinement of concrete.

FURTHER READINGS

ASCE-41-17 (2017) *Seismic Evaluation and Retrofit of Existing Buildings*, American Society of Civil Engineers.

Das, S. and Choudhury, S. (2019) Influence of Effective Stiffness on the Performance of RC Frame Buildings Designed Using Displacement-based Method and Evaluation of Column Effective Stiffness Using ANN, *Engineering Structures,* June 2019, https://doi.org/10.1016/engstruct.2019.109354

Das, S., Mansouri, I., Choudhury, S. Gandomi, A.H. and Hu, J. W. (2021) A Prediction Model for the Calculation of Effective Stiffness Ratios of Reinforced Concrete Columns, *Materials Journal* (SCIE).. V. 14 (7) . https://doi.org/10.3390/ma14071792

Debnath, P.P. and Choudhury S. (2017) Nonlinear Analysis of Shear Wall in Unified Performance-Based Seismic Design of Buildings, *Asian Journal of Civil Engineering*, 18(4), 633–642.

FEMA-356 (2000) Prestandard and Commentary for the Seismic Rehabilitation of Buildings, US Federal Emergency Management Agency.

IS 1893 (Part 1) (2016) *Criteria for Earthquake Resistant Design of Structures*, Bureau of Indian Standard, New Delhi.

Mander, J.B., Priestley, M.J.N. and Park, R. (1988a), Theoretical Stress–Strain Model for Confined Concrete, *Journal of Structural Engineering*, 114(8), 1804–1849.

Mander, J.B., Priestley, M.J.N. and Park, R. (1988b), Observed Stress–Strain Behavior for Confined Concrete, *Journal of Structural Engineering*, 114(8).

Park, R. and Paulay, T. (1974) *Reinforced Concrete Structures*, John Wiley and Sons.

Pettinga, J.D. and Priestley, M.J.N. (2005) Dynamic Behaviour of Reinforced Concrete Frames Designed with Direct Displacement-Based Design, *Journal of Earthquake Engineering*, 9(2), 309–330.

Priestley, M.J.N., Calvi G.M. and Kowwalsky M.J. (2007) *Displacement-Based Seismic Design of Structures*, IUSS Press, Pavia, Italy.

5 Performance Criteria and Performance Levels

5.1 INTRODUCTION

The aim of PBD is to design structures with some intended design objectives to be achieved under some specified hazard level. Performance criteria are quantifiable structural responses, which lead to performance objectives. Performance criteria look like serviceability criteria in some cases, but they are mostly nonlinear in character. The design and acceptance criteria are mainly obtained from experimental results. Most of the existing design philosophies focus on drift as the design target objective. The UPBD method accommodates both drift and performance level in the design methodology. In future, more target objectives may be incorporated.

5.2 PERFORMANCE LEVELS

Performance level (PL) indicates a damage state of an element or whole structure. The PL can be quantified in terms of plastic deformation of the elements of the structure. PLs can be designated in terms of plastic rotations in beams, walls and columns, cracks in members and drift of members. More damage implies lower PL and vice versa. The PLs can be discrete points or a range on a damage scenario. A simple explanation of PL is given in Figure 5.1, where the force–deformation behaviour of an element is demonstrated. The amount of plastic deformation will determine the PL of the element.

PLs have been given some nomenclature along with quantification in FEMA-273 and FEMA-356 documents. The PLs are designated for structural components and non-structural components separately. Finally, the damage state designation of a building is given as a combination of the structural PLs and non-structural PLs. Qualitatively, the PLs are designated in terms of consequence of strong earthquakes on the structural and non-structural components.

The general descriptors of PLs are: (i) immediate occupancy (IO), where the damage in the structural components and non-structural components is so small that the building can be occupied immediately after the earthquake without major repair or retrofitting; (ii) life safety (LS), where the damage in the structural and non-structural components is large but it does not threaten the life of the occupants and (iii) collapse prevention (CP), where the damage in the structural and non-structural components are very high, but the structure does not collapse. There are ranges of performance other than the said discrete point designators (see Figure 5.2).

DOI: 10.1201/9781003441090-5

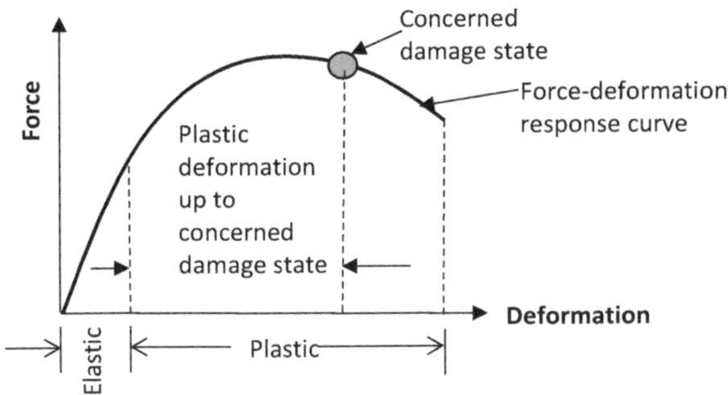

FIGURE 5.1 Illustration of plastic deformation for a PL.

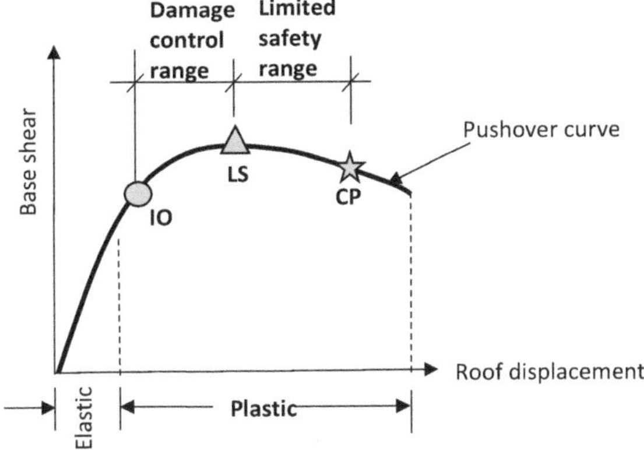

FIGURE 5.2 Performance Levels shown on a pushover curve.

5.2.1 Structural Performance Levels

The structural PLs are expressed as a combination of a letter 'S' and a number. The list of structural PLs is tabulated in Table 5.1 and are illustrated in Figure 5.2.

Figure 5.2 is a plot between base shear and roof displacement of a building. This is called a pushover curve or capacity curve (see details in Chapter 6). A little away from the elastic range (yield point) is the Immediate Occupancy (IO) PL; here damage is limited. Further forward, the damage state is the life safety (LS) PL, where the damage is substantial, but it does not cause a threat to life. At still higher damage, the damage state is called collapse prevention (CP), at which damage is almost complete but the structure has not collapsed. In between IO and LS is the damage control range. In between LS and CP lies the limited safety range. One can opt for building corresponding to the standard discrete PLs such as IO, LS or CP, or one can also have a building on the damage control range or limited safety range. The structural performance levels are tabulated in Table 5.1.

More detailed description of the PLs and acceptance criteria are furnished in Table 5.2.

TABLE 5.1
Structural performance level designators

PL designator	PL name	Description
S1	Immediate Occupancy (IO)	It is defined as the post-earthquake damage state that remains safe to occupy, essentially retaining the pre-earthquake design strength and stiffness of the structure. Here only very limited structural damage occurs. The basic vertical- and lateral-force-resisting systems of the building retain nearly all of their pre-earthquake strength and stiffness.
S2	Damage control *range*	This is defined as the continuous range of damage states between the IO PL and LS PL.
S3	Life Safety (LS)	Structural Performance is defined as the post-earthquake damage state that includes damage to structural components but retains a margin against onset of partial or total collapse. Some structural elements and components are severely damaged, but this does not lead to large falling debris hazards, either within or outside the building. Injuries may occur during the earthquake; however, the overall risk of life-threatening injury as a result of structural damage is expected to be low.
S4	Limited safety *range*	It is a range of damage state lying between LS and CP PLs.
S5	Collapse Prevention (CP)	This is defined as the post-earthquake damage state that includes damage to structural components such that the structure continues to support gravity loads but retains no margin against collapse. Substantial damage to the structure has occurred, potentially including significant degradation in the stiffness and strength of the lateral-force resisting system, large permanent lateral deformation of the structure. However, all significant components of the gravity load resisting system must continue to carry their gravity load demands. Significant risk of injury due to falling hazards from structural debris may exist.
S6	Structural PL *not considered*	Here the structural PL has been considered in design.

5.2.2 Non-structural Performance Levels

Non-structural members in a building include doors, windows, lift, elevator, staircase, heating, ventilation and air conditioning (HVAC) system, oxygen supply pipes, chemical gas supply pipe, infill, building joint, equipment, furniture and contents. The safety of the non-structural items are important for serviceability of the building.

TABLE 5.2
Structural performance levels and acceptance criteria: Vertical elements (after FEMA-356)

Element	Type	IO	LS	CP
Concrete frames	Primary	Minor hairline cracking. Limited yielding possible at a few locations. No crushing (strains below 0.003)	Extensive damage to beams. Spalling of cover and shear cracking (< 3.18 mm width) for ductile columns. Minor spalling in nonductile columns. Joint cracks < 3.18 mm wide.	Extensive cracking and hinge formation in ductile elements. Limited cracking and/or splice failure in some nonductile columns. Severe damage in short columns.
	Secondary	Minor spalling in a few places in ductile columns and beams. Flexural cracking in beams and columns. Shear cracking in joints <1.59 mm width.	Extensive cracking and hinge formation in ductile elements. Limited cracking and/or splice failure in some nonductile columns. Severe damage in short columns.	Extensive spalling in columns (limited shortening) and beams. Severe joint damage. Some reinforcing buckled.
	Drift	1% transient; negligible permanent	2% transient; 1% permanent	4% transient or permanent
Steel moment frames	Primary	Minor local yielding at a few places. No fractures. Minor buckling or observable permanent distortion of members.	Hinges form. Local buckling of some beam elements. Severe joint distortion; isolated moment connection fractures, but shear connections remain intact. A few elements may experience partial fracture.	Extensive distortion of beams and column panels. Many fractures at moment connections, but shear connections remain intact.
	Secondary	-do -	Extensive distortion of beams and column panels. Many fractures at moment connections, but shear connections remain intact.	-do -
	Drift	0.7% transient; negligible permanent	2.5% transient; 1% permanent	5% transient or permanent

(continued)

TABLE 5.2 (Continued)
Structural performance levels and acceptance criteria: Vertical elements (after FEMA-356)

Element	Type	IO	LS	CP
Braced steel frames	Primary	Minor yielding or buckling of braces.	Many braces yield or buckle but do not totally fail. Many connections may fail.	Extensive yielding and buckling of braces. Many braces and their connections may fail.
	Secondary	-do-	-do-	-do-
	Drift	0.5% transient; negligible permanent	1.5% transient; 0.5% permanent	2% transient or permanent
Concrete walls	Primary	Minor hairline cracking of walls, <1.59 mm wide. Coupling beams experience cracking <3.18 mm width.	Some boundary element stress, limited buckling of reinforcement. Some sliding at joints. Damage around openings. Some crushing and flexural cracking. Coupling beams: extensive shear and flexural cracks; some crushing, but concrete generally, remains in place.	Major flexural and shear cracks and voids. Sliding at joints. Extensive crushing and buckling of reinforcement. Failure around openings. Severe boundary element damage. Coupling beams shattered and virtually disintegrated.
Concrete walls	Secondary	Minor hairline cracking of walls. Some evidence of sliding at construction joints. Coupling beams experience cracks <1/8" width. Minor spalling.	Major flexural and shear cracks. Sliding at joints. Extensive crushing. Failure around openings. Severe boundary element damage. Coupling beams shattered and virtually disintegrated.	Panels shattered and virtually disintegrated.
	Drift	0.5% transient; negligible permanent	1% transient; 0.5% permanent	2% transient or permanent
URM Infill Walls	Primary	Minor (<3.18 mm width) cracking of masonry infills and veneers. Minor spalling in veneers at a few corner openings.	Extensive cracking and some crushing but wall remains in place. No falling units. Extensive crushing and spalling of veneers at corners of openings.	Extensive cracking and crushing; portions of face course shed.
	Secondary	-do-	-do-	Extensive crushing and shattering; some walls dislodge.
	Drift	0.6% transient or permanent	0.5% transient; 0.3% permanent	0.1% transient; negligible permanent

URM – other walls	Primary	Minor (<3.18 mm width) cracking of veneers. Minor spalling in veneers at a few corner openings. No observable out-of-plane offsets.	Extensive cracking. Noticeable in-plane offsets of masonry and minor out-of-plane offsets.	Extensive cracking; face course and veneer may peel off. Noticeable in-plane and out-of-plane offsets.
	Secondary	-do-	-do-	Non-bearing panels dislodge.
	Drift	0.3% transient; 0.3% permanent	0.6% transient; 0.6% permanent	1% transient or permanent
RM walls	Primary	Minor (<1/8" width) cracking. No out-of-plane offsets.	Extensive cracking (<6.35 mm) distributed throughout wall. Some isolated crushing.	Crushing; extensive cracking. Damage around openings and at corners. Some fallen units.
	Secondary	-do-	Crushing; extensive cracking; damage around openings and at corners; some fallen units.	Panels shattered and virtually disintegrated.
	Drift	0.2% transient; 0.2% permanent	0.6% transient; 0.6% permanent	1.5% transient or permanent
Foundations	General	Minor settlement and negligible tilting.	Total settlements <152 mm and differential settlements <12.7 mm in 9.1 m.	Major settlement and tilting.

N.B. RM = reinforced masonry; URM = un-reinforced masonry

In some cases, like hospital buildings, the cost of the non-structural items may be up to 85% of the cost of the total cost of the structure. The non-structural PLs are designated as NA, NB etc. as detailed in Table 5.3.

More details of non-structural PLs are given in Table 5.4 and Table 5.5.

TABLE 5.3

Non-structural performance level designators (after FEMA-356)

PL designator	PL name	Description
NA	Operational (N-A)	Non-structural Performance Level N-A, is defined as the post-earthquake damage state in which the non-structural components are able to support the pre-earthquake functions present in the building.
		most non-structural systems required for normal use of the building are functional, although minor clean-up and repair of some items may be required. In addition to assuring that non-structural components are properly mounted and braced within the structure, it is often necessary to provide emergency standby utilities.
NB	Immediate Occupancy (N-B)	Non-structural Performance Level N-B is defined as the post-earthquake damage state that includes damage to non-structural components, but building access and life safety systems including doors, stairways, elevators, emergency lighting, fire alarms, and suppression systems generally remain available and operable, provided that power is available.
		Minor window breakage and slight damage could occur to some components. Presuming that the building is structurally safe, occupants could safely remain in the building, although normal use may be impaired and some clean-up and inspection may be required. In general, components of mechanical and electrical systems in the building are structurally secured and should be able to function if necessary utility service is available. However, some components may experience misalignments or internal damage and be non-operable. Power, water, natural gas, communications lines, and other utilities required for normal building use may not be available. The risk of life-threatening injury due to non-structural damage is very low.
NC	Life Safety (N-C)	Non-structural Performance-Based Level N-C is defined as the post-earthquake damage state that includes damage to non-structural components but the damage is non-life-threatening. Here potentially significant and costly damage has occurred to non-structural components but they have not become dislodged and fallen, threatening life safety

TABLE 5.3 (Continued)
Non-structural performance level designators (after FEMA-356)

PL designator	PL name	Description
		either inside or outside the building. Egress routes within the building are not extensively blocked, but may be impaired by lightweight debris. HVAC, plumbing, and fire suppression systems may have been damaged, resulting in local flooding as well as loss of function. Restoration of the non-structural components may take extensive effort.
ND	Hazard Reduced (N-D)	Non-structural Performance Level N-D is defined as the post-earthquake damage state that includes damage to non-structural components that could potentially create falling hazards. Preservation of egress, protection of fire suppression systems, and similar life safety issues are not addressed in this Non-structural Performance Level. ND represents a post-earthquake damage state in which extensive damage has occurred to non-structural components, but large or heavy items that pose a high risk of falling hazard to a large number of people such as parapets, cladding panels, heavy plaster ceilings, or storage racks are prevented from falling. While isolated serious injury could occur from falling debris, failures that could injure large numbers of persons - either inside or outside the structure, should be avoided.
NE	Non-structural PL not considered	A building design/rehabilitation that does not address non-structural components shall be classified as under this definition. In some cases, the decision to rehabilitate the structure may be made without addressing the vulnerabilities of non-structural components. Also, since many of the most severe hazards to life safety occur as a result of structural vulnerabilities, some municipalities may wish to adopt rehabilitation ordinances that require structural rehabilitation only.

5.2.3 BUILDING PERFORMANCE LEVELS

Building performance levels are expressed in terms of combination of structural PL and non-structural PL. However, a judicious combination is desirable. The building PLs are described in Table 5.6.

$$\text{Building PL} = \text{Structural PL} + \text{Non-structural PL}$$

Any arbitrary combination is not allowed. A high structural PL should not be combined with a low non-structural PL, and vice versa. Because this will lead to damage of the one type before the other. The ideal condition is when both structural and non-structural components get damaged by the amount degree.

TABLE 5.4
Non-structural performance level designators (Architectural components) (after FEMA-356)

Building component	Operational NA	Immediate Occupancy NB	Life Safety NC	Hazards Reduced ND
Cladding	Connections yield; minor cracks (<1.59 mm width) or bending in cladding.	Same as NA	Severe distortion in connections. Distributed cracking, bending, crushing, and spalling of cladding elements. Some fracturing of cladding, but panels do not fall.	Severe distortion in connections. Distributed cracking, bending, crushing, and spalling of cladding elements. Some fracturing of cladding, but panels do not fall in areas of public assembly.
Glazing	Some cracked panes; none broken.	Some cracked panes; none broken.	Extensive cracked glass; little broken glass.	General shattered glass and distorted frames in unoccupied areas. Extensive cracked glass; little broken glass in occupied areas.
Partitions	Cracking to about 1.59 mm width at openings. Minor crushing and cracking at corners.	Same as in NA.	Distributed damage; some severe cracking, crushing, and racking in some areas.	Same as NC
Ceilings	Generally negligible damage. Isolated suspended panel dislocations, or cracks in hard ceilings.	Minor damage. Some suspended ceiling tiles disrupted. A few panels dropped. Minor cracking in hard ceilings.	Extensive damage. Dropped suspended ceiling tiles. Moderate cracking in hard ceilings.	Same as NC
Parapets and ornamentation	Minor damage	Minor damage	Extensive damage; some falling in unoccupied areas.	Same as NC
Chimneys and stacks	Negligible damage.	Minor cracking.	Extensive damage. No collapse.	Same ac NC
Stairs and fire escapes	Negligible damage.	Minor damage.	Some racking and cracking of slabs. Usable.	Extensive racking. Loss of use.
Doors	Minor damage. Doors operable.	Same as NA	Distributed damage. Some racked and jammed doors.	Same as NC

TABLE 5.5
Non-structural performance level designators (Mechanical, Electrical, Plumbing) (after FEMA-356)

Building component	Operational NA	Immediate Occupancy NB	Life Safety NC	Hazards Reduced ND
Elevator	Elevators operate.	Elevators operable; can be started when power available.	Elevators out of service; counterweights do not dislodge.	Elevators out of service; counterweights off rails.
HVAC equipment	Units are secure and operate. Emergency power and other utilities provided, if required.	Units are secure and most operate if power and other required utilities are available.	Units shift on supports, rupturing attached ducting, piping, and conduit, but do not fall.	Most units do not operate; many slide or overturn; some suspended units fall.
Ducts	Negligible damage.	Minor damage at joints, but ducts remain serviceable.	Ducts break loose of equipment and louvers; some supports fail; some ducts fall.	Same as NC
Piping	Negligible damage.	Minor leaks develop at a few joints.	Minor damage at joints, with some leakage. Some supports damaged, but systems remain suspended.	Some lines rupture. Some supports fail. Some piping falls.
Fire Sprinkler Systems	Negligible damage.	Minor leakage at a few heads or pipe joints. System remains operable.	Some sprinkler heads damaged by swaying ceilings. Leaks develop at some couplings.	Some sprinkler heads damaged by collapsing ceilings. Leaks develop at couplings. Some branch lines fail.
Fire Alarm Systems	System is functional.	System is functional.	Ceiling mounted sensors damaged. May not function.	Ceiling mounted sensors damaged. May not function.
Electrical Distribution Equipment	Units are functional. Emergency power is provided, as needed.	Units are secure and generally operable. Emergency generators start, but may not be adequate to service all power requirements.	Units shift on supports and may not operate. Generators provided for emergency power start; utility service lost.	Units slide and/oroverturn, rupturing attached conduit. Uninterruptable Power Source systems fail. Diesel generators do not start.
Plumbing	System is functional. On-site water supply provided, if required.	Fixtures and lines serviceable; however, utility service may not be available.	Some fixtures broken, lines broken; mains disrupted at source.	Some fixtures broken; lines broken; mains disrupted at source.

The allowable combinations of structural PL and non-structural PL are shown in Table 5.6. The standard building nomenclature out of this (see Table 5.7) are as follows:

Operational level building: (OL) 1-A: Applicable for Hospital buildings, other such important buildings.
Immediate occupancy level building: (IO) 1-B: Applicable for important buildings.
Life safety level building: (LS) 3-C: Normal standard buildings, assembly hall other such important buildings.
Collapse prevention level building: (CP) 5-E: Normal buildings.

In a post-earthquake scenario, hospital buildings assume the highest importance. Injured people (likely to be large in number in a major earthquake) need to be hospitalized. The need for a hospital in such a condition is more important than relief and rehabilitation. If a hospital building collapses, it leads to (i) death of indoor patients, (ii) death of doctors, (iii) death of medical staff, (iv) damage of costly equipment and services, (v) disruption of normal services and (vi) decline of admission of injured patients. So, the hospital buildings are to be designed with most stringent requirements and at high seismicity level, may be at MCE level, instead of DBE level.

The relative damage states of buildings are depicted in Figure 5.3. Higher PL corresponds to less damage and lower PL corresponds to high damage. Other combinations of structural PLs and non-structural PLs are shown in Table 5.7. In this table, it is clear that all arbitrary combinations of PLs are not viable. Some combinations are also not economical.

5.3 MEMBER PERFORMANCE LEVELS

The designators of PLs so far discussed depend on the plastic rotations that the members undergo in strong ground motions. The plastic rotations obtained through experiments and refined through consensus are available in FEMA documents sand ASCE-41 series. The plastic rotations for beams, columns and shear walls etc. vary and are given in Tables 5.8 to 5.10. The member PLs in terms of plastic rotations depends on various parameters like reinforcement ratio, shear force etc. It may be noted that there are some differences between FEMA-356 and ASCE-SEI-41-17 values. The later has not differentiated the primary and secondary divisions of members.

In Tables 5.8 to 5.10 the values of force–deformation response parameters (a, b, d, d, e) are not shown. These have been discussed in Chapter 4.

N.B. In Table 5.8, (i) ρ_t is percentage tension steel, ρ_c is percentage compression steel, ρ_{bal} is percentage balanced steel. (ii) 'Avg' means average. (iii) 'C' and 'NC' are abbreviations for conforming and nonconforming transverse reinforcement. A component is conforming if, within the flexural plastic hinge region, hoops are spaced at $\leq d/3$, and if, for components of moderate and high ductility demand, the strength provided by the hoops (Vs) is at least three-fourths of the design shear. Otherwise, the component is considered nonconforming. For India condition, 'C' may be interpreted

TABLE 5.6
Building performance level designators

PL designator	PL name	Structural PL in the combination	Non-structural PL in the combination	Description
OL S1-NA Or 1A	Operational level	S1	NA	In this PL, the structural components of the building shall meet the requirements of Immediate Occupancy Structural Performance Level (S-1) and the non-structural components shall meet the requirements of Operational Non-structural Performance Level (N-A). Buildings meeting this Building PL are expected to sustain minimal or no damage to their structural and non-structural components. The building is suitable for its normal occupancy and use, although possibly in a slightly impaired mode, with power, water, and other required utilities provided from emergency sources, and possibly with some nonessential systems not functioning. Buildings meeting this target Building Performance Level pose an extremely low risk to life safety.
IO S1-NB or 1B	Immediate Occupancy level	S1	NB	It is the combination of IO PL (S-1) for structural N-B PL for non-structural components. Such buildings are expected to sustain minimal or no damage to their structural elements and only minor damage to their non-structural components. While it would be safe to re-occupy a building meeting this target PL immediately following a major earthquake, non-structural systems may not function, either because of the lack of electrical power or internal damage to equipment. Some clean-up and repair may be necessary before the building can function in a normal mode. The risk to life safety at this target Building Performance Level is very low. This level provides most of the protection obtained under the OL Buildings without the cost of providing standby utilities.

(continued)

TABLE 5.6 (Continued)
Building performance level designators

PL designator	PL name	Structural PL in the combination	Non-structural PL in the combination	Description
LS S3-NC or 3C	Life Safety level	S3	NC	LS Building Performance Level is combination of S3 structural PL and NC non-structural PL. Such buildings may experience extensive damage to structural and non-structural components. Repairs may be required before re-occupancy of the building occurs, and repair may be deemed economically impractical I some cases. The risk to life safety in buildings meeting this target Building PL is low.
CP S5-NE or 5E	Collapse Prevention level	S5	NE	This Building PL is a combination of S5 structural PL and NE non-structural PL. Non-structural components are not considered. Buildings meeting this target Building Performance Level may pose a significant hazard to life safety resulting from failure of non-structural components. However, because the building itself does not collapse, gross loss of life may well be avoided. Many buildings meeting this level will be complete economic losses. This level has been sometimes selected as the basis for mandatory seismic rehabilitation ordinances enacted by municipalities, as it results in mitigation of the most severe life-safety hazards at relatively low cost.

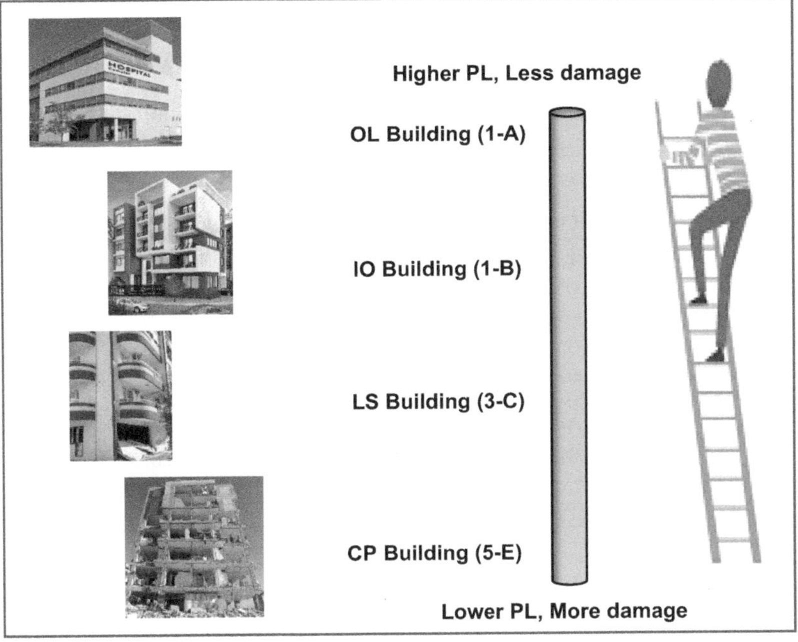

FIGURE 5.3 Standard building nomenclature as per PL.

TABLE 5.7
Viable combinations of building PLs

Building Performance Levels

	Structural PLs					
Non-structural PL ↓	S1 IO	S2 Damage control range	S3 LS	S4 Limited safety range	S5 CP	S6 Not considered
NA Operational	1-A OL building	2-A	NR	NR	NR	NR
NB Immediate Occupancy	1-B IO building	2-B	3-B	NR	NR	NR
NC Life Safety	1-C	2-C	3-C LS building	4C	5-C	6-C
ND Hazard reduced	NR	2-D	3-D	4-D	5-D	6-D
NE Not considered	NR	NR	3-E	4-E	5-E CP building	NA

N.B. NR= not recommended, NA = not applicable.

TABLE 5.8

Plastic rotations for RC beam elements (in radian) (after FEMA-356)

| | | | | | Component type | | | | | | | |
| | | | | | Primary | | | | Secondary | | | |
$\dfrac{\rho_t-\rho_c}{\rho_{bal}}$	Ties conform	$\dfrac{V}{bd\sqrt{f_c}}$	IO	Avg IO	LS	Avg LS	CP	Avg CP	LS	Avg LS	CP	Avg CP
≤ 0.0	C	≤ 3	0.010	**0.00625**	0.02	**0.01125**	0.025	**0.02**	0.02	**0.0187**	0.05	**0.035**
≤ 0.0	C	≥ 6	0.005		0.01		0.02		0.02		0.04	
≥ 0.5	C	≤ 3	0.005		0.01		0.02		0.02		0.03	
≥ 0.5	C	≥ 6	0.005		0.005		0.015		0.015		0.02	
≤ 0.0	NC	≤ 3	0.005	0.00325	0.01	0.007	0.02	0.01125	0.02	0.01125	0.03	0.0175
≤ 0.0	NC	≥ 6	0.0015		0.005		0.01		0.01		0.015	
≥ 0.5	NC	≤ 3	0.005		0.01		0.01		0.01		0.015	
≥ 0.5	NC	≥ 6	0.0015		0.005		0.005		0.005		0.01	

TABLE 5.9

Plastic rotations for RC columns controlled by flexure (after FEMA-356)

| | | | | | Component type | | | | | | | |
| | | | | | Primary | | | | Secondary | | | |
$\dfrac{P}{A_g f_c}$	Ties conform	$\dfrac{V}{bd\sqrt{f_c}}$	IO	Avg IO	LS	Avg LS	CP	Avg CP	LS	Avg LS	CP	Avg CP
≤ 0.0	C	≤ 3	0.005	**0.004**	0.015	**0.01225**	0.02	**0.01575**	0.02	**0.01675**	0.03	**0.02475**
≤ 0.0	C	≥ 6	0.005		0.012		0.016		0.016		0.024	
≥ 0.5	C	≤ 3	0.003		0.012		0.015		0.018		0.025	
≥ 0.5	C	≥ 6	0.003		0.01		0.012		0.013		0.02	
≤ 0.0	NC	≤ 3	0.005	0.0035	0.005	0.003	0.006	0.00725	0.01	0.01125	0.015	0.0112
≤ 0.0	NC	≥ 6	0.005		0.004		0.005		0.008		0.012	
≥ 0.5	NC	≤ 3	0.002		0.002		0.003		0.006		0.01	
≥ 0.5	NC	≥ 6	0.002		0.002		0.002		0.005		0.008	

N.B. Symbols as in Table 5.8.

as those conforming to IS 13920. Here, b is width of beam, d is effective depth of beam, f_c is concrete strength.

In ASCE-SEI-41-17 similar tables are provided. For IO level, the acceptance criteria are same for primary and secondary members. But for LS and CP, the tables give acceptance criteria for *secondary members only*. Those for the primary members are to be computed by the user. This has led to lot of confusions for the users. For example, SAP2000 software has implemented the table values of this document directly. Thus, the *default hinges give properties of secondary members and not primary members*. The user has to generate user-defined hinges for the primary members.

TABLE 5.10

Plastic rotations for RC Shear walls controlled by flexure (after FEMA-356)

$\dfrac{(A_t - A_c)+P}{L_w t_w f_c}$	Confined boundary	$\dfrac{V}{L_w t_w \sqrt{f_c}}$	IO	Avg IO	LS	Avg LS	CP	Avg CP	LS	Avg LS	CP	Avg CP
					— Primary				Secondary			
≤ 0.1	Yes	≤ 3	0.005	**0.003375**	0.010	**0.00675**	0.015	**0.00975**	0.015	**0.00975**	0.020	**0.01425**
≤ 01	Yes	≥ 6	0.004		0.008		0.010		0.010		0.015	
≥ 0.25	Yes	≤ 3	0.003		0.006		0.009		0.009		0.012	
≥ 0.25	Yes	≥ 6	0.0015		0.003		0.005		0.005		0.010	
≤ 0.1	No	≤ 3	0.002	**0.0015**	0.004	**0.00275**	0.008	**0.00475**	0.008	**0.00475**	0.015	**0.0085**
≤ 0.1	No	≥ 6	0.002		0.004		0.006		0.006		0.010	
≥ 0.25	No	≤ 3	0.001		0.002		0.003		0.003		0.005	
≥ 0.25	No	≥ 6	0.001		0.001		0.002		0.002		0.004	

Note: A_t = area of tension steel, A_c = area of compression steel, P = axial force in column, L_w = length of wall, t_w = thickness of wall, f_c = compressive strength of concrete.

TABLE 5.11
Building design/Rehabilitation objectives

		Building Performance Levels			
		OL (1-A)	IO (1-B)	LS (3-C)	CP (5-E)
Earthquake Hazard	50%/50 year	a	b	c	d
Levels	20%/50 year	e	f	g	h
	BSE-1 (~10%/50year)	i	j	k	l
	BSE-2 (~2%/50 year)	m	n	o	p

Notes: Each cell represents a discrete design/rehabilitation objective.
k + p = Basic Safety Objective (BSO)
k + p + (any of a, e, i, h, f, j or n) = Enhanced objective
o, n, m alone = Enhanced objective
k alone or p alone = Limited objectives
c, g, d, h, l = Limited objectives

This gives troubles to the user. ASCE-SEI-17 could easily add columns for primary members in the tables, thus avoiding inconvenience to the users.

5.4 MULTI-OBJECTIVE DESIGN

In Chapter 3, various safety objectives and hazard levels have been described. Designing building for only one objective may not be judicial and economical. A building should satisfy very the condition of less damage at very les seismicity level and moderate damage at moderate hazard level. It may suffer high damage at high hazard level. Combining these concepts, design safety objectives are shown in Table 5.11.

5.4.1 BASIC SAFETY OBJECTIVE (BSO)

The Basic Safety Objective (BSO) is a design/rehabilitation objective that achieves the dual rehabilitation goals of (i) life safety building performance level (3-C) for the BSE-1 earthquake hazard level and (ii) collapse prevention building performance level (5-E) for the BSE-2 earthquake hazard level.

The basic safety objective (BSO) refers to the earthquake risk of life safety category, generally intended by design codes. Buildings meeting the BSO are expected to experience little damage from relatively frequent, moderate earthquakes, but significantly more damage and potential economic loss from the most severe and infrequent earthquakes that could affect them.

5.4.2 ENHANCED DESIGN/REHABILITATION OBJECTIVES

Design/rehabilitation that provides building performance exceeding that of the BSO is termed an enhanced objective (EO). Enhanced objectives shall be achieved using one or a combination of the following two methods:

(i) By designing for target building performance levels that exceed those of the BSO at either the BSE-1 or BSE-2 hazard levels, or both.

(ii) By designing for the target building performance levels of the BSO using an earthquake hazard level that exceeds either the BSE-1 or BSE-2 hazard levels, or both.

So, in a nutshell, enhanced objectives can be obtained by designing for higher target building performance levels (method i) or by designing using higher earthquake hazard levels (method ii) or any combination of these two methods.

5.4.3 Limited Design/Rehabilitation Objectives

Design/rehabilitation that provides building performance less than that of the BSO is termed a limited objective. Limited objectives shall be achieved using the following conditions:

(a) The measures shall not result in a reduction in the performance level of the existing building.

(b) The measures shall not create a new structural irregularity or make an existing structural irregularity more severe.

(c) The measures shall not result in an increase in the seismic forces to any component that is deficient in capacity to resist these forces.

(d) All new or rehabilitated structural components and elements shall be detailed and connected to the existing structure suitably.

5.4.4 Reduced Design/Rehabilitation Objective

Design/rehabilitation that addresses the entire building structural and non-structural systems, but uses a lower seismic hazard or lower target building performance level than the BSO, is termed as reduced rehabilitation objective. Reduced rehabilitation shall be designed for one or more of the following objectives:

(i) Life safety building performance level (3-C) for earthquake demands that are less severe (more probable) than the BSE-1;

(ii) Collapse Prevention building performance level (5-E) for earthquake demands that are less severe (more probable) than the BSE-2;

(iii) Building performance levels 4-C, 4-D, 4-E, 5-C, 5-D, 5-E, 6-D, or 6-E for BSE-1 or less severe (more probable) earthquake demands.

Table 5.11 is to be understood this way – BSO indicates the structure has to satisfy LS criterion under earthquake level BSE-1, and it also has to satisfy CP level at BSE-2. Similarly with other cells. In this table, 'm' is the highest building PL under highest earthquake.

Figure 5.4 shows the design/rehabilitation objectives pictorially.

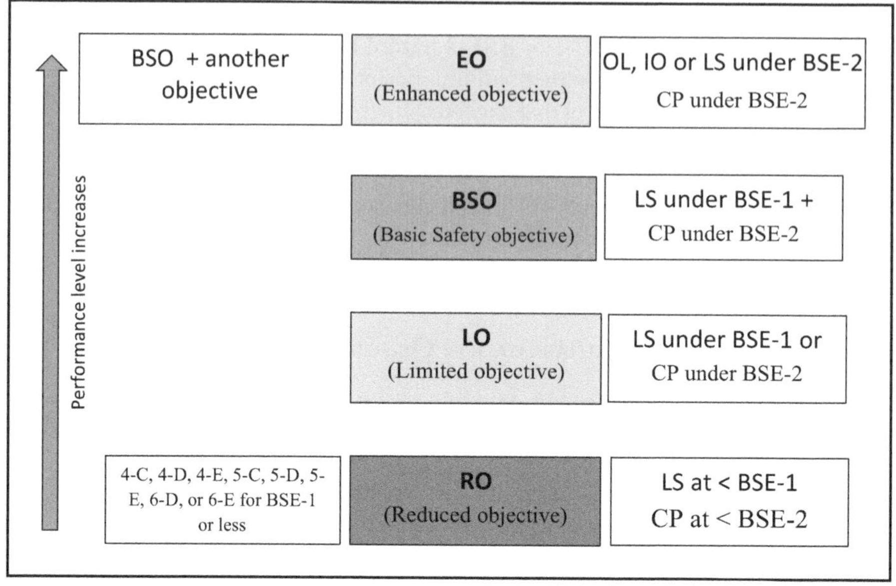

FIGURE 5.4   Design/rehabilitation objectives.

5.5 CLOSURE

In this chapter, the basic performance levels have been described. The safety of structural as well as non-structural components are equally important for overall safety of the structure. In this chapter, special emphasis has been laid down on buildings. The rehabilitation objectives have been discussed.

5.6 EXERCISES

Q 5.6.1 Define performance level.

Q 5.6.2 Describe structural performance level and non-structural performance levels.

Q 5.6.3 Describe building nomenclature with respect to performance levels.

Q 5.6.4 Any arbitrary combination of hazard level and performance level is not justified - discuss.

Q 5.6.5 What do you mean by multi-objective design? Discuss.

FURTHER READINGS

ASCE-SEI-41-17 (2017) Seismic Evaluation and Retrofit of Existing Buildings, ASCE.

FEMA-273 (1996) NEHRP Guidelines for the Seismic Rehabilitation of Buildings, US Federal Emergency Management Agency, Building Seismic Safety Council, Washington DC.

FEMA-356 (2000) Prestandard and Commentary for the Seismic Rehabilitation of Buildings, US Federal Emergency Management Agency.

6 Analysis for the Evaluation of Structures

6.1 INTRODUCTION

In performance-based design, a structure is designed for a set of pre-defined target performance objectives under a specified hazard level. There are engineering techniques to carry out these designs. But the designed structure needs to be checked if it really satisfies the set targets under the pre-set hazard level. This is done through nonlinear analyses of the designed structure. However, linear analysis also may throw some light on the structural behaviours. A classification of analysis types is shown in Figure 6.1.

6.2 LINEAR ANALYSIS

Linear analysis, also known as elastic analysis, is carried out considering the structure to be behaving linearly. This happens when the applied actions are within the corresponding yield strengths of the elements. The analyses that are covered in undergraduate courses like slope deflection method of analysis and matrix method of analysis come under linear analysis. In the realm of dynamics or earthquake engineering, modal analysis and codal response spectrum method of analysis are linear analyses. Linear time history analysis (LTHA) also can be carried out when ground motions are applied on the structure and its response histories are captured without considering nonlinearity in the system. Some insight, like floor spectra of buildings, are obtained through linear dynamic analysis (LDA). Various methods of time history analysis (both linear and nonlinear) are available such as Newmark's method, Wilsons's method, collocation method and Hilber–Hughes–Taylor method. Linear analysis may be applied when higher mode effect is negligible. Linear analyses are applicable when the degree of nonlinearity in the structure is low. Quantitatively, linear procedure is applicable when demand capacity ratio (DCR) is less than 2.0.

6.3 NONLINEAR ANALYSIS

Nonlinearity arises out of (i) material nonlinearity and (ii) P-delta effect and large deflection. Here the elements yield and undergo further plastic deformation. In the computer model, nonlinear hinges are to be assigned to appropriate locations of the

DOI: 10.1201/9781003441090-6

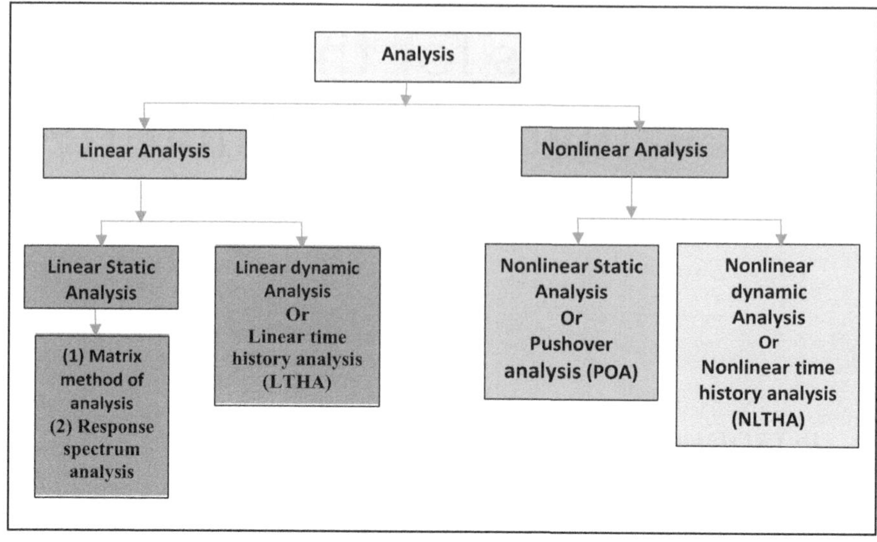

FIGURE 6.1　Analysis types.

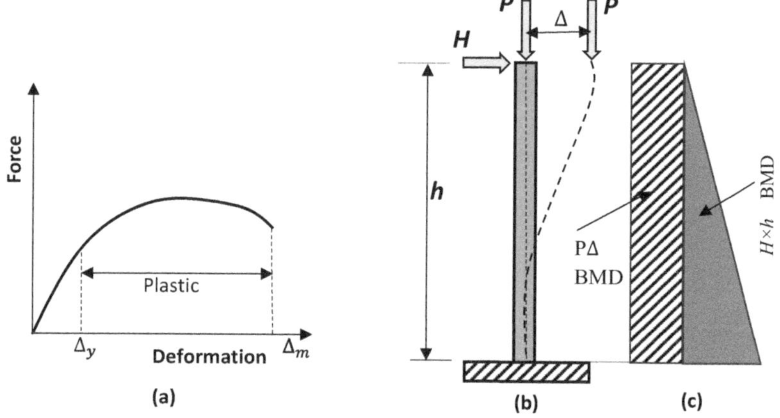

FIGURE 6.2　(a) Nonlinear response of structure/element and (b) P-Δ effect.

members. Members will show nonlinear force–deformation response (Figure 6.2). The nonlinear actions may be in flexure, shear, torsion or axial force. Accordingly, we can have flexural hinge, shear hinge and axial hinge in the element.

Nonlinear analyses may be (i) nonlinear static analysis (NSA) or (ii) nonlinear dynamic analysis (NDA). Nonlinear static analysis is called also pushover analysis (POA). Nonlinear dynamic analysis is also called nonlinear time history analysis (NLTHA). POA uses monotonically increasing static load of some specified pattern. POA results approximately simulates the dynamic conditions. The NLTHA is supposed to give most accurate results as the response. From NLTHA, the performance level of the structure can be judged, the drift can be computed and variation of any response parameter over the tie duration of applied ground motion can be obtained.

6.3.1 P-Delta Effect and its Consideration

P-delta effect (or P-Δ effect) has been partly described in Section 4.7. It arises due to sway of columns (Figure 6.2b). There is initial bending moment at the base of the column equal to $H \times h$. Now as the column sways laterally, new moment arises whose magnitude is $P \times \Delta$. As the deflection increases, the new component of the moment also increases. This may eventually lead to failure of the column.

P-delta effect can be of two types: (i) P-delta effect for linear analysis and (ii) dynamic P-delta effect.

6.3.1.1 P-Delta Effect for Static Linear Procedure (FEMA-356)

For determining the P-delta effect in linear procedure (linear analysis), Eq. (6.3.1.1) is used to evaluate the stability coefficient θ_i.

$$\theta_i = \frac{P_i \delta_i}{V_i h_i} \tag{6.3.1.1}$$

where P_i = portion of the total weight of the structure including dead, permanent live, and 25% of transient live loads acting on the columns and bearing walls within story level i

δ_i = lateral drift in story i, in the direction under consideration, at its centre of rigidity, using the same units as for measuring h_i

V_i = the total calculated lateral shear force in the direction under consideration at storey i due to earthquake response to the selected ground shaking level, as indicated by the selected linear analysis procedure

h_i = height of story i

Action to be taken is as follows:

If $\theta_i < 0.1$ in all stories: P-delta effect need not be taken.

If $0.1 < \theta_i < 0.33$: seismic forces and deformations in storey i is to be increased by the factor $\dfrac{1}{1-\theta_i}$ for the purpose of rehabilitation/redesign

If $\theta_i > 0.33$, the building is not safe. Re-design/rehabilitation to the full extent is needed till θ_i becomes less than 0.33.

6.3.1.2 P-Delta Effect for Static Nonlinear Procedure

For nonlinear procedures, static P-delta effects shall be incorporated in the analysis by including in the mathematical model the nonlinear force–deformation relationship of all elements and components subjected to axial forces.

6.3.1.3 P-Delta Effect for Dynamic Nonlinear Procedure

Dynamic P-delta effect shall be considered using coefficient C_3.

If θ_i <0.1 in all stories, take $C_3 = 1.0$.

Else, take $C_3 = 1 + 5\dfrac{\theta_{i,max} - 1}{T}$, T being the first mode time period.

The maximum value of θ_i is taken for all the stories.
For further understanding, refer to FEMA-356.

6.4 PUSHOVER ANALYSIS

6.4.1 GENERAL

Pushover analysis (POA) is a nonlinear static analysis. The structure is at first loaded with a gravity load. Now, some pattern of lateral loads of small magnitude is applied. This pushes the structure in the direction of the applied loads. The base shear and displacement at the top of the structure is noted. The lateral loads are increased by some percentage. The new displacement is noted. This process is continued till some desired displacement level is attained or till the structure becomes unstable. The lateral load applied is called the monotonically increasing load. The curve that we get by joining the response points (base shear vs. roof displacement) is called pushover curve (POC) or capacity curve. The curve is supposed to simulate the backbone curve of hysteresis. That is why the static procedure simulates the dynamic earthquake effects partially.

To understand the process, let us consider a portal frame as in Figure 6.3. Suppose the flexural strengths of members 1, 2, and 3 are S_1, S_2 and S_3, respectively, such that $S_1 < S_2 < S_3$. The frame is loaded with a gravity load. Now a small lateral load H is applied at top. The generated base shear is $V_{b1} = H$. This produces displacement at top by amount Δ_1. We note down the point (V_{b1}, Δ_1). Now, let us increase the lateral load by 2% (i.e. new lateral load is $1.02H$). Let the top displacement be Δ_2. Note the point (V_{b2}, Δ_2). These steps will be repeated and a series of points (V_{bi}, Δ_i). will be obtained for increasing values of i. Joining these points, we get the POC. Note that

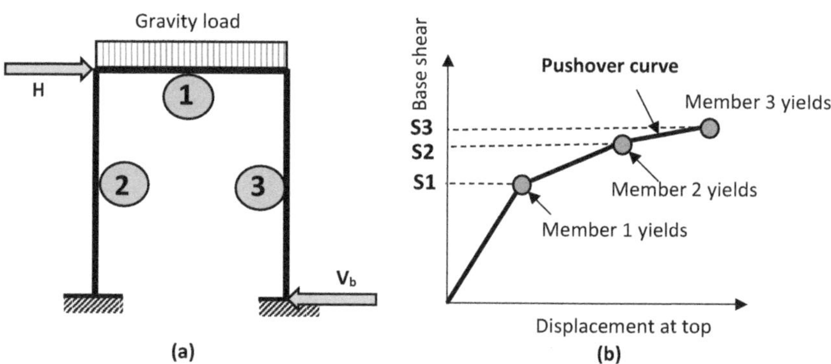

FIGURE 6.3   (a) A portal frame and (b) pushover curve.

during the process the member with the least strength yields first. Gradually, other members also yield.

For multilevel structures like buildings, the lateral load of the chosen pattern is placed at each floor level. Then the question arises – what lateral load is to be taken for a structure? The lateral load can be constant over the height and may vary in a triangular fashion or parabolic fashion. FEMA-356 has given the flowing two series of lateral load patterns: pattern A and pattern B. At least, one pattern from A and one pattern from B is to be taken for POA.

6.4.2 LATERAL LOAD PATTERNS

Pushover analysis (POA) is carried out on a structure already subjected to gravity load; a step-by-step lateral load is applied to push the system in one side. As the lateral load is applied, the opposite base shear equal in magnitude with the total lateral load comes into play. Initially, some steps will undergo the process in the elastic condition. After some pushing steps, members will start forming plasticity, partially or fully. Gradually, more members will fail. At each step of lateral load increment, the base shear (V_i) and the roof displacement ($\Delta_{roof\,i}$) are noted. Pushover curve (also called capacity curve) is the line joining all such points ($\Delta_{roof\,i}, V_i$).

The question arises – what type of lateral load is to be applied? Some simplified lateral loads are: (i) triangular distribution, (ii) rectangular distribution and (iii) inverted triangular distribution. Some arbitrary lateral load patterns are shown in Figure 6.4. Such lateral loadings may be called as arbitrary. Our aim is to get a simulation of backbone curve through a simplified analysis (backbone curve is the average path line of hysteretic damage cycles). So, a judicious choice of lateral load is necessary. A first-mode proportional lateral load pattern seems to be rational, as the structure will bend in the first mode shape. But the first mode use will exclude the higher mode effect. Also, the masses in the system play an important role in the nonlinear behaviour. So, a mass-proportional lateral load distribution is also a rational one. Considering all these, FEMA has given some specified lateral load pattern, which is highlighted below.

The lateral loads are given in two parts, which are shown here (Section 6.4.2.1 and Section 6.4.2.2) as part A and part B. At least one loading from each part is to be taken in POA.

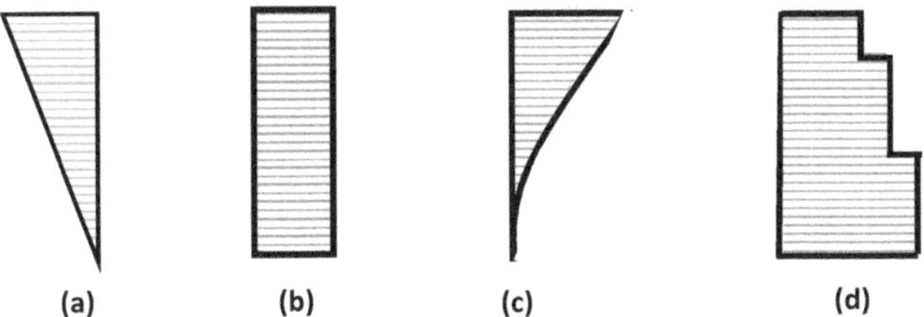

FIGURE 6.4 Lateral load patterns: (a) triangular, (b) constant, (c) mode-proportional and (d) mass proportional.

6.4.2.1 Lateral Load Patterns A

A.1 A vertical distribution *proportional to the values of* C_{vx} given in Equation (6.4.2.1). The use of this distribution shall be permitted only when more than 75% of the total mass participates in the fundamental mode in the direction under consideration, and the uniform distribution is also used.

A.2 A vertical distribution proportional to the shape of the *fundamental mode* in the direction under consideration. The use of this distribution shall be permitted only when more than 75% of the total mass participates in this mode.

A.3 A vertical distribution *proportional to the storey shear distribution* is calculated by combining modal responses from a response spectrum analysis of the building, including sufficient modes to capture at least 90% of the total building mass and using the appropriate ground motion spectrum. This distribution shall be used when the period of the fundamental mode exceeds 1.0 sec.

$$C_{vx} = \frac{W_x h_x^k}{\sum_i^n W_i h_i^k}$$

(6.4.2.1)

where C_{vx} = vertical distribution factor at x-th floor
   W_x = weight of x-th floor
   W_i = weight of i-th floor
   h_i = height of i-th floor from the base of the building
   n = number of stories
   h_x = height of x-th floor from the base of the building
   k = 2.0 for $T \geq 2.5$ s
    = 1.0 for $T \leq 0.5$ (linear interpolation permitted)
   T = fundamental time period of building.

6.4.2.2 Lateral Load Patterns B

B.1 A *uniform distribution* consisting of lateral forces at each level is proportional to the total mass at each level.

B.2 An *adaptive load distribution* that changes as the structure is displaced. The adaptive load distribution shall be modified from the original load distribution using a procedure that considers the properties of the yielded structure.

The uniform load distribution (B.1) means the lateral load at any floor is applied in proportion to the mass of that floor.

The adaptive load distribution (B.2) has to change as per the current displacement of the structure. This is more accurate but much more difficult to implement.

Out of the necessary two to be taken from two groups, A.2 (mode proportional) and B.1 (mass proportional) are easy to apply. It may be noted that mode proportional loading can be applied if the mode under consideration (first mode in the direction concerned) has at least 75% mass participation. It is obvious that higher mode effects are left out. This has been taken care of in multimodal POA (MPOA), discussed in Section 6.4.3. As such, mode-proportional (A.2) loading represents a lower bound of capacity of the structure. Since all masses are involved, uniform loading (B.1) represents the upper bound of the capacity.

6.4.2.3   How Many Pushover Curves?

Pushover analysis is to be done for a structure under two lateral load patterns as described above. The pushover curves (POC) are to be obtained in both the orthogonal directions of the structural plan. So, we need a minimum of four POCs. Again, if there is irregularity in the structure, POC along positive and negative directions (i.e. load applied from the other side) will be different. So, you can have 2×2×2 or 8 POCs for a structure.

A POC is also called capacity curve, as it depicts the capacity of the structure at various displacement levels. A POC is typically drawn between the roof displacement and base shear. The base shear at any lateral load application step is equal to the sum of the applied lateral forces. A typical POC is show in Figure 6.5.

6.4.3   MULTIMODE PUSHOVER ANALYSIS

We have seen in Section 6.4.2 that mode-proportional POA with one mode cannot represent the structural behaviours correctly. So, the need for multimodal POA (MPOA) was felt. Sasaki et al. (1998) reported a MPOA. Here, the load pattern is taken based on the modes considered. The load at a level is the mode shape times the mass at that level. For each load pattern, the POA is carried out and the response of interest is found out. In the final diagram, the demand curve and as many pushover curves as the number of modes considered will appear. From the graph, the study of the structure in different modes and the total structural behaviour becomes clear.

Chopra and Goel (2001, 2002, 2003a, 2003b) also reported the MPOA procedure. Here, the pushover curve is produced based on a particular mode of interest and then it is bilinearized. It is now converted to force–displacement relation for equivalent single degree of freedom (ESDOF) system. The peak response of interest is evaluated. Similarly, the steps are repeated for higher modes and finally the peak responses are combined by square root of sum of square (SRSS) rule. The limitations of modal pushover analysis (MPA) are that they consider one mode at a time and hence the combination result remains doubtful' and that the P-delta effect is ignored.

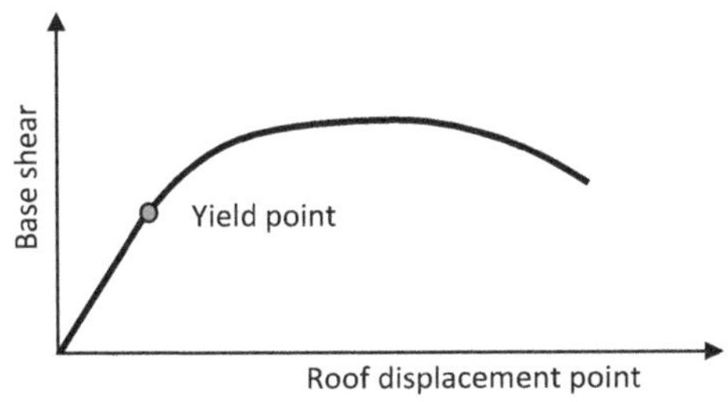

FIGURE 6.5   A typical pushover curve.

6.4.4 Modified Modal Pushover Analysis

Chopra and Goel (2004) reported a modified modal pushover analysis (MMPA) procedure. It was argued that this procedure is more realistic than the MPOA. In the MMPA procedure, the response contributions of higher vibration modes are computed by assuming the building to be linearly elastic, thus reducing the computational efforts. The natural frequencies and modes for linearly elastic building are computed. For the first mode, the base shear versus roof displacement pushover curve is obtained. The pushover curve is idealized as a bilinear curve. The idealized pushover curve is converted to the force–displacement relation. The peak roof displacement associated with the first-mode inelastic single degree of freedom (SDOF) system is calculated for the combined gravity load and lateral load. Now response is calculated for gravity load alone. The dynamic response due to various modes is computed. The total response is obtained by using the SRSS rule.

6.4.5 Other Types of Pushover Analyses

Gupta and Kunnath (2000) developed an *adaptive pushover analysis* where the pushover load changes with the steps. A damped elastic response spectrum is used. Modal analysis is done and storey forces for each of mode considered is computed. The storey forces are scaled with modal base shear. Nonlinear static analysis is now done. Member forces are now computed. If any member yields, the stiffness matrix is updated.

Aydinoğlu (2007) reported an *incremental response spectrum analysis* (IRSA) for multi-mode pushover analysis. It is a piecewise linear response spectrum analysis applied at each step of NSP. The equal displacement principle is utilized. P-delta effect is also considered in this method. Goel and Chopra (2005) extended the modal pushover analysis to compute member forces.

Kuramoto and Matsumoto (2004) reported a *mode adaptive pushover analysis*. The curves from CSM are used here. Stiffness-dependent lateral load distribution proportional to arbitrary mode of vibration at each loading step is applied. Higher mode effect is considered in the sequential analysis.

6.5 EVALUATION OF STRUCTURES: PERFORMANCE POINT

The buildings designed through the philosophy of PBD need to be evaluated to check if it can really attain the target design objectives under the given hazard level. We need to compare the target with achieved values (performance level and drift). In this context, a point on the capacity curve called *performance point* (PP) is very important. The performance point is a unique point where the capacity of the structure is equal to the (reduced) demand on the structure. The demand imposed from earthquake gets changed as the structure suffers damages. The damage induces hysteresis, which dissipates energy. As a result, the demand reduces. This is called *reduced demand*.

We can now define: *Performance point is a unique point where the capacity of the structure equals the reduced demand in the structure.*

The performance point can be described in terms of real response values ($\Delta_{roof, PP}$, V_b) or in terms of spectral values ($S_{d,PP}$, S_a). The displacement corresponding to the PP ($\Delta_{roof, PP}$ or, $S_{d,PP}$) is called target displacement (Δ_t) or design displacement (Δ_d). We have to determine the design displacement so that we can design the structure for the target displacement.

The methods of determination of the displacement at PP are as follows:

   (i)   Displacement coefficient method (FEMA-356)
   (ii)   Capacity spectrum method (ATC-40)
   (iii) Equivalent linearization method (FEMA-440)
   (iv) Displacement modification method (FEMA-440).

These methods are discussed in the sections to follow.

6.6   DISPLACEMENT COEFFICIENT METHOD

The displacement coefficient method was originally given in FEMA-273 (1997) and later on incorporated in FEMA-356 (2000) with a little modification. Here the target displacement is obtained theoretically based on some coefficients and effective time period. The modal participation factor for first mode may be used in the process. The coefficients used accommodate the effects of equivalent single degree of freedom system, inelastic action, hysteresis and P-Δ influence. For each of these influencing components, a separate coefficient is given. The method is straightforward, as it does not require construction of any diagram. However, with time, it was found that the method gave inconsistent results in many cases and this led to the modification of the coefficients as given in FEMA-440 (2004).

The method gives a formula for the computation of target displacement as given by Eq. (6.6.1).

$$\Delta_t = C_0 C_1 C_2 C_3 S_a \frac{T_e^2}{4\pi^2} g \tag{6.6.1}$$

where Δ_t is the target displacement

      g is acceleration due to gravity

      C_0 = Modification factor to relate spectral displacement of an equivalent SDOF system to the roof displacement of the MDOF building system, is calculated using one of the following procedures:

- The first modal participation factor at the level of the control node (that is, roof) is as follows:
- The modal participation factor at the level of the control node calculated using a shape vector corresponding to the deflected shape of the building at the target displacement. This procedure shall be used if the adaptive load pattern defined is used.
- The appropriate value is taken from Table 6.1.

TABLE 6.1
Values of coefficient C_0 (FEMA-356)

| | Shear building | | Other buildings |
No. of stories	Triangular load pattern	Uniform load pattern	For ay load pattern
1	1.0	1.0	1.0
2	1.2	1.15	1.2
3	1.2	1.2	1.3
5	1.3	1.2	1.4
10 or more	1.3	1.2	1.5

Note: (1) Linear interpolation is permitted. (2) Shear building is that in which the drift decreases with increase of height.

TABLE 6.2
Values of C_m (FEMA-356)

No. of Stories	Concrete Moment Frame	Concrete Shear Wall	Concrete Pier-Spandrel	Steel Moment Frame	Steel Concentric Braced Frame	Steel Eccentric Braced Frame	Others
1 to 2	1.0	1.0	1.0	1.0	1.0	1.0	1.0
> 3	0.9	0.8	0.8	0.9	0.9	0.9	1.0

C_1 = Modification factor related to expected maximum inelastic displacements to displacements calculated for linear elastic response = 0 for $T_e \geq T_s$; T_e is effective time period, T_s is characteristic time period that refers to the time period in the design spectrum as a junction of constant acceleration and constant velocity segments

$$= \frac{\left[1.0 + \frac{(R-1)T_s}{T_e}\right]}{R} \text{ for } T_e < T_s; \ R \text{ is strength factor given as ratio of elastic strength}$$

demand to calculated yield strength coefficient, calculated by following expression:

$$R = \frac{S_a W}{V_y} C_m$$

C_m = effective mass factor to account for higher mode effects as given in Table 6.2. The effective mass factor to account for higher mode mass participation effects obtained from Table 6.2. C_m shall be taken as 1.0 if the fundamental period, T, is greater than 1.0 sec.

NB. If fundamental time period $T > 1.0$ sec, $C_m = 1.0$.

TABLE 6.3
Values of C_2 (FEMA-356)

Structural PL	$T \geq 0.1$ sec		$T \leq T_s$ sec	
	Framing type 1	Framing type 2	Framing type 1	Framing type 2
IO	1.0	1.0	1.0	1.0
LS	1.3	1.0	1.1	1.0
CP	1.5	1.0	1.2	1.0

Notes: (1) Linear interpolation to be applied for other values of T; (2) Type 1 structure: Structures in which more than 30% of the storey shear at any level is resisted by any combination of the following components, elements, or frames: ordinary moment-resisting frames, concentrically braced frames, frames with partially restrained connections, tension-only braces, unreinforced masonry walls, shear-critical, piers, and spandrels of reinforced concrete or masonry. Rest is taken as type 2 structures.

C_2 = Modification factor represents the effect of pinched hysteretic shape, stiffness degradation and strength deterioration on maximum displacement response. The values of C_2 for different framing systems and structural performance levels are obtained from Table 6.3. Alternatively, the use of C_2 = 1.0 shall be permitted for nonlinear procedures.

$$C_3 = 1.0 + \frac{|\alpha|(R-1)^{1.5}}{T_e}$$

α = Ratio of post-yield stiffness to effective elastic stiffness where the nonlinear force–displacement relation shall be characterized by a bilinear relation as shown in Figure 6.6.

Some of the symbols used above can be read by considering Figure 6.6.

For bilinearization of POC with respect to any PP, a line from the PP is extended backwards, while a line from the origin is drawn inclined upwards such that two lines meet at point A where the magnitude of base the shear force is V_y in such a way that (i) the line from origin cuts the capacity curve at $0.6\,V_y$ and (ii) area bounded by POC and first line is equal to the area bounded by second line and POC.

6.7 CAPACITY SPECTRUM METHOD

Capacity spectrum method (CSM) was given in ATC-40 document. It is an iterative procedure based on the geometrical construction for finding the target displacement. The method visually determines the performance point (PP). A general feel of the process is given in Figure 6.7. More details will follow. The standard ordinates of the

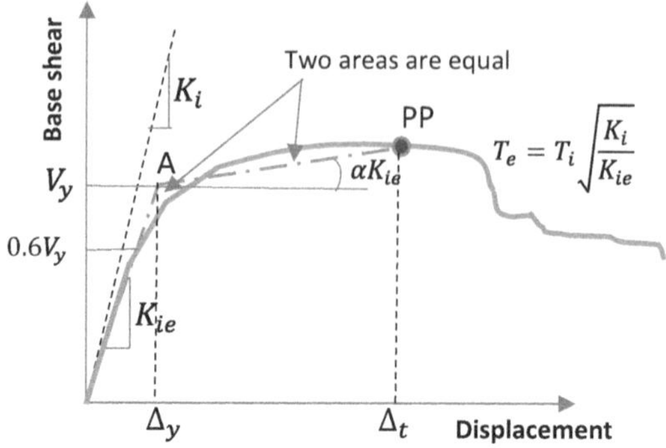

FIGURE 6.6   Bilinearization of capacity curve.

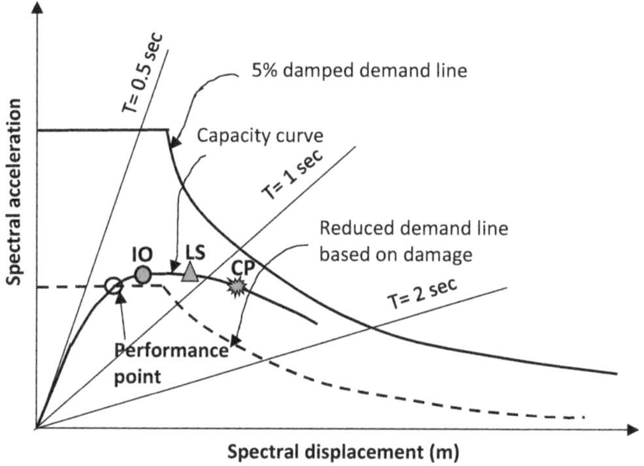

FIGURE 6.7   A general picture of CSM showing performance point.

capacity curve are roof displacement and base shear. Making some transformation of the ordinate values, the same can be converted to spectral displacement and spectral acceleration. Because of the use of spectral values, the method is called the capacity spectrum method.

Figure 6.7 shows a capacity curve superimposed with a demand curve, on $S_d - S_a$ plot. In design codes, the demand curve is 5% damped demand curve. The structure receives energy from ground motion, which gets dissipated through hysteresis, as damage sets in. As seismic action keeps rising to higher level, the structures undergo increasing damages. The damage reduces the stiffness and hence increases the time period. This is called *period lengthening*. If the lengthened period corresponds to the falling parabolic part of the demand diagram, the demand on the structure decreases. The damage is analogous to the increase in damping. So, we can interpret the reduction in demand to be due to increased damping. The decreased demand is called *reduced demand*, shown by dotted line in Figure 6.7. We shall discuss how to arrive at the reduced demand corresponding to a particular damage state, in subsequent sections. The intersection point of the reduced demand curve with the capacity curve is the *performance point* (PP). The significance of PP is this that here the (reduced) demand in the structure is equal to the capacity of the structure. So, corresponding to this point is the ideal displacement for which the structure should be designed for a desired PL (damage state). This is the *target or design displacement*.

Note that the radial lines from the centre of the Figure 6.7 indicate constant period lines. The reason for this is that the relationship between S_d and S_a is given as $S_a = \omega^2 S_d$, where $\omega = 2\pi/T$. If T is constant, then the relationship becomes analogous to $y = mx$ style of a straight line passing through the origin. Hence, the constant period lines are radial lines from the origin in the plot.

6.7.1 ADRS FORMAT OF CAPACITY SPECTRUM AND DEMAND SPECTRUM

Traditional POC is roof displacement (say, in m) versus base shear (say, in kN). The traditional demand spectrum is time period (s) versus spectral acceleration ratio (S_a/g). If we wish to superimpose the two diagrams in a single plot, it is not possible as the axes values are all different. We can convert the axes parameters by using dynamical equations so that the x-axis parameters are the same for both demand diagram and capacity diagram. A similar treatment can be applied for y-axis too. Now the two diagrams can be superimposed in the same plot. The conversion equations are Eqs. (6.7.1.1) to (6.7.1.3). Such converted axes parameters are called acceleration-displacement response spectrum (ADRS) format. ADRS format is actually $S_d - S_a$ format.

$$\text{Convert } T - \text{axis to } S_d - \text{axis: } S_{d,i} = \frac{T_i^2}{4\pi^2} \tag{6.7.1.1}$$

$$\text{Convert } \Delta_{roof} - \text{axis to } S_d - \text{axis: } S_{d,i} = \frac{\Delta_{roof,i}}{(PF)_1 \, \phi_{1,roof}} \tag{6.7.1.2}$$

$$\text{Convert } V_b - \text{axis to } S_a - \text{axis: } S_{a,i} = \frac{V_{b,i}}{\alpha_1 W} \tag{6.7.1.3}$$

where
$(PF)_1$ = Modal participation factor for first mode

$$= \left[\frac{\sum_{i=1}^{N} W_i \phi_{i1}}{\sum_{i=1}^{N} W_i \phi_{i1}^2} \right] \tag{6.7.1.4}$$

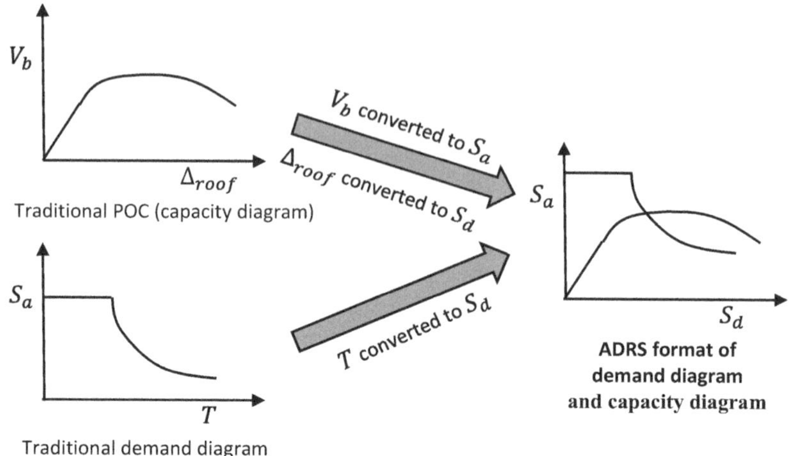

FIGURE 6.8   Demand diagram and capacity diagram in ADRS format.

α_1 = modal mass coefficient for the first mode

$$= \frac{\left[\sum_{i=1}^{N} W_i \phi_{i1}\right]^2}{\left[\sum_{i=1}^{N} W_i\right]\left[\sum_{i=1}^{N} W_i \phi_{i1}^2\right]} \qquad (6.7.1.5)$$

W_i = seismic weight at floor i

ϕ_{i1} = mode shape coefficient for first mode a floor i.

In Eq. (6.7.1.1) to (6.7.1.3) i indicates i-th point. The conversion is explained in Figure 6.8.

6.7.2   DAMPING CORRESPONDING TO A DAMAGE STATE

We have discussed the process of bilinearization of POC in Section 6.6. Here we find the *damping* corresponding to a damage state that is designated as the PP. The displacement corresponding to the PP is target displacement Δ_t (or, design displacement Δ_d). Consider Figure 6.9 where the bilinearized POC is shown. The bilinearization gives us the yield base shear V_y and yield displacement Δ_y. The POC shown as dark curve has been bilinearized by the dotted lines OA and AB. After bilinearization, the hysteretic loop is given by parallelogram GBEF and area under the POC is equal to area of quadrantal parallelogram OABH. For the sake of brevity, we shall represent the PP by (d_{pi}, a_{pi}) and yield point A by (d_y, a_y).

The energy dissipated in one complete hysteresis cycle (E_D) is equal to the area under the parallelogram GBEF, which is four times the area of one parallelogram

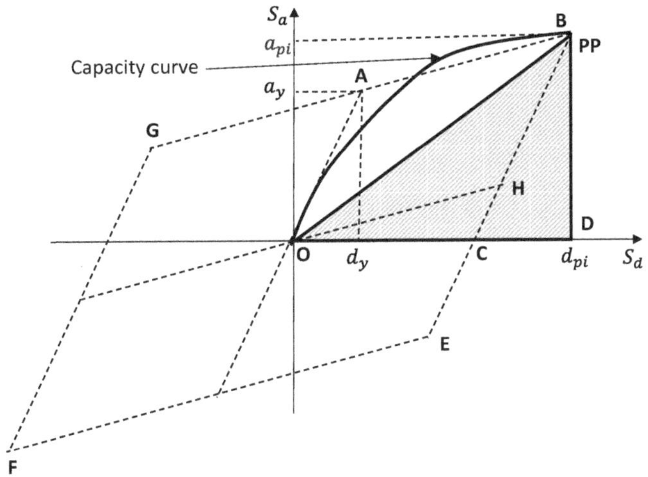

FIGURE 6.9 Hysteresis behaviour under POC.

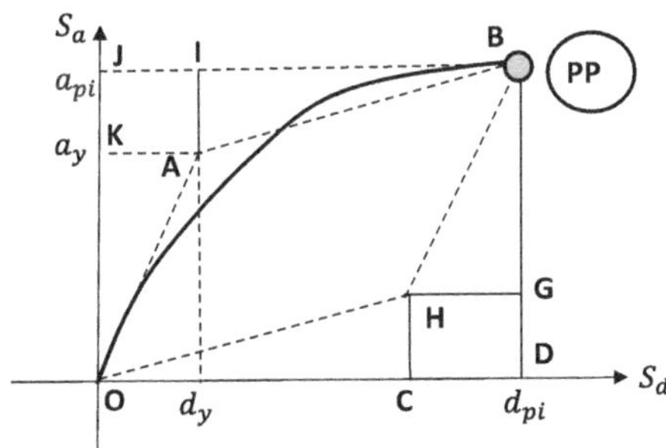

FIGURE 6.10 One quadrant of hysteresis behaviour under POC.

OABH. The maximum strain energy (E_{so}) stored is given by the area of triangle OBD = (1/2) $d_{pi}a_{pi}$.

$$E_{so} = \frac{d_{pi}a_{pi}}{2} \tag{6.7.2.1}$$

The damping associated in the hysteresis is given by Eq. (6.7.2.2).

$$\xi_h = \frac{1}{4\pi}\frac{E_D}{E_{so}} \tag{6.7.2.2}$$

Let us now calculate the value of E_D from the geometry. For this purpose, we consider the first quadrant of the hysteresis loop as shown in Figure 6.10.

Referring to Figure 6.10, by symmetry, area KAIJ = area CDGH

Area OAK = area HGB
Area ABI = area OCH
Area of parallelogram OABH
$\quad\quad$ = area of rectangle ODBJ − 2(area KAIJ + area OAK + area ABI)

$$= a_{pi}d_{pi} - 2\left[\left(a_{pi} - a_y\right)d_y + 0.5a_y d_y + 0.5\left(d_{pi} - d_y\right)\left(a_{pi} - a_y\right)\right]$$

$$= a_y d_{pi} - d_y a_{pi}$$

Total hysteretic energy is area of bigger parallelogram = $E_D = 4(a_y d_{pi} - d_y a_{pi})$.

Hysteretic damping

$$\xi_h = \frac{1}{4\pi}\frac{E_D}{E_{so}} = \frac{1}{4\pi}\frac{4(a_y d_{pi} - d_y a_{pi})}{0.5 a_{pi} d_{pi}}$$

$$\xi_h = \frac{1}{\pi}\frac{a_y d_{pi} - d_y a_{pi}}{a_{pi} d_{pi}} \tag{6.7.2.3}$$

Total damping is the summation of hysteretic damping and material damping.

$$\xi_t = \xi_h + \xi_m \tag{6.7.2.4}$$

Material damping for concrete is 5% and that for steel, it is 2%.

The reduction factor to get demand spectrum at other damping may be given by Eq. (6.7.2.5).

$$\text{Reduction factor} = \sqrt{\frac{10}{5 + \xi_h}} \tag{6.7.2.5}$$

However, ATC-40 gives reduction factors for three types of buildings. The reduction factors for getting reduced demand are shown in Table 6.4.

TABLE 6.4
Acceleration reduction factors for reduced demand

ξ_h (%)	As per formula (6.7.2.5)	For new building as per ATC-40
0	1.41	1.00
5	1.00	0.78
10	0.81	-
15	0.70	0.55
20	0.63	-
25	0.58	0.44
30	0.53	-
35	0.50	0.38
40	0.47	-
45	0.45	0.33

6.7.3 DETERMINATION OF PERFORMANCE POINT

There are three methods (named as procedure A, B, C) given in ATC-40 for finding the PP. Here we shall discuss the first method, which is assumed to be the most accurate one. The steps involved are described below.

Data given are: Designed/existing building; 5% damped demand diagram as hazard level.

(1) Generate the POC for the building and convert it to ADRS format. Convert the demand diagram to ADRS format. Superimpose both in the same plot.

(2) Consider an arbitrary PP say (d_{p1}, a_{p1}). The selection of the point may depend on the judgement of how much damage the building may suffer under MCE level of ground shaking, or in other words, its PL.

(3) Bilinearize the capacity curve and obtain yield point (d_y, a_y).

(4) Obtain damping from Eq. (6.7.2.3).

(5) Obtain the reduced demand diagram by multiplying the ordinates of demand diagram by factor given by Eq. (6.7.2.5).

(6) Note the intersection point of reduced demand diagram with capacity curve, say (d_{p2}, a_{p2}).

if, $0.95\, d_{p1} \le d_{p2} \le 1.05 d_{p1}$

acceptable PP is (d_{p2}, a_{p2});

Else, go to step (2) with new PP (d_{p3}, a_{p3}).

The process is explained in Figure 6.11.

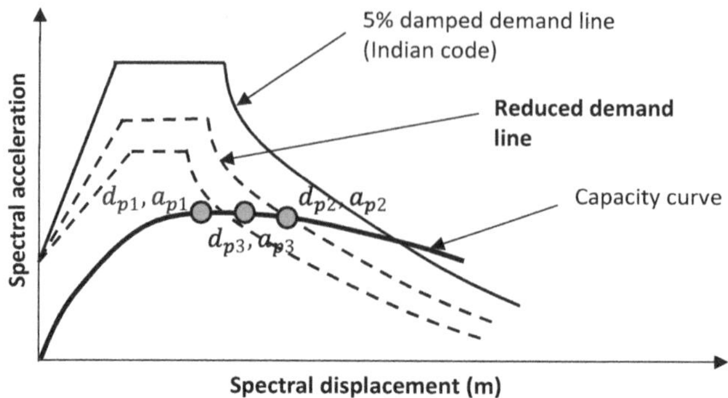

FIGURE 6.11   Iteration for obtaining PP.

6.8   DISPLACEMENT MODIFICATION METHOD (FEMA-440)

FEMA-440 has improved coefficients C_1 and C_2 in FEMA-356 coefficient method.

$$\Delta_t = C_0 C_1 C_2 C_3 S_a \frac{T_e^2}{4\pi^2} g \tag{6.8.1}$$

$$C_1 = 1 + \frac{R-1}{aT_e^2} \tag{6.8.2}$$

If $T < 0.2$ sec, C_1 value corresponding to 0.2 sec may be taken
   If $T > 1.0$ sec, $C_1 = 1.0$.

$$a = 130 \text{ for Class B site}$$
$$= 90 \text{ for Class C site}$$
$$= 60 \text{ for Class D site}$$

$$C_2 = 1 + \frac{1}{800}\left(\frac{R-1}{T}\right)^2 \tag{6.8.3}$$

If $T < 0.2$ sec, C_2 value of 0.2 sec may be taken
If $T > 0.7$ sec, $C_2 = 1.0$.

Other coefficients are same as those of FEMA-356.

Displacement modification procedures estimate the total maximum displacement of the oscillator by multiplying the elastic response, assuming initial linear properties and damping by one or more coefficients. The coefficients are typically derived empirically from series of nonlinear response history analyses of oscillators with varying periods and strengths (FEMA-440).

6.9 EQUIVALENT LINEARIZATION METHOD (FEMA-440)

Equivalent linearization technique assumes that the maximum total displacement (elastic plus inelastic) of an SDOF oscillator can be estimated by the elastic response of an oscillator with a larger period and damping than the original. These procedures use estimates of ductility to compute effective period and effective damping. The steps in the method are highlighted below.

(1) Consider the demand curve and capacity curve. Convert them to the ADRS format.

(2) Consider an arbitrary initial PP, say (d_{p1}, a_{p1}). The selection of the point may depend on the judgement of how much damage the building may suffer under MCE level of ground shaking, or in other words, its PL.

(3) Bilinearize the capacity curve and obtain yield point (d_y, a_y).

(4) Obtain post yield stiffness ratio, ductility and effective damping as follows:

$$\alpha = \frac{a_{pi} - a_y}{d_{pi} - d_y} \frac{d_y}{a_y}$$

$$\mu = \frac{d_{pi}}{d_y}$$

Effective time period T_e and effective damping ξ_e are obtained from set of equations (FEMA-440).

(5) Determine the estimated maximum displacement, d_i, using the intersection of the radial effective period, T_e, with the ADRS for ξ_e. The estimated

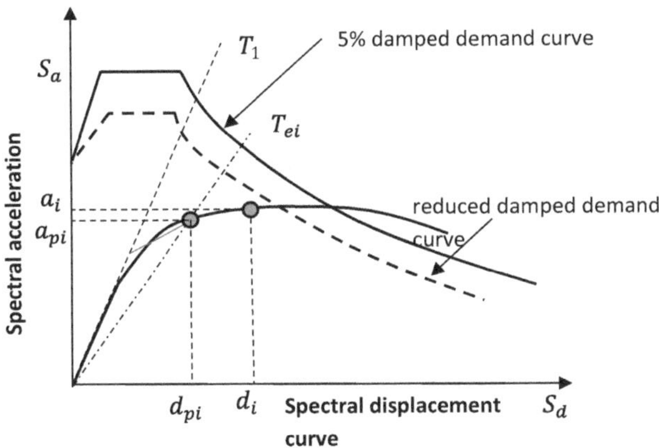

FIGURE 6.12 Equivalent linearization procedure.

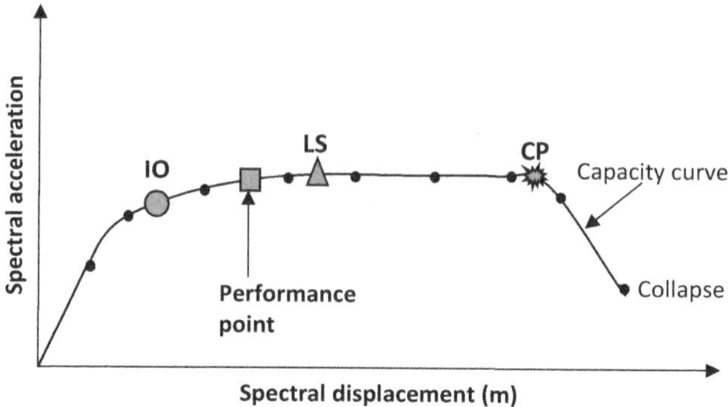

FIGURE 6.13 PL of a building.

maximum acceleration, a_i, is that corresponding to d_i on the capacity curve (see Figure 6.12).

(6) Compare the estimated maximum displacement, d_j, with the initial (or previous) assumption. If it is within acceptable tolerance, the performance point corresponds to a_i and d_i. If it is not within acceptable tolerance, then repeat the process from Step 2 using a_i and d_i or some other selected point, as a starting point.

6.10 PERFORMANCE POINT AND PERFORMANCE LEVEL OF STRUCTURES

After carrying out the nonlinear static procedure (NSP) we get the PP. This will help us know the PL of the building as a whole. Consider Figure 6.13. The PP has been obtained by using any of the NSPs. The damage states IO, LS and CP are also shown. Point IO corresponds to the displacement beyond which members exhibit LS PL. Similarly, the concept applies to LS and CP states. Now, the PP here is beyond IO and with LS states. So, it is building of LS PL.

6.11 TIME HISTORY ANALYSIS

The dynamic evaluation of the structure is carried by POA and time history analysis. Time history analysis may be linear time history analysis (LTHA) or nonlinear time history analysis (NLTHA). Time history analysis is a step-by-step analysis. The record interval of ground motion is further split into finer segments and analysis is done for satisfying dynamic equilibrium at each step for each node. On analysis, we get the displacement response history, velocity response history and acceleration response history. The time history analysis enables us to obtain quantities as (but not limited to) drift, floor and roof response histories, hinge results, variation of base shear and other quantities and floor spectra. From the hinge results at last step of time history analysis, we get the performance level of the structure. The calculation of interstorey drift ratio (IDR) is explained below.

Let $u_i(t)$ denote displacement history of i-th floor and $u_{i+1}(t)$ that of $(i + 1)$-th floor.

$$\text{Drift vector} = \left| u_{i+1}(t) - u_i(t) \right|$$

$$\text{IDR vector} = \text{drift vector}/\, h_s .$$

where, h_s is interstorey height between i-th and $(i +1)$-th floor.

It may be mentioned that floor spectra are the average floor response spectra of floors when any floor is subjected to acceleration time history of that floor. Floor mounted equipment are to be designed for average floor spectra and not design spectrum of code.

6.11.1   How Many Earthquakes?

For evaluation a structure through time history analysis, the number of earthquakes to be considered in the analysis is to be decided. For incremental dynamic analysis (IDA), again a different suite of ground motions to be considered. More ground motions lead to more accurate prediction of response covering record-to-record variability.

6.11.1.1   ASCE-7-16 Provisions

Prior to 2016 version, this document prescribed three ground motions and later on it prescribed seven ground motions. When three ground motions were to be taken, the maximum response was to be adopted for engineering purposes. When seven grounds motions were prescribed, the average response could be used. In ASCE-7-16 version, 11 ground motions have been prescribed, and the average response is to be reported.

6.11.1.2   FEMA P-695 Provisions

These provisions are used for incremental dynamic analysis leading to fragility evaluation. The ground motions are divided into two categories, namely, near field records and far field records. In far field records, there are 22 recorded ground motions. The magnitude of the earthquake varies from 6.5 to 7.6. The earthquakes are recorded earthquakes in the United States, Japan, Turkey, Italy, Taiwan and Iran.

In near field records, there are 28 recorded ground motions. The earthquake magnitude varies from 6.5 to 7.6. The earthquakes in the set are recorded earthquakes in the United States, Turkey, Italy, Taiwan and Canada. There is no strict consensus about the distance of fault up to which it can be called a near field. Generally, it is taken as 10–15 km from fault source. Near field structures are prone to wave action of earthquakes. The tall structures in near field distance may show resonance during earthquake.

Near field records have higher ground velocity which may lead to higher damages.

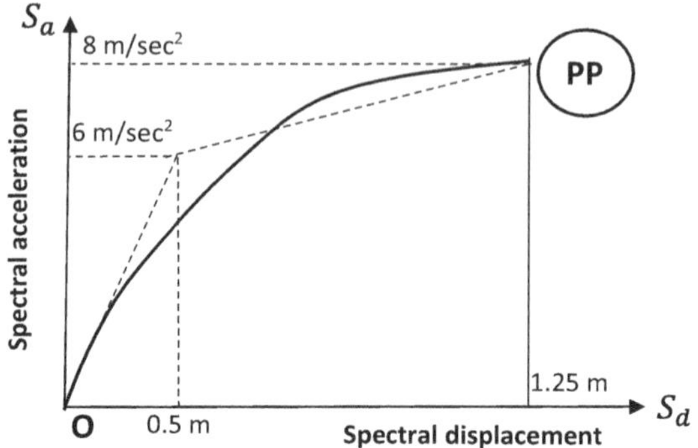

FIGURE 6.14   Figure on Example Ex6.12.1.

6.12   EXAMPLE

Ex6.12.1 A hysteresis curve for an RC section is shown in Figure 6.14. Find the damping associated with the performance point.

From the figure given. we observe that yield point, (d_y, a_y) is (0.5, 6); the performance point, (d_p, a_p) is (1.25, 8), in metre unit.

$$\xi_h = \frac{1}{\pi}\frac{a_y d_p - d_y a_p}{a_p d_p} = \frac{1}{\pi} \times \frac{6 \times 1.25 - 0.5 \times 8}{8 \times 0.95} = 0.11.$$

Total damping is summation of hysteretic damping and material damping.

$$\xi_t = \xi_h + \xi_m = 0.11 + 0.05 = 0.16 \text{ or } 16\%.$$

6.13   CLOSURE

The results of a pushover analysis can sometimes be misleading unless they are interpreted correctly. It must be emphasized that the pushover analysis is approximate in nature and is based on static loading. As such, it cannot represent dynamic phenomena with a high degree of accuracy. It may not detect some important deformation modes that can occur in a structure subjected to severe earthquakes, and it may exaggerate others. Inelastic dynamic response may differ significantly from predictions based on invariant or adaptive static load patterns, particularly if higher mode effects become important. The influence of higher modes, which is neglected in the standard pushover analysis, varies for different response parameters (FEMA-440).

6.14   EXERCISES

Ex6.14.1 A hysteresis curve for an RC section shows yield point as (0.6 m, 7.3 m/s^2) and the performance point as (1.32, 9.4 m/s^2). Find the effective damping and demand reduction factor.

Ex6.14.2 Why do we analyse the structures?

Ex6.14.3 Explain what is meant by P-delta effect. How it is accounted for?

Ex6.14.4 Give a brief description of pushover analysis.

Ex6.14.5 What is modal pushover analysis? What are its different types?

Ex6.14.6 Give an account of displacement coefficient method.

Ex6.14.7 How is bilinearization of POC done? What is its use?

Ex6.14.8 Give detailed steps in capacity spectrum method.

Ex6.14.9 What is ADRS format?

Ex6.14.10 How do you decide the performance point of a structure?

Ex6.14.11 Distinguish between performance point and performance level.

FURTHER READINGS

ATC-40 (1996) Seismic Evaluation and Retrofit of Existing Concrete Buildings, Applied Geochronology Council.

Aydınoğlu, M.N. (2007) A Response Spectrum-Based Nonlinear Assessment Tool For Practice: Incremental Response Spectrum Analysis (IRSA), Indian Society of Earthquake Technology (ISET), *Journal of Earthquake Technology*, Paper No. 481, 44(1), March, pp. 169–192.

Chopra, A.K. and Goel, R.K. (1999a) Capacity-Demand-Diagram Methods Based on Inelastic Design Spectrum, *Earthquake Spectra*, 15(4), 637–656.

Chopra, A.K. and Goel, R.K. (1999b) Capacity-Demand Diagram Methods Based on Inelastic Design Spectrum, PEER Report No. 1999-10, p. 43–67.

Chopra, A.K. and Goel, R.K. (1999c) Capacity-Demand-Diagram Methods for Estimating Seismic Deformation of Inelastic Structures: SDF Systems, Report No. PEER-1999/02, April 1999.

Chopra, A.K. and Goel, R.K. (2000) Evaluation of NSP to Estimate Seismic Deformation: SDF Systems, *Journal of Structural Engineering, ASCE*, 126(4), 482–490.

Chopra, A.K. and Goel, R.K. (2001) A Modal Pushover Analysis Procedure to Estimate Seismic Demands for Buildings: Theory and Preliminary Evaluation, PEER Report No. 2001-3, January 2001.

Chopra, A.K. and Goel, R.K. (2002) A Modal Pushover Analysis Procedure for Estimating Seismic Demands for Buildings, *Earthquake Engineering and Stress Dynamics*, 31, 561–582.

Chopra, A.K. and Goel, R.K. (2003a) A Modal Pushover Analysis Procedure to Estimate Seismic Demands for Unsymmetrical-Plan Buildings: Theory and Preliminary Evaluation, Report No. EERC/2003-08, Sept. 2003.

Chopra, A.K. and Goel, R.K. (2003b) A Modal Pushover Analysis Procedure to Estimate Seismic Demands for Buildings: Summary and Evaluation, 5th National Conference on Earthquake Engineering, May, Istanbul, Turkey.

Chopra, A.K. and Goel, R.K. (2004) Modal Pushover Analysis: Symmetric- and Unsymmetric-Plan Buildings, International Workshop on Performance-Based Seismic Design, Bled, Slovenia, 28 June–1 July 2004.

FEMA-273 (1997) NEHRP Guidelines for the Seismic Rehabilitation of Buildings, US Federal Emergency Management Agency, Building Seismic Safety Council, Washington, DC.

FEMA-349 (2000) Action Plan for Performance-Based Seismic Design, US Federal Emergency Management Agency, Earthquake Engineering Research Institute.

FEMA-356 (2000) Prestandard and Commentary for the Seismic Rehabilitation of Buildings, US Federal Emergency Management Agency.

FEMA-368 (2001) NEHRP Recommended Provisions for Seismic Regulations for New Buildings and other Structures, US Federal Emergency Management Agency.

FEMA-389 (2004) Premier for Design Professionals, US Federal Emergency Management Agency.

FEMA-396 (2003) Incremental Seismic Rehabilitation of Hospital Buildings, December, US Federal Emergency Management Agency.

FEMA-440 (2004) Improvement of Nonlinear Static Seismic Analysis Procedures, ATC-55 Project, Applied Technology Council and US Federal Emergency Management Agency.

FEMA P695 (2009) Quantification of Building Seismic Performance Factors, Federal Emergency Management Agency.

Goel, R.K. and Chopra, A.K. (2005) Extension of Modal Pushover Analysis to Compute Member Forces", *Earthquake Spectra*, 21(1), Feb. pp. 125–139.

Gupta, B. and Kunnath, S.K. (2000) Adaptive Spectra-Based Pushover Procedure for Seismic Evaluation of Structures", *Earthquake Spectra*, May, pp. 367–391.

Kunnath, S.K. and Kalkan, E. (2004) Evaluation of Seismic Deformation Demands Using Nonlinear Procedures in Multistory Steel and Concrete Moment Frames, *ISET Journal of Earthquake Technology*, Paper No. 445, Vol. 41, No. 1, March 2004, pp. 159–181.

Kuramoto, H. and Matsumoto, K. (2004) Mode-Adaptive Pushover Analysis for Multi-Story RC Buildings, 13th World Conference on Earthquake Engineering, Canada, 1–6 August, Paper No. 2500.

Sasaki, K.K., Freeman, S.A. and Paret, T.F. (1998) Multimode Pushover Procedure (Mmp)—A Method to Identify the Effects of Higher Modes in a Pushover Analysis, Proceedings of the 6th US National Conference on Earthquake Engineering.

7 Direct Displacement-Based Design of RC Frame Buildings

7.1 INTRODUCTION

The displacement-based design in which the target design displacement is evaluated directly through the application of mechanics, may be called as direct displacement-based design (DDBD).

Though there are some variants of DDBD, the one developed by Pettinga and Priestley (2005) for RC frame buildings is handy and easy to use. This method has been described in this chapter. It is to be noted that the methodology of Pettinga and Priestley (2005) addresses *interstorey drift* $\left(\theta_d\right)$ as the only design performance objective.

For designating a point on nonlinear part of force–deformation curve (Figure 7.1a), secant stiffness or effective stiffness (k_e) is used. The target displacement corresponds to some intended damage level or damage state of the structure. Such damage state corresponds to a particular displacement value. This is the target displacement (Δ_d). The initial stiffness of the structure k and the initial time period of the structure is initially T (elastic time period); but these change as the structure reaches the nonlinear state. The time period corresponding to any point on the nonlinear part of the force–deformation curve is designated as effective time period (T_e). The elastic time period relationship in general is given by Eq. (7.1.1).

$$T = 2\pi\sqrt{\frac{m}{k}} \tag{7.1.1}$$

Eq. (7.1.1.) may be written as

$$k = 4\pi^2 \frac{m}{T^2} \tag{7.1.2}$$

For the target point (or any other point on the nonlinear part of the force–deformation curve), we can rewrite Eq. (7.1.2) as Eq. (7.1.3).

DOI: 10.1201/9781003441090-7

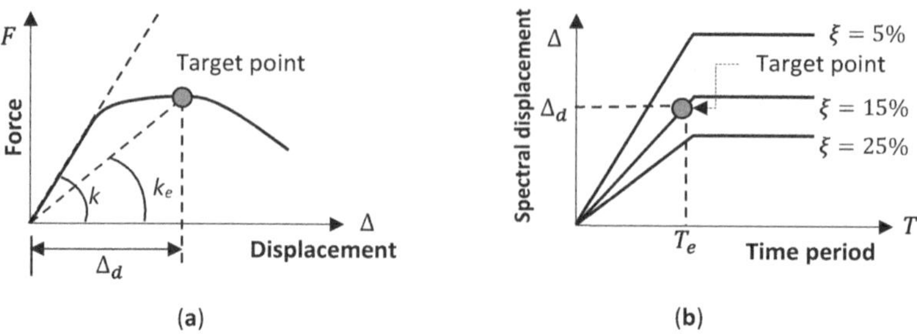

FIGURE 7.1 (a) Force–deformation curve and (b) displacement spectra.

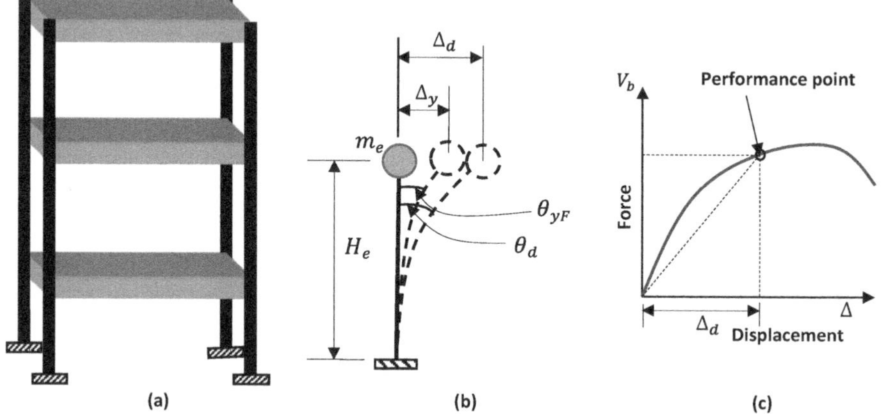

FIGURE 7.2 (a) MDOF building, (b) equivalent SDOF system and (c) force–deformation behaviour.

$$k_e = 4\pi^2 \frac{m_e}{T_e^2} \tag{7.1.3}$$

In Eq. (7.1.3), k_e is effective stiffness, m_e is effective mass (related to equivalent SDOF system), T_e is effective time period corresponding to the target point. The value of T_e is obtained from displacement spectra (constructed from design spectra) corresponding to the design displacement and system effective damping (Figure 7.1(b)).

From the basic stiffness relationship, with V_b as base shear, we can write,

$$V_b = k_e \Delta_d \tag{7.1.4}$$

Consider the 3D frame building in Figure 7.2(a). It has masses and at different floor levels and stiffnesses of different storeys. From dynamical relations, it is possible to convert the 3D building to an equivalent single degree of freedom (ESDOF) system (Fig. 7.2b).

The ESDOF system shall have an equivalent single mass (m_e) and equivalent height (H_e). The equivalent height corresponds to an equivalent overturning moment replicating the overturning moment of the real building. The ESDOF system shall undergo a displacement compatible to and arising out of the dynamic properties of the real building. The displacement pattern of the real building depends upon its shape profile (generally first mode profile) and the amount of drift it undergoes. The total angular displacement of the ESDOF system is the angular design drift (θ_d).

As per Figure 7.2(b), the yield displacement (Δ_y) of the ESDOF system is given by Eq. (7.1.5).

$$\Delta_y = \theta_{yF} H_e \tag{7.1.5}$$

As per Priestley (2003), the frame yield rotation (θ_{yF}) is given by Eq. (7.1.6), in which ε_y is yield strain of rebar steel ($= f_y / E_s$), l_b is length of beam in the direction of displacement considered, and h_b is depth of the beam.

$$\theta_{yF} = 0.5\frac{\varepsilon_y l_b}{h_b} \tag{7.1.6}$$

In the above equations, θ_{yF} is yield rotation (i.e. maximum elastic rotation) of frame, H_e is effective height of ESDOF system, l_b is length of beam, h_b is depth of beam, Δ_y is yield displacement of ESDOF system.

The ESDOF system properties are given as follows:

$$\Delta_d = \frac{\sum_{i=1}^{n} m_i \Delta_i^2}{\sum_{i=1}^{n} m_i \Delta_i} \tag{7.1.7}$$

$$m_e = \frac{\sum_{i=1}^{n} m_i \Delta_i}{\Delta_d} \tag{7.1.8}$$

$$H_e = \frac{\sum_{i=1}^{n} m_i \Delta_i h_i}{\sum_{i=1}^{n} m_i \Delta_i} \tag{7.1.9}$$

In Eqs. (7.1.7) to (7.1.9), Δ_d is target design displacement, m_i is mass of i-th floor, n is total number stories in the building, Δ_i is displacement of i-th floor as per shape profile considered and, h_i is height of i-th floor above the base.

Note that for building with beams of unequal lengths, Eq. (7.1.6) may be applied for each beam length separately, and then the average value of yield rotation may be obtained.

7.2 SHAPE PROFILE AND DISPLACEMENT PROFILE

The first mode shape profiles for buildings have been suggested by researchers. Let us denote ϕ_i as shape coefficient and Δ_i as displacement for i-th floor. The shape profile suggested by Priestley and Calvi (1997) for RC frame buildings is given by Eqs. (7.2.1) to (7.2.3).

$$\text{For } n \le 4: \qquad \phi_i = \frac{h_i}{H} \tag{7.2.1}$$

$$\text{For } 4 \le n < 20: \ \phi_i = \frac{h_i}{H}\left(\frac{16 - 0.5\dfrac{h_i}{H}(n-4)}{16 - 0.5(n-4)}\right) \tag{7.2.2}$$

$$\text{For } n \ge 20 \qquad \phi_i = \frac{2h_i}{H}\left(1 - 0.5\frac{h_i}{H}\right) \tag{7.2.3}$$

In the above equations, h_i is the height of i-the floor from base and H is overall height of the building.

Pettinga and Priestley (2005) suggested the following shape profile for RC frame buildings:

$$\text{For} \le 4: \ \phi_i = \frac{h_i}{H} \tag{7.2.4}$$

$$\text{For} > 4: \ \phi_i = \frac{4}{3}\frac{h_i}{H}\left(1 - 0.25\frac{h_i}{H}\right) \tag{7.2.5}$$

Shape profile has been depicted in Figure 7.3(a). The displacement profile of the building (designated by Δ_i) is a function of shape profile and displacement due to drift. The effect of drift is taken into account through the displacement of the storey with largest storey height, or largest mass (called critical storey). The critical storey has been explained in Figure 7.3.

The critical storey is defined as the storey having critical (maximum) mass and/or critical interstorey height (maximum interstorey height). Denoting the storey height of critical storey as h_{sc}, the lateral drift of critical storey is given by

$$\Delta_c = \theta_d h_{sc} \tag{7.2.6}$$

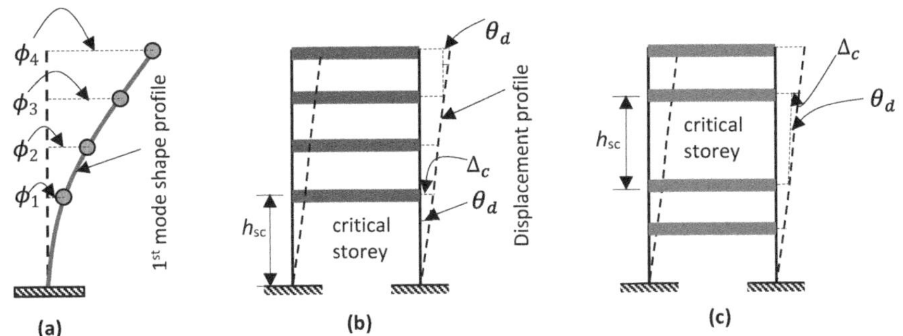

FIGURE 7.3 (a) First mode shape profile. (b) and (c) Frame with critical storey and displacement profile.

Denoting the shape coefficient of critical storey as ϕ_c, the lateral displacement of i-th floor is taken as proportional to the critical storey parameters and is given as Eq. (7.2.7).

$$\Delta_i = \phi_i \frac{\Delta_c}{\phi_c} \tag{7.2.7}$$

7.2.1 Distribution of the Base Shear over the Floors

The base shear computed from displacement criteria may be distributed over floors as per the following equations suggested by Pettinga and Priestley (2005).

$$\text{For } n \leq 10 \quad F_i = V_b \frac{m_i \Delta_i}{\sum_{i=1}^{n} m_i \Delta_i} \tag{7.2.8}$$

$$\text{For } n > 10 \begin{cases} F_i = 0.9V_b \dfrac{m_i \Delta_i}{\sum_{i=1}^{n} m_i \Delta_i} \\[3mm] F_{roof} = F_{i,\,roof} + 0.1V_b \end{cases} \tag{7.2.9}$$

In the above two equations, n is the total number of stories and F_{roof} is the lateral force put at roof.

7.3 LOAD COMBINATION AND DESIGN

The loads considered are those of gravity loads and seismic loads. As the structure is in nonlinear stage, the expected strengths of materials are to be used in design. The load combinations are:

$$DL + LL$$

$$DL + LL \pm F_x$$

$$DL + LL \pm F_y$$

where F_x and F_y are forces at floor levels in two orthogonal directions of the building and as given in Eqs. (7.2.8) and (7.2.9).

The expected material strengths for materials are used in design. The expected material strengths for concrete (f_{ce}) and rebar (f_{se}) are given by FEMA-356 as,

$$f_{ce} = 1.5 f_{ck} \tag{7.3.1a}$$

$$f_{se} = 1.25 f_y \tag{7.3.1b}$$

Example 7.3.1 *For the building plan shown in Figure 7.4, compute the frame yield rotation. The grade of rebar steel is Fe415 and overall depth of all beams is 500 mm.*

Solution

Yield strength of rebar = 415 MPa.

Yield strain $\varepsilon_y = \dfrac{f_y}{E_s} = 415/(2 \times 10^5) = 0.002075$.

Along the short direction: $l_b = 5$ m, $h_b = 0.5$ m.

$$\theta_{yF} = 0.5 \frac{\varepsilon_y l_b}{h_b} = 0.5 \frac{0.002075 \times 5}{0.5} = \mathbf{0.010375} \text{ radian.}$$

Along the long direction:
For $l_b = 4.5$ m, $h_b = 0.5$ m.

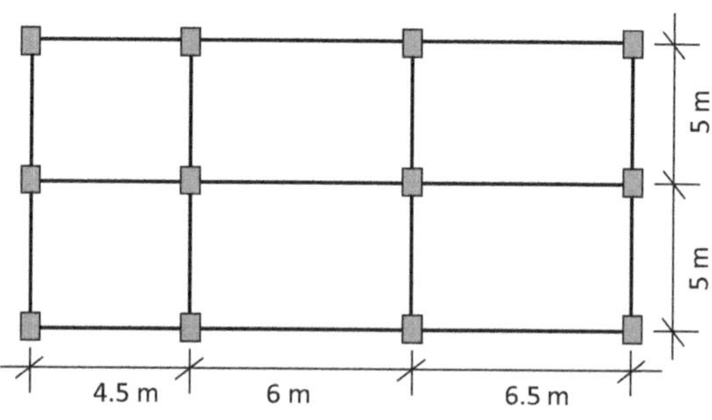

FIGURE 7.4   Plan of building in Example 7.3.1.

$$\theta_{yF,1} = 0.5\frac{\varepsilon_y l_b}{h_b} = 0.5\frac{0.002075 \times 4.5}{0.5} = 0.009338 \text{ radian.}$$

For $l_b = 6$ m, $h_b = 0.5$ m.

$$\theta_{yF,2} = 0.5\frac{\varepsilon_y l_b}{h_b} = 0.5\frac{0.002075 \times 6}{0.5} = 0.01245 \text{ radian.}$$

For $l_b = 6.5$ m, $h_b = 0.5$ m.

$$\theta_{yF,3} = 0.5\frac{\varepsilon_y l_b}{h_b} = 0.5\frac{0.002075 \times 6.5}{0.5} = 0.013488 \text{ radian.}$$

Average frame yield rotation $= \frac{1}{3}\left(\theta_{yF,1} + \theta_{yF,2} + \theta_{yF,3}\right) = \mathbf{0.011758}$ radian.

7.4 EQUIVALENT DAMPING OF THE BUILDING SYSTEM

The equivalent damping corresponding to the target point of the building comprises of material damping and hysteretic damping. The material damping for concrete is taken as 5% of the critical value. The hysteretic damping is a function of ductility and has been suggested by Pettinga and Priestley (2005) as Eq. (7.4.1).

$$\xi_h = 120\left(\frac{1 - \dfrac{1}{\sqrt{\mu}}}{\pi}\right)\% \tag{7.4.1}$$

Here, displacement ductility (μ) is given as

$$\mu = \frac{\Delta_d}{\Delta_y} \tag{7.4.2}$$

Hence, the equivalent damping (ξ_{eq}) for RC buildings corresponding to target point is given by Eq. (7.4.3):

$$\xi_{eq} = 5 + 120\left(\frac{1 - \dfrac{1}{\sqrt{\mu}}}{\pi}\right)\% \tag{7.4.3}$$

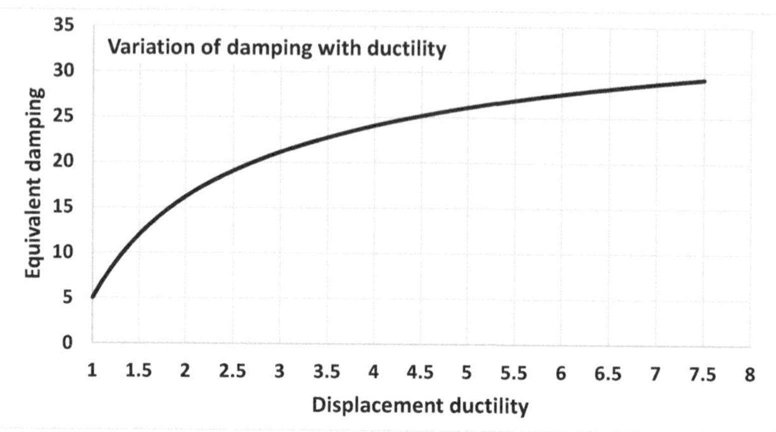

FIGURE 7.5 Variation of equivalent damping with displacement ductility.

7.4.1 EVALUATION OF THE PERFORMANCE OF THE DESIGNED BUILDING

The building designed as per DDBD needs to be evaluated for its actual performance and has to be compared with the target objectives. The evaluation of the building shall be carried out with (i) nonlinear static analysis (pushover analysis), (ii) linear dynamic analysis and (iii) nonlinear dynamic analysis. The drifts are computed from the displacement histories at various floors. Drift can be linear or nonlinear. If the achieved drift differs by more than 20 from target drift, redesign is necessary.

Eq. (7.4.3) is plotted in Figure 7.5. The figure shows the nature of variation of equivalent damping with displacement ductility. The figure may be used for the computation of equivalent damping from a given displacement ductility. It may be noted that Eq. (7.4.3) and Figure 7.5 are valid for RC structures.

7.5 NUMERICAL EXAMPLES

For the computation of ESDOF system properties, the data can be put in the tabular form shown in Table 7.1.

Example 7.5.1 *Find the design forces at each floor level for the 8-storey building, the plan of which is shown in Figure 7.6. The ground storey height is 4 m and other storey height is 3.5 m. The seismic weight in floors is 10 kN/m² and that in roof is 6 kN/m². The grade of rebar steel is Fe415 and overall depth of all beams is 600 mm. Design drift is 2.2%. The building is to be designed at 0.36g level of Indian seismic code.*

Solution

The seismic weight in each floor $= (3 \times 6) \times (2 \times 5) \times 10 = 1800$ kN.
Seismic mass in each floor $= 1800 \times 10^3/9.81 = 183500$ kg.
The seismic weight in roof $= (3 \times 6) \times (2 \times 5) \times 6 = 1080$ kN.
Seismic mass in roof $= 1080 \times 10^3/9.81 = 110100$ kg.

TABLE 7.1
Computing relevant values

Floor no. (from bottom)	Height of floor (h_i) (m)	Mass (m_i) (kg)	Shape coefficient	Displacement profile (Δ_i)	$m_i\Delta_i$	$m_i\Delta_i^2$	$m_i\Delta_i h_i$
1							
2							
3							
...							
					Sum	Sum	Sum

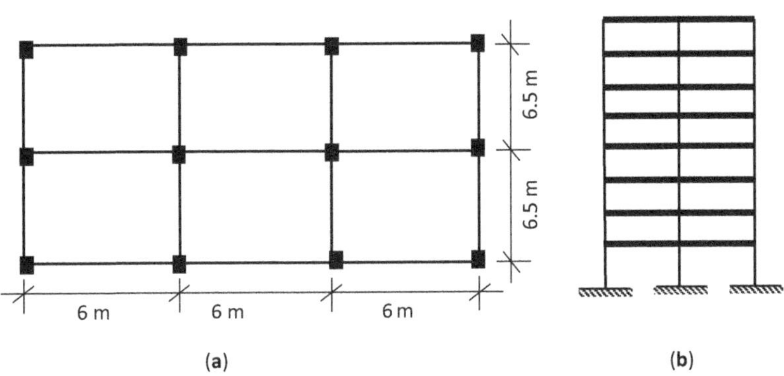

(a)

(b)

FIGURE 7.6   (a) Plan of building and (b) side elevation of the frame in Example 7.5.1.

Using Eq. (7.2.5), shape profile is given by $\phi_i = \dfrac{4}{3}\dfrac{h_i}{H}\left(1 - 0.25\dfrac{h_i}{H}\right)$,

$H = 4 + 7 \times 3.5 = 28.5$ m. The computed values are tabulated in Table Ex7.5.1.1. Critical storey is ground storey. Hence, $\phi_c = 0.1806$ [from Eq. (7.2.5) when hi = 4 m]. The critical storey displacement $= \Delta_c = \theta_d h_{sc} = 0.022 \times 4 = 0.088$ m. Now, ESDOF system properties are computed as follows:

$$\Delta_d = \frac{\sum_{i=1}^{n} m_i\Delta_i^2}{\sum_{i=1}^{n} m_i\Delta_i} = \frac{144240}{411710} = 0.350\,\text{m}.$$

$$m_e = \frac{\sum_{i=1}^{n} m_i\Delta_i}{\Delta_d} = \frac{411710}{0.3503} = 117500\,\text{kg}.$$

$$H_e = \frac{\sum_{i=1}^{n} m_i\Delta_i h_i}{\sum_{i=1}^{n} m_i\Delta_i} = \frac{7791010}{411710} = 18.923\,\text{m}.$$

TABLE EX7.5.1.1
Solution to Example 7.5.1

Floor no. (from bottom)	Ht. of floor (h_i)(m)	Mass (m_i)(kg)	Shape coefficient (ϕ_i) (Eq. 7.2.5)	Displacement profile (Δ_i) (m) (Eq. 7.2.7)	$m_i\Delta_i$	$m_i\Delta_i^2$	$m_i\Delta_i h_i$
1	4	183500	0.1806	0.0880	16148	1421	64592
2	7.5	183500	0.3278	0.1598	29314	4683	219856
3	11	183500	0.4650	0.2266	41581	9422	457392
4	14.5	183500	0.5921	0.2886	52949	15278	767759
5	18	183500	0.7091	0.3456	63418	21917	1141517
6	21.5	183500	0.8161	0.3978	72987	29031	1569223
7	25	183500	0.9131	0.4450	81658	36338	2041438
8	28.5	110100	1	0.4874	53657	26150	1529231
Sum					411710	144240	7791010

Design forces along the long direction

Yield strain of rebar = $415/(2 \times 10^5) = 0.002075$.

The length of beam along long direction is 6 m. The yield rotation of frame by

$$\theta_{yF} = 0.5\frac{\varepsilon_y l_b}{h_b} = 0.5\frac{0.002075 \times 6}{0.6} = 0.010375 \text{ radian.}$$

Yield displacement,

$$\Delta_y = \theta_{yF}H_e = 0.010375 \times 18.923 = 0.196 \text{ m.}$$

Displacement ductility,

$$\mu = \frac{\Delta_d}{\Delta_y} = 0.350/0.196 = 1.786.$$

Equivalent damping is given by Eq. (7.4.3):

$$\xi_{eq} = 5 + 120\left(\frac{1-\dfrac{1}{\sqrt{\mu}}}{\pi}\right) = 5 + 120\left(\frac{1-\dfrac{1}{\sqrt{1.786}}}{\pi}\right) = \mathbf{14.6\%}$$

Consider displacement spectra of Figure 7.7, prepared for acceleration spectra of IS 1892-2016. From the graph, corresponding to $\Delta_d = 0.350$ m, and $\xi_{eq} = 14.6\%$, $T_e = 4$ s.

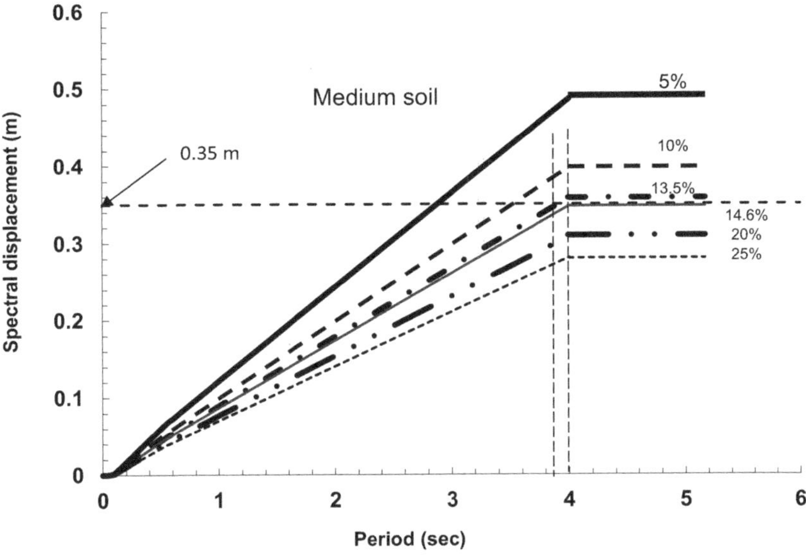

FIGURE 7.7 Displacement spectra for Example 7.5.1.

The equivalent stiffness of ESDOF system,

$$k_e = 4\pi^2 \frac{m_e}{T_e^2} = 4\pi^2 \frac{117500}{4^2} = 289920 \text{ N/m} = 289.9 \text{ kN/m.}$$

$$V_b = k_e \Delta_d = 289.9 \times 0.3503 \text{ kN} = 101.6 \text{ kN.}$$

Distribution of base shear over floors

Now, $F_i = V_b \dfrac{m_i \Delta_i}{\sum_{i=1}^n m_i \Delta_i}$. The values are tabulated in Table Ex7.5.1.2.

Design forces along the short direction

Length of the beam is 6.5 m. The yield rotation of frame is

$$\theta_{yF} = 0.5 \frac{\varepsilon_y l_b}{h_b} = 0.5 \frac{0.002075 \times 6.5}{0.6} = 0.0112 \text{ radian.}$$

Yield displacement,

$$\Delta_y = \theta_{yF} H_e = 0.0112 \times 18.923 = 0.212 \text{ m.}$$

TABLE EX7.5.1.2
Design forces in floor in the *long* direction in Example 7.5.1

Floor no. (from bottom)	m_i (kg)	Δ_i (m)	$m_i\Delta_i$ (kg-m)	V_b (kN)	F_i (kN)
1	183500	0.088	16148		4.0
2	183500	0.1598	29314		7.2
3	183500	0.2266	41581	101.6	10.3
4	183500	0.2886	52949		13.1
5	183500	0.3456	63418		15.6
6	183500	0.3978	72987		18.0
7	183500	0.445	81658		20.2
8	110100	0.4874	53657		13.2
		Sum	411712		101.6

Displacement ductility,

$$\mu = \frac{\Delta_d}{\Delta_y} = 0.350/0.213 = 1.65.$$

Equivalent damping is given by

$$\xi_{eq} = 5 + 120\left(\frac{1 - \dfrac{1}{\sqrt{\mu}}}{\pi}\right) = 5 + 120\left(\frac{1 - \dfrac{1}{\sqrt{1.65}}}{\pi}\right) = 13.5\%.$$

Consider the displacement spectra of Figure 7.7, corresponding to $\Delta_d = 0.350$ m and $\xi_{eq} = 13.5\%$, $T_e = 3.85$ s.

The equivalent stiffness of ESDOF system,

$$k_e = 4\pi^2\frac{m_e}{T_e^2} = 4\pi^2\frac{117500}{3.85^2} = 312950 \text{ N/m} = 312.9 \text{ kN/m}.$$

$$V_b = k_e\Delta_d = 312.9 \times 0.350 = 109.5 \text{ kN}$$

Distribution of forces over floors

$F_i = V_b\dfrac{m_i\Delta_i}{\sum_{i=1}^{n}m_i\Delta_i}$. The values are tabulated in Table Ex7.5.1.3.

TABLE EX7.5.1.3
Design forces in floor in the *short* direction in Example 7.5.1

Floor no. (from bottom)	m_i (kg)	Δ_i (m)	$m_i\Delta_i$ (kg-m)	V_b (kN)	F_i (kN)
1	183500	0.088	16148	109.5	4.3
2	183500	0.1598	29314		7.8
3	183500	0.2266	41581		11.1
4	183500	0.2886	52949		14.1
5	183500	0.3456	63418		16.9
6	183500	0.3978	72987		19.4
7	183500	0.445	81658		21.7
8	110100	0.4874	53657		14.3
		Sum	411712		109.5

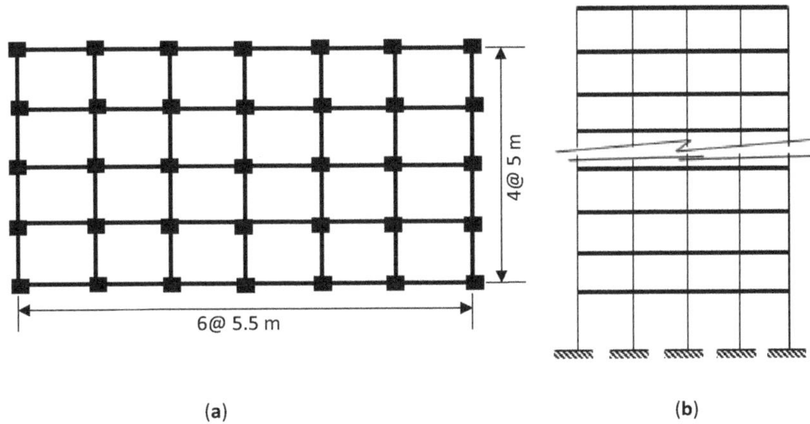

FIGURE 7.8   (a) Plan of the building and (b) elevation of short frame in Example 7.5.2.

Example 7.5.2 *Find the design forces for the 18-storey building, the plan of which is shown in Figure 7.8. The ground storey height is 4.5 m and other storey height is constant to 3.4 m. The seismic weight in floors is 20 kN/m² and that in roof is 10 kN/ m². The grade of rebar steel is Fe500 and overall depth of all beams is 800 mm in the long direction and 700 mm in the short direction. Design drift is 2%. The building is to be designed at 0.55g level of Euro-Code.*

Solution
The seismic weight in each floor = (6 × 5.5) × (4 × 5) × 20 = 13200 kN.
Seismic mass in each floor = 13200 × 10³/9.81 = 1345566 kg = 1345.566 ton-mass.
The seismic weight in roof = (6 × 5.5) × (4 × 5) × 10 = 6600 kN.
Seismic mass in roof = 6600 × 10³/9.81 = 672783 kg = 672.783 ton-mass

TABLE EX7.5.2.1
ESDOF system properties computation for Example 7.5.2

Floor no. (from bottom)	Ht. of floor (h_i)(m)	Mass (m_i)(ton)	Shape coefficient (ϕ_i)	Displacement profile (Δ_i) (m)	$m_i\Delta_i$	$m_i\Delta_i^2$	$m_i\Delta_i h_i$
1	4.5	1345.566	0.0946	0.0900	121.101	10.9	545.0
2	7.9	1345.566	0.1637	0.1558	209.645	32.7	1656.2
3	11.3	1345.566	0.2309	0.2197	295.647	65.0	3340.8
4	14.7	1345.566	0.2960	0.2817	379.106	106.8	5572.9
5	18.1	1345.566	0.3592	0.3419	460.023	157.3	8326.4
6	21.5	1345.566	0.4204	0.4001	538.397	215.4	11575.5
7	24.9	1345.566	0.4797	0.4565	614.228	280.4	15294.3
8	28.3	1345.566	0.5369	0.5109	687.517	351.3	19456.7
9	31.7	1345.566	0.5921	0.5635	758.262	427.3	24036.9
10	35.1	1345.566	0.6454	0.6142	826.466	507.6	29008.9
11	38.5	1345.566	0.6967	0.6630	892.126	591.5	34346.9
12	41.9	1345.566	0.7460	0.7099	955.244	678.1	40024.7
13	45.3	1345.566	0.7933	0.7549	1015.819	766.9	46016.6
14	48.7	1345.566	0.8386	0.7981	1073.852	857.0	52296.6
15	52.1	1345.566	0.8819	0.8393	1129.342	947.9	58838.7
16	55.5	1345.566	0.9233	0.8787	1182.289	1038.8	65617.0
17	58.9	1345.566	0.9626	0.9161	1232.694	1129.3	72605.6
18	62.3	672.783	1	0.9517	640.278	609.3	39889.3
				Sum	**13012**	**8774**	**528449**

Using Eq. (7.2.5), shape profile is given by $\phi_i = \dfrac{4}{3}\dfrac{h_i}{H}\left(1 - 0.25\dfrac{h_i}{H}\right)$,

$H = 4.5 + 17 \times 3.4 = 62.3$ m. The computed values are tabulated in Table Ex7.5.2.1.

Critical storey is ground storey. Hence, $\phi_c = 0.0.09457$ (from Table Ex7.5.2.1). The critical storey displacement $= \Delta_c = \theta_d h_{sc} = 0.02 \times 4.5 = 0.09$ m.

Now, ESDOF system properties are computes as below:

$$\Delta_d = \frac{\sum_{i=1}^{n} m_i\Delta_i^2}{\sum_{i=1}^{n} m_i\Delta_i} = \frac{8774}{13012} = 0.674 \text{ m.}$$

$$m_e = \frac{\sum_{i=1}^{n} m_i\Delta_i}{\Delta_d} = \frac{13012.0}{0.674} = 19298 \text{ ton-mass.}$$

$$H_e = \frac{\sum_{i=1}^{n} m_i \Delta_i h_i}{\sum_{i=1}^{n} m_i \Delta_i} = \frac{528449}{13012} = 40.612 \text{ m.}$$

Yield strain of rebar $= 500/(2 \times 10^5) = 0.0025$.

Design forces along *long* direction
The length of beam along long direction is 5.5 m. The yield rotation of frame by,

$$\theta_{yF} = 0.5 \frac{\varepsilon_y l_b}{h_b} = 0.5 \times \frac{0.0025 \times 5.5}{0.8} = 0.008594 \text{ radian.}$$

Yield displacement,

$$\Delta_y = \theta_{yF} H_e = 0.008594 \times 40.612 = 0.349 \text{ m.}$$

Displacement ductility,

$$\mu = \frac{\Delta_d}{\Delta_y} = 0.674/0.349 = 1.93.$$

Equivalent damping is given by,

$$\xi_{eq} = 5 + 120 \left(\frac{1 - \frac{1}{\sqrt{\mu}}}{\pi} \right) = 5 + 120 \left(\frac{1 - \frac{1}{\sqrt{1.93}}}{\pi} \right) = 15.7\%.$$

Consider displacement spectra of Figure 7.9, corresponding to $\Delta_d = 0.674$ m and $\xi_{eq} = 15.7\%$, $T_e = 4.55$ sec.
The equivalent stiffness of ESDOF system,

$$k_e = 4\pi^2 \frac{m_e}{T_e^2} = 4\pi^2 \frac{19298 \times 1000}{4.55^2} = 3680001 \text{ N/m} = 36800 \text{ kN/m.}$$

$$V_b = k_e \Delta_d = 36800 \times 0.674 = 24803 \text{ kN.}$$

Distribution of force over floors

Number of storey being more than 10, $F_i = 0.9 V_b \dfrac{m_i \Delta_i}{\sum_{i=1}^{n} m_i \Delta_i}$. Finally, $0.1 V_b$ is added
to roof level force. The values are tabulated in Table Ex7.5.2.2.

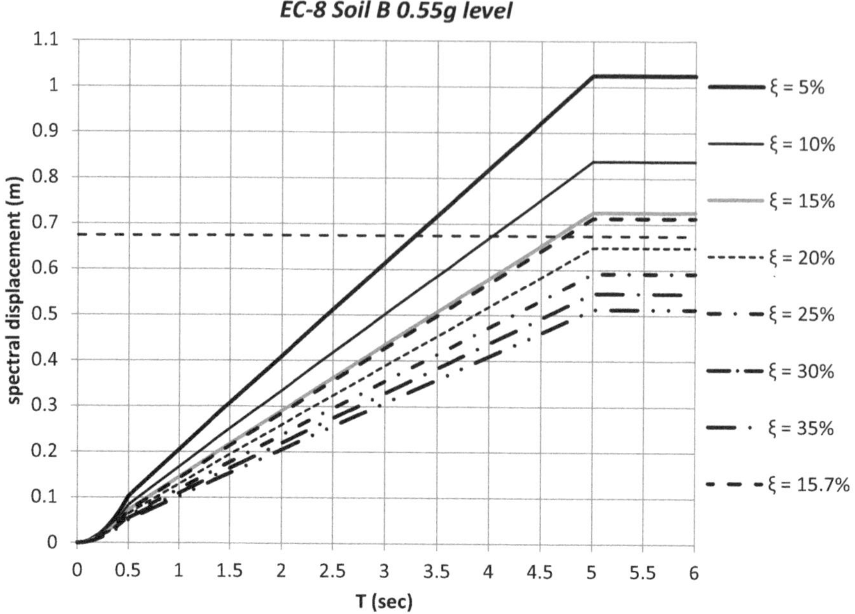

FIGURE 7.9 Displacement spectra used in the *long* direction for Example 7.5.2.

Roof level force $= 1098.4 + 0.1V_b = 1098.4 + 0.1 \times 24803 = 3578$ kN.

Design forces along short direction

Length of beam is 5 m and depth of beam is 0.7 m. The yield rotation of frame by,

$$\theta_{yF} = 0.5\frac{\varepsilon_y l_b}{h_b} = 0.5\frac{0.0025 \times 5}{0.7} = 0.00893\,\text{radian}.$$

Yield displacement,

$$\Delta_y = \theta_{yF}H_e = 0.00893 \times 40.612 = 0.363\,\text{m}.$$

Displacement ductility,

$$\mu = \frac{\Delta_d}{\Delta_y} = 0.674/0.363 = 1.86.$$

TABLE EX7.5.2.2
Forces in the *long* direction in Example 7.5.2

Floor no. (from bottom)	m_i (ton)	Δ_i (m)	$m_i\Delta_i$ (ton-m)	$0.9V_b$ (kN)	F_i (kN)
1	1345.57	0.0900	121.10		207.8
2	1345.57	0.1558	209.65		359.7
3	1345.57	0.2197	295.65		507.2
4	1345.57	0.2817	379.11	22323	650.4
5	1345.57	0.3419	460.02		789.2
6	1345.57	0.4001	538.40		923.7
7	1345.57	0.4565	614.23		1053.7
8	1345.57	0.5109	687.52		1179.5
9	1345.57	0.5635	758.26		1300.8
10	1345.57	0.6142	826.47		1417.9
11	1345.57	0.6630	892.13		1530.5
12	1345.57	0.7099	955.24		1638.8
13	1345.57	0.7549	1015.82		1742.7
14	1345.57	0.7981	1073.85		1842.3
15	1345.57	0.8393	1129.34		1937.5
16	1345.57	0.8787	1182.29		2028.3
17	1345.57	0.9161	1232.69		2114.8
18	672.783	0.9517	640.28		1098.4
		Sum	13012		22323.0

Equivalent damping is given by,

$$\xi_{eq} = 5 + 120\left(\frac{1-\dfrac{1}{\sqrt{\mu}}}{\pi}\right) = 5 + 120\left(\frac{1-\dfrac{1}{\sqrt{1.86}}}{\pi}\right) = 15.2\%.$$

Consider displacement spectra of Figure 7.10, corresponding to $\Delta_d = 0.674$ m and $\xi_{eq} = 15.2\%$, $T_e = 4.54$ sec.

The equivalent stiffness of ESDOF system,

$$k_e = 4\pi^2\frac{m_e}{T_e^2} = 4\pi^2\frac{19298\times1000}{4.54^2} = 36962414 \text{ N/m} = 36962 \text{ kN/m}.$$

$$V_b = k_e\Delta_d = 36962\times0.674 = 24913 \text{ kN}$$

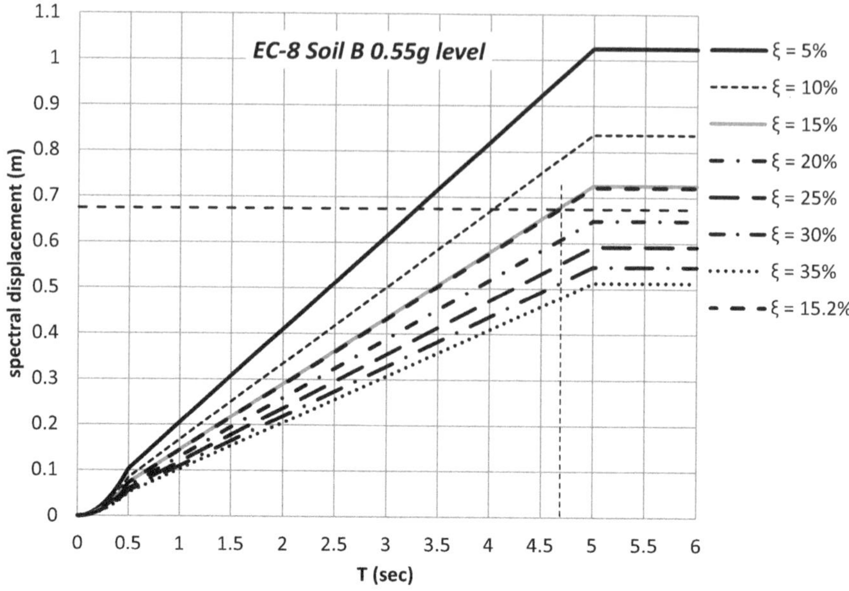

FIGURE 7.10 Displacement spectra for *short* direction for Example 7.5.2.

Distribution of forces over floors

$$F_i = 0.9V_b \frac{m_i \Delta_i}{\sum_{i=1}^{n} m_i \Delta_i}$$. Remaining $0.1V_b$ is added at roof level force. The values are

tabulated in Table Ex7.5.2.3.

The roof level force = 1103.3 + $0.1V_b$ = 1103.3 + 0.1 × 24913 = 3594 kN.

7.6 TWO EXPRESSIONS OF EQUIVALENT DAMPING

Two formulae of equivalent damping based on ductility, for RC frame buildings, are available: Eq. (7.6.1) given by Pettinga and Priestley (2005) and Eq. (7.6.2) given by Priestley et al. (2007). These two equations are plotted in Figure 7.11.

$$\xi_e = 0.05 + 1.2\left(\frac{1 - \mu^{-0.5}}{\pi}\right) \tag{7.6.1a}$$

$$\xi_e = 5 + 120\left(\frac{1 - \mu^{-0.5}}{\pi}\right)\% \tag{7.6.1b}$$

$$\xi_e = 0.05 + 0.71\left(\frac{\mu - 1}{\mu\pi}\right) \tag{7.6.2a}$$

TABLE EX7.5.2.3
Forces in the *short* direction in Example 7.5.2

Floor no. (from bottom)	m_i (kg)	Δ_i (m)	$m_i\Delta_i$ (kg-m)	$0.9V_b$ (kN)	F_i (kN)
1	1345.57	0.0900	121.10		208.7
2	1345.57	0.1558	209.65		361.3
3	1345.57	0.2197	295.65		509.4
4	1345.57	0.2817	379.11	22422	653.3
5	1345.57	0.3419	460.02		792.7
6	1345.57	0.4001	538.40		927.7
7	1345.57	0.4565	614.23		1058.4
8	1345.57	0.5109	687.52		1184.7
9	1345.57	0.5635	758.26		1306.6
10	1345.57	0.6142	826.47		1424.1
11	1345.57	0.6630	892.13		1537.3
12	1345.57	0.7099	955.24		1646.0
13	1345.57	0.7549	1015.82		1750.4
14	1345.57	0.7981	1073.85		1850.4
15	1345.57	0.8393	1129.34		1946.0
16	1345.57	0.8787	1182.29		2037.3
17	1345.57	0.9161	1232.69		2124.1
18	a	0.9517	640.28		1103.3
		Sum	13012		22421.7

$$\xi_e = 5 + 71\left(\frac{\mu - 1}{\mu\pi}\right)\%$$
(7.6.2b)

Figure 7.11 shows that the Eq. (7.6.2) gives lower values of equivalent damping.

7.7 VARIATION OF Δ_d WITH θ_d

It is worthwhile to have an idea about how the design displacement changes with design drift. The design displacement increases with target drift and height of building. It may so happen that the design displacement is so high that we do not get a solution for effective time period for the computed effective damping; the horizontal line through Δ_d remains above the displacement spectrum line corresponding to ξ_e.

For the purpose of drawing a plot between θ_d and Δ_d, we consider a 12-storey building with ground storey height 4.3 m and other storey height as 3.3 m. The floor mass is taken as 10,000 kg and roof mass as 7000 kg. The value of θ_d is varied from 0.5% to 4% increasing it at the rate 0.5%. One sample table is shown in Table 7.2. Other tables are not shown for brevity. Finally, all the points between θ_d and Δ_d are plotted.

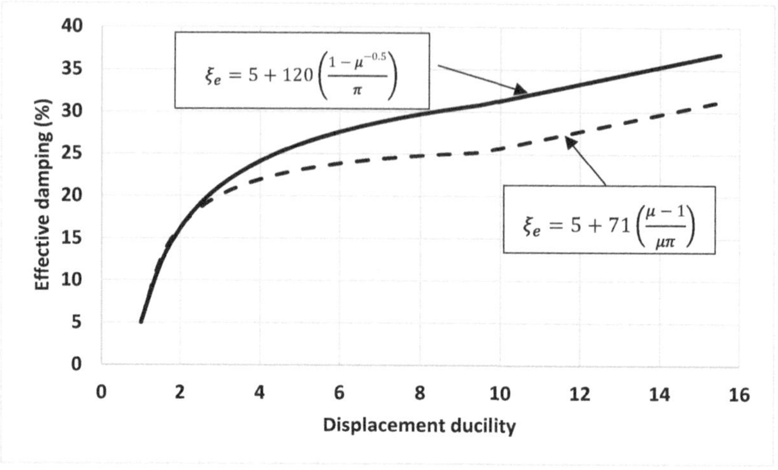

FIGURE 7.11   Variation of equivalent damping with displacement ductility.

TABLE 7.2
ESDOF system properties for $\theta_d = 0.5\%$

Floor no. (from bottom)	Ht. of floor (h_i)(m)	Mass (m_i)(ton)	Shape coefficient (ϕ_i)	Displacement profile (Δ_i) (m)	$m_i\Delta_i$	$m_i\Delta_i^2$	$m_i\Delta_i h_i$
1	4.5	1345.566	0.0946	0.0900	121.101	10.9	545.0
2	7.9	1345.566	0.1637	0.1558	209.645	32.7	1656.2
3	11.3	1345.566	0.2309	0.2197	295.647	65.0	3340.8
4	14.7	1345.566	0.2960	0.2817	379.106	106.8	5572.9
5	18.1	1345.566	0.3592	0.3419	460.023	157.03	8326.4
6	21.5	1345.566	0.4204	0.4001	538.397	215.4	11575.5
7	24.9	1345.566	0.4797	0.4565	614.228	280.4	15294.3
8	28.3	1345.566	0.5369	0.5109	687.517	351.3	19456.7
9	31.7	1345.566	0.5921	0.5635	758.262	427.3	24036.9
10	35.1	1345.566	0.6454	0.6142	826.466	507.6	29008.9
11	38.5	1345.566	0.6967	0.6630	892.126	591.5	34346.9
12	41.9	1345.566	0.7460	0.7099	955.244	678.1	40024.7
				Sum	**11222**	**1294**	**310539**

$$\Delta_d = \frac{\sum_{i=1}^{n} m_i \Delta_i^2}{\sum_{i=1}^{n} m_i \Delta_i} = \frac{1294}{11222} = 0.115 \text{ m.}$$

Similarly, computations are carried for other drifts. Finally, the values are tabulated in Table 7.3. The values are plotted in Figure 7.12. It is a rising straight line. This

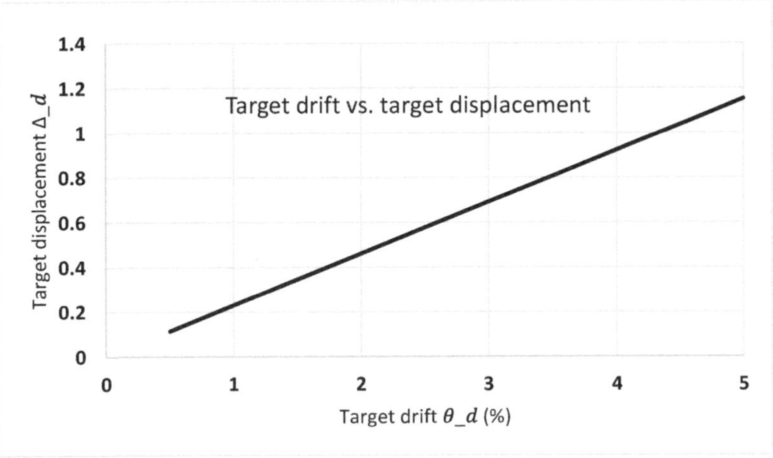

FIGURE 7.12 θ_d vs. Δ_d relationship.

TABLE 7.3
Values Δ_d for different θ_d

Design drift (θ_d in %)	Target design displacement (Δ_d in m)
0.5	0.115
1	0.231
1.5	0.346
2	0.461
2.5	0.577
3	0.692
3.5	0.807
4	0.923
4.5	1.038
5	1.153

is because of Eqs. (7.2.6) and (7.2.7), which are linear. However, the slope of the straight line will vary with different buildings.

The equivalent mass or equivalent height of the building does not change with θ_d, as θ_d has no effect on mass, and equivalent height arises out of overturning moment in real building and ESDOF system.

7.8 CASES WHEN DESIGN DISPLACEMENT IS MORE THAN THE SPECTRAL DISPLACEMENT

It may so happen that the design displacement (Δ_d) is more than the corner period (T_c) displacement (Δ_c) of spectral displacement line corresponding to equivalent

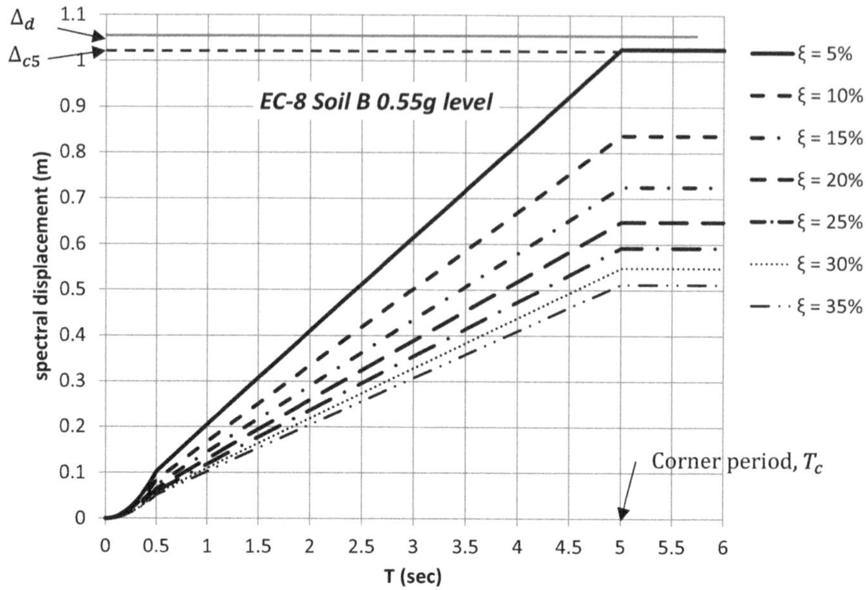

FIGURE 7.13   Displacement spectrum for Case I.

damping. Two cases may arise. These two cases are highlighted here after Priestley et al. (2007).

Case I: *When Δ_d is more than 5% damped elastic displacement at corner point* (Δ_{c5})

This case is shown in Figure 7.13. In this figure, the corner period displacement at 5% damped spectrum is 1.04 m, whereas the design displacement is 1.06 m. This actually indicates that the system is elastic.

In this case the design base shear is given by Eq. (7.8.1).

$$V_b = K_{elastic}\Delta_{c5} \qquad\qquad (7.8.1)$$

In Eq. (7.8.1),

$K_{elastic}$ = elastic stiffness of the system

Δ_{c5} = corner period spectrum displacement corresponding to 5% damping.

Case II: *When Δ_d is more than displacement at corner point (Δ_c) corresponding to equivalent damping*

The case is described through Figure 7.14. The design displacement is Δ_d (= 0.85 m) and equivalent damping is 20%. The displacement corresponding to corner period in 20% damping line is Δ_c (= 0.65 m). The steps in solution are as below:

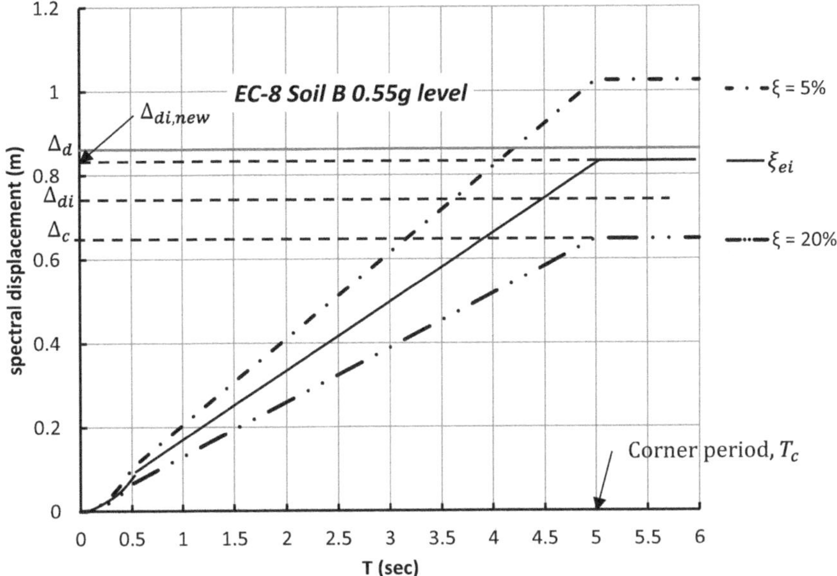

FIGURE 7.14   Displacement spectrum for Case II.

(i)   Take an intermediate design displacement Δ_{di}.

(ii) New ductility $\mu_i = \Delta_{di}/\Delta_y$. Find new equivalent damping

$$\xi_{ei} = 5 + 120\left(\frac{1-\mu_i^{-0.5}}{\pi}\right)\%.$$

(iii) Find the corner period displacement $\Delta_{di,new}$ corresponding to spectrum line ξ_{ei}.

(iv) Compare Δ_{di} and $\Delta_{di,new}$. If these are sufficiently close, stop. Else, go to step (i) with a fresh intermediate spectral displacement value.

Example 7.8.1 *Find the design base shear for a 12-storey RC frame building with target drift of 2.5%. The floor mass is 10,000 kg and the roof mass is 7000 kg. The ground storey height is 4 m and other storey height is 3.3 m. The building is to be designed for EC-8 design spectrum for type B soil and 0.36g seismicity level. f_y = 500 MPa; length of beam is 5 m and beam depth is 600 mm. In the initial pre-design model of the building, a force of 10,000 kN was applied laterally at roof level which caused a displacement of 0.45 m at roof level.*

Solution:

Ground storey is the critical storey.

The critical storey displacement $= \Delta_c = \theta_d h_{sc} = 0.025 \times 4 = 0.1$ m

TABLE EX7.8.1.1
ESDOF system properties in Example 7.8.1

Floor no. (from bottom)	Ht. of floor (h_i)(m)	Mass (m_i)(ton)	Shape coefficient (ϕ_i)	Displacement profile (Δ_i) (m)	$m_i\Delta_i$	$m_i\Delta_i^2$	$m_i\Delta_i h_i$
1	4	10000	0.1291	0.1	1000	100	4000
2	7.3	10000	0.2306	0.17867	1786.7	319.2	13043
3	10.6	10000	0.3276	0.25387	2538.7	644.5	26911
4	13.9	10000	0.4202	0.32562	3256.2	1060.3	45261
5	17.2	10000	0.5083	0.39389	3938.9	1551.5	67750
6	20.5	10000	0.5920	0.45871	4587.1	2104.1	94035
7	23.8	10000	0.6712	0.52006	5200.6	2704.6	123774
8	27.1	10000	0.7459	0.57794	5779.4	3340.2	156623
9	30.4	10000	0.8161	0.63237	6323.7	3998.9	192239
10	33.7	10000	0.8819	0.68333	6833.3	4669.3	230281
11	37	10000	0.9432	0.73082	7308.2	5341.0	270404
12	40.3	7000	1	0.77485	5424.0	4202.8	218586
				Sum	53976.7	30036.4	1442904

$H = 4 + 11 \times 3.3 = 40.3$ m.

$\phi_c = 0.1291$ (from Table Ex7.8.1.1).

Shape profile is given by $\phi_i = \dfrac{4}{3}\dfrac{h_i}{H}\left(1 - 0.25\dfrac{h_i}{H}\right)$.

Table Ex7.8.1.1 is prepared for computation of ESDOF system properties.

$$\Delta_d = \frac{\sum_{i=1}^{n} m_i\Delta_i^2}{\sum_{i=1}^{n} m_i\Delta_i} = \frac{30036.4}{53976.7} = 0.556 \text{ m.}$$

$$m_e = \frac{\sum_{i=1}^{n} m_i\Delta_i}{\Delta_d} = \frac{53976.7}{0.556} = 97079 \text{ kg.}$$

$$H_e = \frac{\sum_{i=1}^{n} m_i\Delta_i h_i}{\sum_{i=1}^{n} m_i\Delta_i} = \frac{1442904}{53976.7} = 26.73 \text{ m.}$$

Yield strain of rebar = $500/(2 \times 10^5) = 0.0025$.

The length of beam is 5 m. The yield rotation of frame is

$$\theta_{yF} = 0.5\frac{\varepsilon_y l_b}{h_b} = 0.5 \times \frac{0.0025 \times 5}{0.6} = 0.0104 \text{ radian.}$$

Yield displacement,

$$\Delta_y = \theta_{yF} H_e = 0.0104 \times 26.73 = 0.278 \text{ m}.$$

Displacement ductility,

$$\mu = \frac{\Delta_d}{\Delta_y} = 0.556/0.278 = 1.997.$$

Equivalent damping is given by,

$$\xi_{eq} = 5 + 120 \left(\frac{1 - \dfrac{1}{\sqrt{\mu}}}{\pi} \right) = 5 + 120 \left(\frac{1 - \dfrac{1}{\sqrt{1.997}}}{\pi} \right) = 16.2\%.$$

Figure 7.15 shows that the design displacement is higher than the displacement at corner period for displacement spectrum at 16.2% damping. So, the usual solution does not exist.

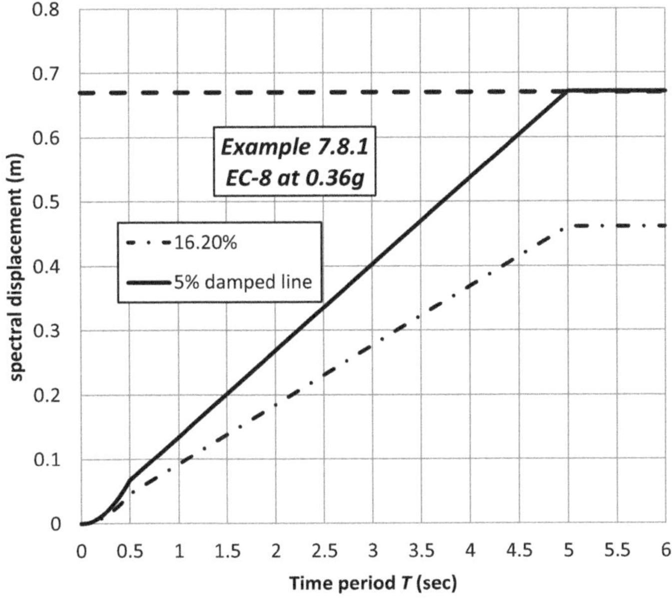

FIGURE 7.15 EC-8 displacement spectrum for Example 7.8.1.

Using Eq. (7.8.1),

$$V_b = K_{elastic}\Delta_{c5}$$

Here, $\Delta_{c5} = 0.67$ m.

A force of 10,000 kN produces a displacement of 0.45 m at the roof level.

The elastic stiffness = 10000/0.45 = 22222 kN/m.

Displacement corresponding to 5% damped line at corner period = $\Delta_{c5} = 0.67$ m.

$$V_b = K_{elastic}\Delta_{c5} = 22222 \times 0.67 = 14889 \text{ kN}.$$

Example 7.8.2 *Find the design base shear for a 14-storey RC frame building with target drift of 2.5%. The floor mass is 12,000 tons and roof mass is 8000 tons. The ground storey height is 4.1 m and other storey height is 3.3 m; the building is to be designed for EC-8 design spectrum for type B soil and 0.4g seismicity level. $f_y = 500$ MPa; length of beam is 5.0 m, beam depth is 700 mm.*

Solution:

Ground storey is the critical storey.

The critical storey displacement = $\Delta_c = \theta_d h_{sc} = 0.025 \times 4.1 = 0.1025$ m.

$H = 4.1 + 13 \times 3.3 = 47$ m.

$\phi_c = 0.1138$ (from Table Ex7.8.2.1).

TABLE EX7.8.2.1

ESDOF system properties in Example 7.8.2

Floor no. (from bottom)	Ht. of floor (h_i) (m)	Mass (m_i) (ton)	Shape coefficient (ϕ_i)	Displacement profile (Δ_i) (m)	$m_i\Delta_i$	$m_i\Delta_i^2$	$m_i\Delta_i h_i$
1	4.1	12000	0.1138	0.1025	1230.0	126	5043
2	7.4	12000	0.2017	0.1817	2180.2	396	16133
3	10.7	12000	0.2863	0.2579	3094.8	798	33114
4	14	12000	0.3676	0.3312	3973.9	1316	55635
5	17.3	12000	0.4456	0.4015	4817.5	1934	83342
6	20.6	12000	0.5204	0.4688	5625.5	2637	115886
7	23.9	12000	0.5918	0.5332	6398.0	3411	152913
8	27.2	12000	0.6600	0.5946	7135.0	4242	194072
9	30.5	12000	0.7249	0.6530	7836.5	5118	239012
10	33.8	12000	0.7865	0.7085	8502.4	6024	287381
11	37.1	12000	0.8448	0.7611	9132.8	6951	338826
12	40.4	12000	0.8998	0.8106	9727.6	7886	392996
13	43.7	12000	0.9515	0.8572	10287.0	8818	449540
14	47	8000	1.0000	0.9009	7207.2	6493	338737
				Sum	87148	56150	2702631

Shape profile is given by $\phi_i = \dfrac{4}{3}\dfrac{h_i}{H}\left(1-0.25\dfrac{h_i}{H}\right)$.

Table Ex7.8.2.1 is prepared for computation of ESDOF system properties.

$$\Delta_d = \frac{\sum_{i=1}^{n} m_i \Delta_i^2}{\sum_{i=1}^{n} m_i \Delta_i} = \frac{56150}{87148} = 0.644 \text{ m.}$$

$$m_e = \frac{\sum_{i=1}^{n} m_i \Delta_i}{\Delta_d} = \frac{87148}{0.644} = 135259 \text{ ton.}$$

$$H_e = \frac{\sum_{i=1}^{n} m_i \Delta_i h_i}{\sum_{i=1}^{n} m_i \Delta_i} = \frac{2702631}{87148} = 31.012 \text{ m.}$$

Yield strain of rebar $= 500/(2 \times 10^5) = 0.0025$.
The length of beam is 5 m. The yield rotation of frame is given by,

$$\theta_{yF} = 0.5\frac{\varepsilon_y l_b}{h_b} = 0.5 \times \frac{0.0025 \times 5}{0.7} = 0.00893 \text{ radian.}$$

Yield displacement,

$$\Delta_y = \theta_{yF} H_e = 0.0104 \times 31.012 = 0.277 \text{ m.}$$

Displacement ductility,

$$\mu = \frac{\Delta_d}{\Delta_y} = 0.644/0.277 = 2.325.$$

Equivalent damping is given by,

$$\xi_{eq} = 5 + 120\left(\frac{1-\dfrac{1}{\sqrt{\mu}}}{\pi}\right) = 5 + 120\left(\frac{1-\dfrac{1}{\sqrt{2.325}}}{\pi}\right) = 18.1\%.$$

Figure 7.16 shows that the design displacement is higher than the displacement at the corner period for displacement spectrum at 18.1% damping; it is also below the

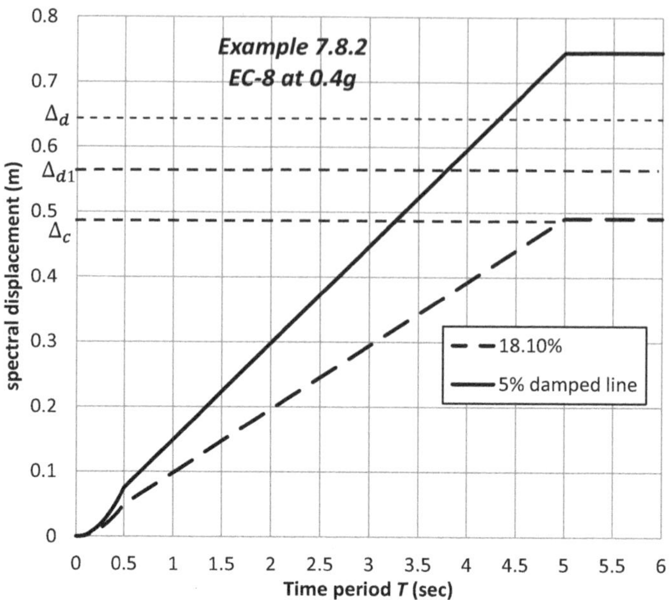

FIGURE 7.16   EC-8 displacement spectrum for Example Ex7.8.2.

corner period displacement for 5% damped spectral line. So, the usual solution does not exist. Refer to case II of Section 7.8.

The case is described through Figure 7.16. The design displacement is Δ_d (= 0.644 m) and equivalent damping is 18.1%. The displacement corresponding to corner period in 18.1% damping line is Δ_c (= 0.49 m). The steps in solution are as follows:

(i) Take an intermediate design displacement $\Delta_{d1} = (0.644 + 0.49)/2 = 0.567$ m.

(ii) New ductility $\mu_1 = \dfrac{\Delta_{d1}}{\Delta_y} = \dfrac{0.567}{0.277} = 2.05$. Find new equivalent damping

$$\xi_{e1} = 5 + 120\left(\frac{1 - \mu_1^{-0.5}}{\pi}\right)\% = 16.5\%.$$

This is shown in Figure 7.17.

(iii) At 16.5% damping, the corner period displacement Δ_c is 0.5 m. Select the new design displacement Δ_{d2} as $(0.5 + 0.567)/2 = 0.534$ m.

(iv) New ductility $\mu_2 = \dfrac{\Delta_{d2}}{\Delta_y} = \dfrac{0.534}{0.277} = 1.93$. Find new equivalent damping

$$\xi_{e1} = 5 + 120\left(\frac{1 - \mu_2^{-0.5}}{\pi}\right)\% = 15.7\%.$$

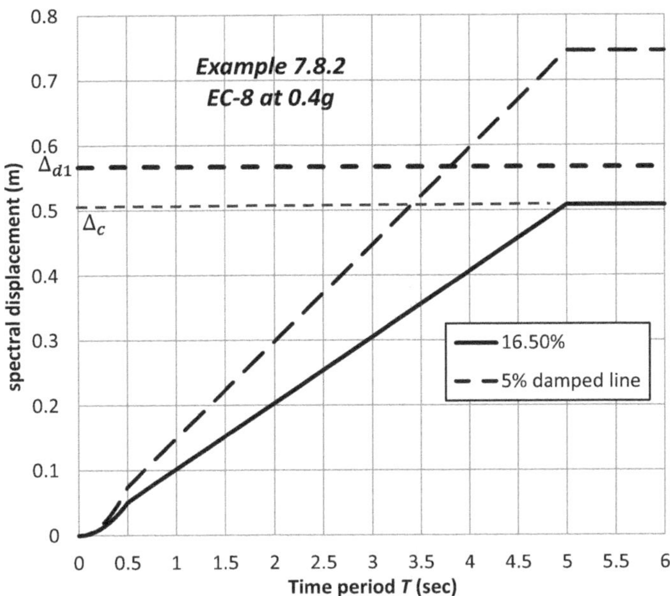

FIGURE 7.17   EC-8 displacement spectrum for Example 7.8.2 (second case).

This stage is shown in Figure 7.18.

From Figure 7.18, it is seen that the design displacement Δ_{d2} (= 0.534 m) and new Δ_c (= 0.52 m) are very close to each other. So, no more iteration is carried out. The corner period is 5 seconds.

$$k_e = 4\pi^2 \frac{m_e}{T_e^2} = 4\pi^2 \frac{135259}{5^2} = 213592 \,\text{kN/m}.$$

$$V_b = k_e \Delta_d = 213592 \times 0.534 = 114058 \text{ kN}.$$

7.9   BASE SHEAR CONSIDERING P-DELTA EFFECT

The basics of P-delta effect have been discussed in Section 4.7 of Chapter 4. Here we discuss the calculation of base shear considering the P-delta effect. Consider Figure 7.19.

Let

V_b = base shear as per DDBD method

$V_{b\Delta}$ = additional base shear caused by P-delta effect.

It is evident that P-delta effect causes overturning moment equal to $P\Delta$. This overturning moment is counterbalanced by $V_{b\Delta}$.

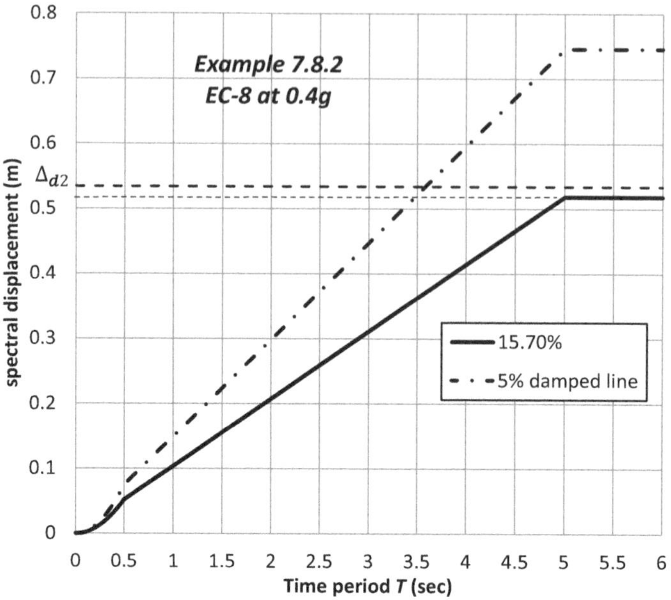

FIGURE 7.18 EC-8 displacement spectrum for Example 7.8.2.

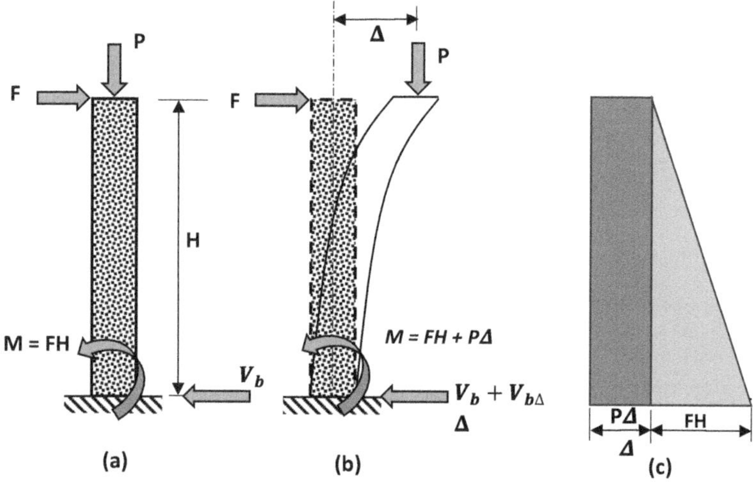

FIGURE 7.19 P-delta effect on base shear: (a) P-delta model and (b) additional base shear (c) BMD.

So, we can write

$$P\Delta = V_{b\Delta} H \tag{7.9.1}$$

where H is the height of the system.

From Eq. (7.9.1),

$$V_{b\Delta} = \frac{P\Delta}{H}$$

(7.9.2)

Now, the displacement Δ in DDBD method is the design displacement Δ_d.
So, Eq. (7.9.2) is written as Eq. (7.9.3).

$$V_{b\Delta} = \frac{P\Delta_d}{H}$$

(7.9.3)

$$V_b = K_e \Delta_d \text{ as usual.}$$

So, total base shear formula can be written as

$$V_b = K_e \Delta_d + \frac{P\Delta_d}{H}$$

(7.9.4)

Sometimes, a modifier is used in the second term of the base shear.
   It may be noted that design base moment is given by Eq. (7.9.5).

$$M_{base} = K_e \Delta_d H + P\Delta_d$$

(7.9.5)

7.10   CLOSURE

The DDBD method as applied to RC frame buildings has been highlighted in this chapter. The method is simple to understand and apply. The DDBD method can be applied to other structures as well, where the methodology slightly differs. It may be noted that DDBD method takes only one target objective, namely, drift. Also, it does not give any insight on how to get the initial member sizes.

7.11   EXERCISES

Q7.11.1 Give a brief account of DDBD method.
Q7.11.2 Give design steps in DDBD method for the design of RC frame buildings.
Q7.11.3 Discuss how the unequal beam lengths are tackled in the DDBD method.
Q7.11.4 Discuss the effect of P-delta effect on base shear and base moment in a structure.
Q7.11.4 Explain the way of tackling the case when design displacement remains above the corner period displacement.
Q7.11.5 A 12-storey RC frame building plan is shown in Figure 7.20. The seismic load in the plan is 10 kN/m^2 in floors and 8 kN/m^2 in roof level. Find the design forces along two directions of the building. The ground storey height is 4.2 m and other storey height is 3.3 m. The building is to be designed for a drift of

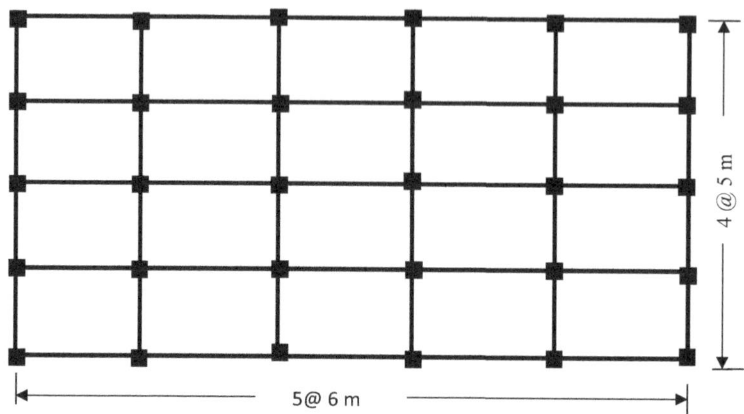

FIGURE 7.20　Plan of building for **Q** 7.11.5.

1.8%. Use EC-8 spectrum for soil type B at 0.45g seismic level. The rebar yield strength is 500 MPa. The depth of the beam in the short direction is 600 mm and that in the long direction is 700 mm. Find the design forces in both directions of the plan.

FURTHER READINGS

Pettinga, J.D. and Priestley, M.J.N. (2005) Dynamic Behaviour of Reinforced Concrete Frames Designed with Direct Displacement-Based Design, *Journal of Earthquake Engineering*, 9(2), 309–330.

Priestley, M.J.N. (2003) Myths and Fallacies in Earthquake Engineering, Revisited, *European School for Advanced Studies in Reduction of Seismic Risk*, 9th Mallet-Milne Lecture.

Priestley, M.J.N. and Calvi, G.M. (1997) Concepts and Procedures for Direct Displacement-based Design and Assessment, Proc. of International Workshop on Seismic Design Methodologies for Next Generation of Codes, Bled, Slovenia.

Priestley, M.J.N., Calvi G.M. and Kowwalsky M.J. (2007) *Displacement-Based Seismic Design of Structures*, IUSS Press, Pavia, Italy.

8 DDBD for Dual System

8.1 INTRODUCTION

A dual system building consists of shear walls and frame elements. It is also known as frame-shear wall system. Figure 8.1 shows a dual system building. Shear wall is a structural RC wall which carries the lateral forces arising out of earthquake or wind, in its own plane. Frame elements consist of beams and columns. Shear wall takes the major portion of the lateral base shear. Frame elements are primarily designed for carrying the vertical loads. However, as the frame is monolithically cast with the walls, the frame also carries a portion of the lateral base shear. Typically, 25–35% of the base shear is allowed to be carried by the frame, and the remaining portion is to be carried by the shear walls. For example, if the frame carries 30% of the base shear, the shear wall will carry 70% of the base shear.

Shear walls are like cantilever deep beams. The horizontal length of wall is the depth of such a deep beam. Shear wall can dissipate a large amount of energy absorbed from the ground motion through hysteresis. In the preceding chapters, we have discussed the DDBD method. In this chapter, the DDBD method primarily developed by Sullivan et al. (2006) for dual system is discussed. The thickness of shear wall is small, generally in the range of 150–350 mm. The length of shear wall may be many metres and the height is equal to the height of the storey. It may be noted that the length of wall means the horizontal dimension of the wall. Shear wall is continuously provided along the height of the building. The placement of shear wall in plan should be symmetrical and each wall needs another wall as a shadow on the opposite side of the plan. Shear wall provided at the periphery of the plan can resist more torsion.

8.2 FRAME–SHEAR WALL INTERACTION

The frame and wall are monolithically cast. As such, they have to deflect together. But the frame and wall bending behaviours are not identical. A frame system with rigid diaphragm and beams bends in a shear mode. This is depicted in Figure 8.2. The shear wall being a cantilever deep beam bends in a cantilever bending mode. The curvatures of bending for wall and frame are opposite to each other. If the frame has convexity of curvature outside, the wall will have the convexity of curvature inside. During earthquake shaking, such curvature will change sides. The wall and the frame

DOI: 10.1201/9781003441090-8

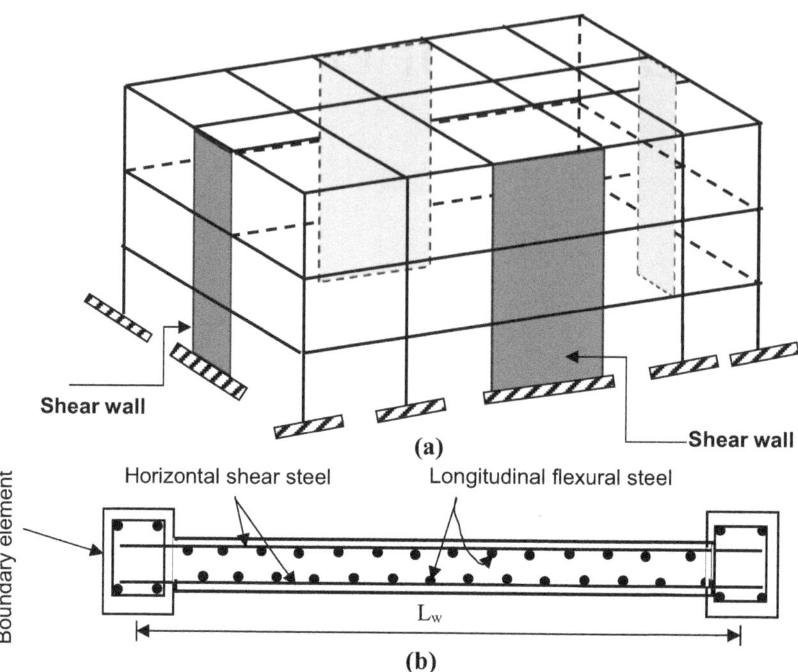

FIGURE 8.1 Dual system: (a) 3D view of building and (b) sectional plan of a shear wall.

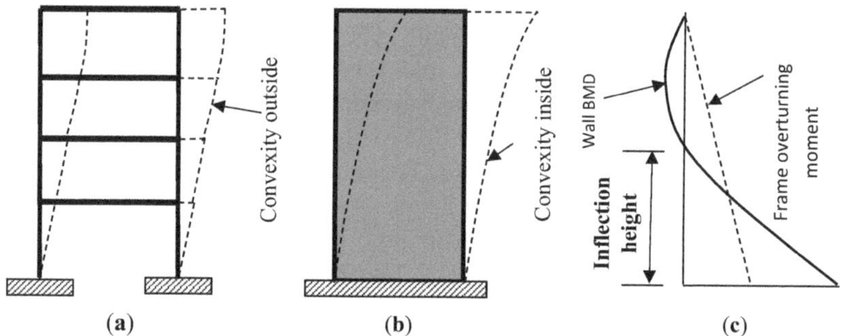

FIGURE 8.2 Frame–shear wall interaction: (a) frame bending in shear mode, (b) wall bending as cantilever and (c) BMD.

thus undergo a tug of war like situation while responding to external actions. This is called frame–shear wall interaction. Because of this interaction, the wall develops a point of zero bending moment at some height, which is called the inflection height.

In Figure 8.2, the frame displays the shear bending mode while the wall is showing cantilever bending. The convexity of deflected curve for frame is on the right side while that for the wall is on the left side. The interaction of frame–wall makes a bending moment diagram, which changes sign for wall bending. The point where the BMD changes sign is the inflection point. The height of inflection is shown in

Figure 8.2(c). Above the inflection height, the frame pushes the wall towards the left, while below the inflection point, the wall pulls the frame to the right. In analysis and design, it is necessary to compute the height of inflection (h_{inf}).

8.3 DDBD FOR DUAL SYSTEM

The DDBD for a dual system after Sullivan et al. (2006a, 2006b) is discussed below. The design method takes drift (θ_d) as the only target design criterion.

In RC frame building, we started with shape profile of the building (Chapter 7). Here, we shall consider the yield displacement of walls first. The final displacement of any floor is the sum of yield displacement of wall at that floor and additional displacement caused by drift. In Figure 8.3(a), an MDOF dual system is shown. The MDOF system can be converted to an ESDOF system as shown in Figure 8.3(b). The equivalent system properties have been discussed in the previous chapters. The yield curvature of the wall (ϕ_{yw}) is gven by Eq. (8.3.1). Here, ε_y is yield strain of rebar of wall, L_w is horizontal length of wall.

$$\phi_{yW} = \frac{2\varepsilon_y}{L_W} \tag{8.3.1}$$

The displacement profile of the wall is obtained from the yield displacement of walls (Δ_{iy}) in different floors as expressed by Eqs. (8.3.2) and (8.3.3). The frame yield rotation is given by Eq. (8.3.4), as was discussed in the previous chapter. The final deflections at floor levels (Δ_i) is given by Eq. (8.3.5).

$$\Delta_{iy} = \frac{\phi_{yW} h_{inf} h_i}{2} - \frac{\phi_{yW} h_{inf}^2}{6} \quad \text{when } h_i \geq h_{inf} \tag{8.3.2}$$

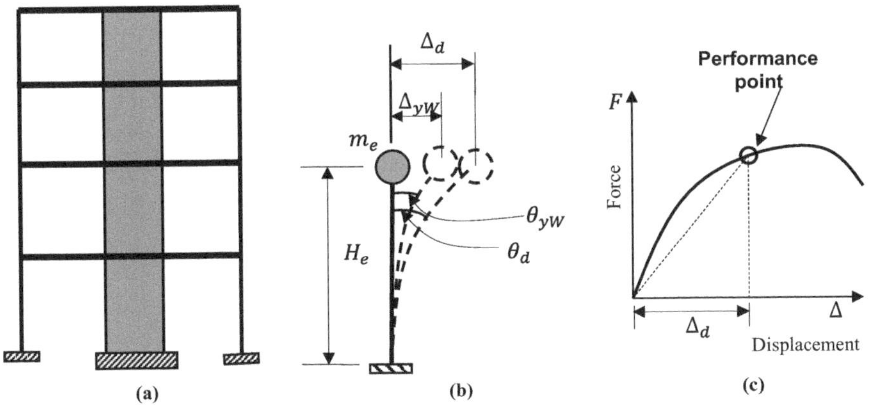

FIGURE 8.3 (a) MDOF dual system, (b) ESDOF system and (c) force–deformation response.

$$\Delta_{iy} = \frac{\phi_{yW} h_i^2}{2} - \frac{\phi_{yW} h_i^3}{6h_{inf}} \quad \text{when } h_i < h_{inf} \tag{8.3.3}$$

$$\theta_{yF} = \frac{0.5\varepsilon_y l_b}{h_b} \tag{8.3.4}$$

$$\Delta_i = \Delta_{iy} + \left(\theta_d - \frac{\phi_{yW} h_{inf}}{2}\right) h_i \tag{8.3.5}$$

where Δ_{iy} is the wall yield displacement at i-th floor level, h_i is the height of i-th floor from base, h_{inf} is the inflection height of wall, θ_{yF} is the yield rotation of frame, θ_d is the design drift, Δ_i is the lateral deflection of the floor at the i-th level.

The ESDOF system properties are same as discussed in previous chapters. However, these are reproduced here for ready reference.

$$\Delta_d = \frac{\sum_{i=1}^{n} m_i \Delta_i^2}{\sum_{i=1}^{n} m_i \Delta_i} \tag{8.3.6}$$

$$m_e = \frac{\sum_{i=1}^{n} m_i \Delta_i}{\Delta_d} \tag{8.3.7}$$

$$H_e = \frac{\sum_{i=1}^{n} m_i \Delta_i h_i}{\sum_{i=1}^{n} m_i \Delta_i} \tag{8.3.8}$$

where m_e is the effective mass, Δ_d is design displacement and H_e is the effective height of ESDOF system, respectively.

The ductility of the frame is given by Eq. (8.3.9). The ductility of the frame varies with storey. If the beam depth is constant over the floors, then the average value of frame ductility may be taken. Ductility of wall is given by Eq. (8.3.10). Here, L_p is the plastic hinge length of wall, d_b the diameter of rebar in wall. The *smaller* value of L_p out of Eqs. (8.3.10b) and (8.3.10c) is to be taken.

$$\mu_{F,i} = \left(\frac{\Delta_i - \Delta_{i-1}}{h_i - h_{i-1}}\right) \frac{1}{\theta_{yF}} \tag{8.3.9}$$

$$\mu_W = 1 + \frac{1}{L_p \phi_{yW}}\left(\theta_d - \frac{\phi_{yW} h_{inf}}{2}\right) \qquad (8.3.10a)$$

$$L_p = 0.022 f_y d_b + 0.054 h_{inf} \qquad (8.3.10b)$$

$$L_p = 0.2 L_W + 0.03 h_{inf} \qquad (8.3.10c)$$

The wall hysteretic damping (ξ_W) and frame damping (ξ_F) are given by Eqs. (8.3.11) and (8.3.12), respectively. In these equations, r is post yield stiffness ratio (ratio of stiffness after yield to initial stiffness), and it generally varies from 0.05 to 0.1. $T_{e,trial}$ is trial effective time period and is given by Eq. (8.3.13a). Here, n is total number of stories in the building. μ_{ESDOF} is system ductility and is given by Eq. (8.3.13b).

$$\xi_W = \frac{95}{1.3\pi}\left(1 - \mu_W^{-0.5} - 0.1 r \mu_W\right)\left(1 + \left(T_{e,trial} + 0.85\right)^{-4}\right) \qquad (8.3.11)$$

$$\xi_F = \frac{120}{1.3\pi}\left(1 - \mu_F^{-0.5} - 0.1 r \mu_F\right)\left(1 + \left(T_{e,trial} + 0.85\right)^{-4}\right) \qquad (8.3.12)$$

$$T_{e,trial} = \frac{n}{6}\sqrt{\mu_{ESDOF}} \qquad (8.3.13a)$$

$$\mu_{ESDOF} = \frac{M_W \mu_W + M_{OTF} \mu_F}{M_W + M_{OTF}} \qquad (8.3.13b)$$

The final ESDOF effective damping is given by Eq. (8.3.14).

$$\xi_{ESDOF} = \frac{M_W \xi_W + M_{OTF} \xi_F}{M_W + M_{OTF}} \qquad (8.3.14)$$

The design base shear is given by Eq. (8.3.15).

$$V_b = k_e \Delta_d \qquad (8.3.15a)$$

$$k_e = 4\pi^2 \frac{m_e}{T_e^2} \qquad (8.3.15b)$$

The base shear is distributed over the floors as per Eq. (8.3.16):

$$\text{For } n \leq 10 \quad F_i = V_b \frac{m_i \Delta_i}{\sum_{i=1}^{n} m_i \Delta_i} \tag{8.3.16a}$$

$$\text{For } n > 10 \quad F_i = 0.9V_b \frac{m_i \Delta_i}{\sum_{i=1}^{n} m_i \Delta_i} \tag{8.3.16b}$$

$$F_{roof} = F_{roof} \text{ from Eq. } (8.3.16b) + 0.1V_b \tag{8.3.16c}$$

Eqs. (8.3.16b) and (8.3.16c) take care of higher mode effects.

8.4   EXAMPLES

Example 8.4.1 *Find the lateral design forces in floors for a 10-storey dual system, which is to be designed for a 2.5% drift. Take f_y as 500 MPa, diameter of longitudinal rebar as 20 mm, interstorey height as 3.4 m and ground storey height as 4.4 m. Floor mass is 10,000 kg in each floor and the roof mass is 6000 kg. Assume that the frame takes 35% of the base shear. Length of the wall is 5 m, length of the beam is 5 m and depth of the beam is 0.6 m. Take post-yield stiffness ratio as 0.1. Use EC-8 spectrum for soil type B at 0.4g seismicity level.*

Solution: Table Ex8.4.1.1 is prepared for the computation of height of inflection. For this purpose, absolute values of forces and moments are not necessary. The relative values in proportion of mass and height of floors is sufficient for the purpose. In this table, column (6) is the cumulative sum of column (5) items. Putting these lateral forces at respective floors, the moments produced by the forces are put in column (7). Column (7) is obtained from Eq. (8.4.1), where, MOT is frame overturning moment.

$$M_{OTi} = M_{OT,i+1} + V_{t,i+1}\left(h_{i+1} - h_i\right) \tag{8.4.1}$$

As frame is designed to take 35% of base shear, the frame force remains constant at 0.35 in column (8). As wall shear is the total shear less frame shear, column (9) is obtained by subtracting column (8) from column (6). Column (10) is the moment produced by forces at different heights in column (9) and it can be computed using Eq. (8.4.2).

$$M_{Wi} = M_{W,i+1} + V_{W,i+1}\left(h_{i+1} - h_i\right) \tag{8.4.2}$$

A graph is drawn between wall moment and height of the building, which is shown in Figure 8.4. From the graph, inflection height (h_{inf}) is 21.6 m. M_W = 11.491 kN-m, M_{OTF} = 23.47 kN-m.

TABLE EX8.4.1.1
Computation of height of inflection for Example 8.4.1

Floor m_i (kg)	m_i (kg)	h_i (kg)	$m_i h_i$	$m_i h_i$/sum $(m_i h_i)$	V_{ti} relative	M_{OTi} relative	V_F relative	V_{wi} relative	M_w relative
								(9)=	
(1)	(2)	(3)	(4)	(5)	(6)	(7)	(8)	(6)–(8)	(10)
10	6000	35	210000	0.115	0.115	0.000	0.35	-0.235	0.000
9	10000	31.6	316000	0.173	0.287	0.390	0.35	-0.063	-0.800
8	10000	28.2	282000	0.154	0.442	1.367	0.35	0.092	-1.013
7	10000	24.8	248000	0.136	0.577	2.869	0.35	0.227	-0.701
6	10000	21.4	214000	0.117	0.694	4.831	0.35	0.344	0.071
5	10000	18	180000	0.098	0.792	7.190	0.35	0.442	1.240
4	10000	14.6	146000	0.080	0.872	9.884	0.35	0.522	2.744
3	10000	11.2	112000	0.061	0.933	12.849	0.35	0.583	4.519
2	10000	7.8	78000	0.043	0.976	16.023	0.35	0.626	6.503
1	10000	4.4	44000	0.024	1.000	19.341	0.35	0.650	8.631
		0				23.741			11.491
	sum	1830000	1						

N.B. In the table V_{ti} is total shear in i-th floor, V_F is frame shear, V_{Wi} is wall shear in i-th floor.

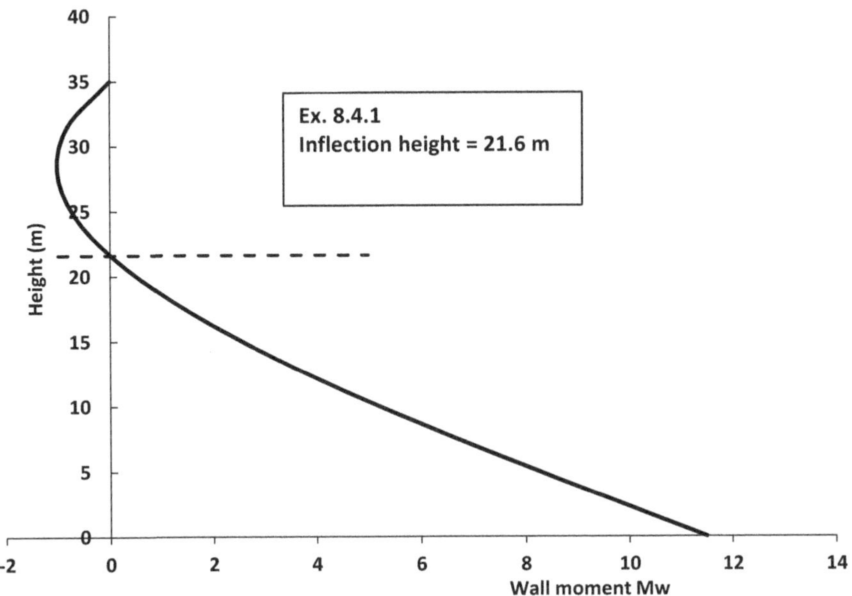

FIGURE 8.4 Height of inflection for Example 8.4.1.

TABLE EX8.4.1.2
ESDOF system properties computation for Example 8.4.1

Floor	h_i	m_i	Δ_{iy}	Δ_i	$m_i\Delta_i$	$m_i\Delta_i^2$	$m_i\Delta_i h_i$
10	35	6000	0.3002	0.7972	4783	3814	167420
9	31.6	10000	0.2635	0.7122	7122	5073	225068
8	28.2	10000	0.2268	0.6272	6272	3934	176882
7	24.8	10000	0.1901	0.5422	5422	2940	134476
6	21.4	10000	0.1534	0.4572	4572	2091	97849
5	18	10000	0.1170	0.3726	3726	1388	67068
4	14.6	10000	0.0826	0.2899	2899	840	42323
3	11.2	10000	0.0519	0.2109	2109	445	23623
2	7.8	10000	0.0268	0.1375	1375	189	10726
1	4.4	10000	0.0090	0.0715	715	51	3146
0	0			SUM	**38997**	**20765**	**948582**

Now, we prepare Table Ex8.4.1.2 by using Eqs. (8.3.2) to (8.3.5).

Note that the yield curvature of wall $\phi_{yW} = \dfrac{2\varepsilon_y}{L_W} = (2 \times 500/200000)/5 = 0.001$ per m.

Note that the inflection height can also be found out by interpolation without drawing the graph.

$$\Delta_d = \frac{\sum_{i=1}^{n} m_i\Delta_i^2}{\sum_{i=1}^{n} m_i\Delta_i} = \frac{20765}{38997} = 0.532 \text{ m.}$$

$$m_e = \frac{\sum_{i=1}^{n} m_i\Delta_i}{\Delta_d} = \frac{38997}{0.532} = 73302 \text{ kg.}$$

$$H_e = \frac{\sum_{i=1}^{n} m_i\Delta_i h_i}{\sum_{i=1}^{n} m_i\Delta_i} = \frac{948582}{38997} = 24.324 \text{ m.}$$

$$\theta_{yF} = \frac{0.5\varepsilon_y l_b}{h_b} = 0.5 \times 0.0025 \times 5/0.6 = 0.0104 \text{ radian.}$$

For ductility of frame as per Eq. (8.3.9), we prepare Table Ex8.4.1.3.

From this Table Ex8.4.1.3, average frame ductility μ_F is 2.29.

Now we find the wall ductility. Plastic hinge length of wall is smaller out of:

$$L_p = 0.022 f_y d_b + 0.054 h_{inf} = 0.022 \times 500 \times 20/1000 + 0.054 \times 21.6 = 1.386 \text{ m.}$$

and, $L_p = 0.2L_W + 0.03h_{inf} = 0.2 \times 5 + 0.03 \times 21.6 = 1.648 \text{ m}$

TABLE EX8.4.1.3
Computation of frame ductility for Example 8.4.1

Floor	h_i	Δ_i	μ_F
10	35	0.8302	2.400
9	31.6	0.7452	2.400
8	28.2	0.6602	2.400
7	24.8	0.5752	2.400
6	21.4	0.4902	2.400
5	18	0.4052	2.398
4	14.6	0.3202	2.361
3	11.2	0.2366	2.258
2	7.8	0.1566	2.087
1	4.4	0.0827	1.805
0	0	0.0000	-
		sum	22.909
		Average	**2.29**

That is, $L_p = 1.386$ m.

$$\phi_{yW} = 0.001$$

Wall ductility is given by, $\mu_W = 1 + \dfrac{1}{L_p \phi_{yW}}\left(\theta_d - \dfrac{\phi_{yW} h_{inf}}{2}\right) =$

$$1 + \frac{1}{1.386 \times 0.001}\left(0.025 - \frac{0.001 \times 21.6}{2}\right) = 11.24.$$

System ductility is given by,

$$\mu_{ESDOF} = \frac{M_W \mu_W + M_{OTF} \mu_F}{M_W + M_{OTF}} = \frac{11.491 \times 11.24 + 23.47 \times 2.29}{11.24 + 23.47} = 5.233.$$

Trial effective time period $T_{e,trial} = \dfrac{n}{6}\sqrt{\mu_{ESDOF}} = \dfrac{10}{6}\sqrt{5.233} = 3.813$ s. Given, $r = 0.1$.

$$\xi_W = \frac{95}{1.3\pi}\left(1 - \mu_W^{-0.5} - 0.1r\mu_W\right)\left(1 + \left(T_{e,trial} + 0.85\right)^{-4}\right)$$

$$= \frac{95}{1.3\pi}\left(1 - 11.24^{-0.5} - 0.1 \times 0.1 \times 11.24\right)\left(1 + \left(3.816 + 0.85\right)^{-4}\right) = 13.7.$$

TABLE EX8.4.1.4
Lateral design forces in Example 8.4.1

Floor	h_i (m)	m_i (kg)	$m_i h_i$	V_b (kN)	F_i (kN)
10	35	6000	210000	**75958**	8716
9	31.6	10000	316000		13116
8	28.2	10000	282000		11705
7	24.8	10000	248000		10294
6	21.4	10000	214000		8883
5	18	10000	180000		7471
4	14.6	10000	146000		6060
3	11.2	10000	112000		4649
2	7.8	10000	78000		3238
1	4.4	10000	44000		1826
		sum	1830000		**75958**

$$\xi_F = \frac{120}{1.3\pi}\left(1-\mu_F^{-0.5}-0.1r\mu_F\right)\left(1+\left(T_{e,trial}+0.85\right)^{-4}\right)$$

$$=\frac{120}{1.3\pi}\left(1-2.29^{-0.5}-0.1\times0.1\times2.29\right)\left(1+\left(3.816+0.85\right)^{-4}\right)=9.32.$$

$\xi_{ESDOF} = \dfrac{M_W\xi_W + M_{OTF}\xi_F}{M_W + M_{OTF}} = 10.7\%$. From EC-8 spectra at 0.4g level (Figure 8.5), $T_e = 4.5$ s.

$$k_e = 4\pi^2\frac{m_e}{T_e^2} = 142778 \text{ kN/m.}$$

$$V_b = k_e\Delta_d = 75958 \text{ kN.}$$

The base shear is distributed over the floors as per Eq. (8.3.16) and shown in Table Ex8.4.1.4.

$$\text{For } n \leq 10\, F_i = V_b\frac{m_i\Delta_i}{\sum_{i=1}^{n}m_i\Delta_i}$$

Example 8.4.2 *Find the lateral design forces in floors for a 20-storey dual system, which is to be designed for a 2.2% drift. Take f_y as 500 MPa, diameter of rebar as 25 mm, interstorey height as 3.2 m, ground storey height as 4.2 m. Floor mass is*

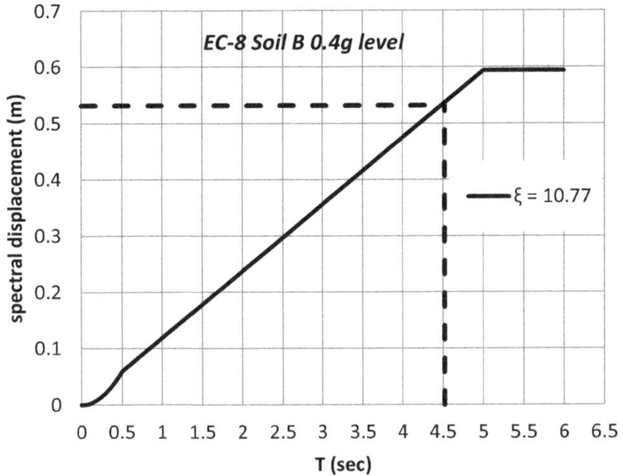

FIGURE 8.5 EC-8 spectrum for Example 8.4.1.

TABLE EX8.4.2.1
Computation of height of inflection for Example 8.4.2

Floor	m_i (kg)	h_i (m)	m_ih_i	$m_ih_i/$ sum (m_ih_i)	V_{ti} relative	M_{OTi} relative	V_F relative	V_{wi} relative	M_w relative
20	9000	65	585000	0.078	0.078	0.000	0.3	-0.222	0.000
19	11000	61.8	679800	0.091	0.169	0.250	0.3	-0.131	-0.71
18	11000	58.6	644600	0.086	0.255	0.791	0.3	-0.045	-1.129
17	11000	55.4	609400	0.081	0.337	1.608	0.3	0.037	-1.272
16	11000	52.2	574200	0.077	0.413	2.685	0.3	0.113	-1.155
15	11000	49.0	539000	0.072	0.485	4.008	0.3	0.185	-0.792
14	11000	45.8	503800	0.067	0.553	5.561	0.3	0.253	-0.199
13	11000	42.6	468600	0.063	0.615	7.330	0.3	0.315	0.610
12	11000	39.4	433400	0.058	0.673	9.299	0.3	0.373	1.619
11	11000	36.2	398200	0.053	0.727	11.454	0.3	0.427	2.814
10	11000	33.0	363000	0.049	0.775	13.779	0.3	0.475	4.179
9	11000	29.8	327800	0.044	0.819	16.259	0.3	0.519	5.699
8	11000	26.6	292600	0.039	0.858	18.88	0.3	0.558	7.360
7	11000	23.4	257400	0.034	0.892	21.625	0.3	0.592	9.145
6	11000	20.2	222200	0.030	0.922	24.481	0.3	0.622	11.041
5	11000	17.0	187000	0.025	0.947	27.431	0.3	0.647	13.031
4	11000	13.8	151800	0.020	0.967	30.462	0.3	0.667	15.102
3	11000	10.6	116600	0.016	0.983	33.558	0.3	0.683	17.238
2	11000	7.4	81400	0.011	0.994	36.703	0.3	0.694	19.423
1	11000	4.2	46200	0.006	1	39.883	0.3	0.700	21.643
0	0	0	0	0	1	**44.083**			**24.583**
	sum		7482000						

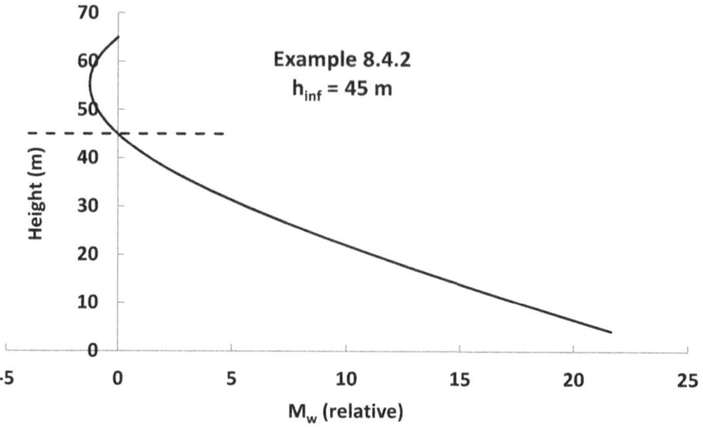

FIGURE 8.6   Height of inflection for Example 8.4.2.

11,000 kg and the roof mass is 9000 kg. Assume that the frame takes 30% of base shear. Length of the wall is 10 m, length of the beam is 5 m and depth of the beam is 0.8 m. Take EC-8 design spectrum at 0.65g level.

Solution: Table Ex8.4.2.1 is prepared for the computation of height of inflection, as explained in Ex. 8.4.1. From this Table $M_W = 24.583$ kN-m (relative), $M_{OTF} = 44.083$ kN-m (relative).

From this Table, Figure 8.6 is drawn, wherefrom, $h_{inf} = 45$ m.

Now, ESDOF system properties are found out by preparing Table Ex8.4.2.2.

$$\Delta_d = \frac{\sum_{i=1}^{n} m_i \Delta_i^2}{\sum_{i=1}^{n} m_i \Delta_i} = \frac{111461}{132928} = 0.839 \text{ m}$$

$$m_e = \frac{\sum_{i=1}^{n} m_i \Delta_i}{\Delta_d} = \frac{132928}{0.839} = 158,528 \text{ kg}$$

$$H_e = \frac{\sum_{i=1}^{n} m_i \Delta_i h_i}{\sum_{i=1}^{n} m_i \Delta_i} = \frac{6050140}{132928} = 45.515 \text{ m.}$$

$$\theta_{yF} = \frac{0.5\varepsilon_y l_b}{h_b} = 0.5 \times 0.0025 \times 5/0.8 = 0.0078 \text{ radian.}$$

$$\phi_{yW} = \frac{2\varepsilon_y}{L_W} = (2 \times 500/200000)/10 = 0.0005 \text{ per m.}$$

TABLE EX8.4.2.2
ESDOF system properties computation in Example 8.4.2

Floor	h_i	m_i	Δ_{iv}	Δ_i	$m_i\Delta_i$	$m_i\Delta_i^2$	$m_ih_i\Delta_i$
20	65	9000	0.5626	1.2611	11350	14314	737770
19	61.8	11000	0.5266	1.1907	13098	15597	809469
18	58.6	11000	0.4906	1.1203	12324	13807	722175
17	55.4	11000	0.4546	1.0499	11549	12126	639837
16	52.2	11000	0.4186	0.9795	10775	10555	562455
15	49	11000	0.3826	0.9091	10001	9092	490030
14	45.8	11000	0.3466	0.8387	9226	7738	422560
13	42.6	11000	0.3106	0.7684	8452	6494	360059
12	39.4	11000	0.2749	0.6983	7681	5363	302632
11	36.2	11000	0.2398	0.6288	6917	4349	250394
10	33	11000	0.2057	0.5604	6164	3454	203409
9	29.8	11000	0.173	0.4933	5426	2676	161692
8	26.6	11000	0.142	0.4279	4707	2014	125205
7	23.4	11000	0.1132	0.3646	4011	1463	93858
6	20.2	11000	0.0868	0.3038	3342	1015	67511
5	17	11000	0.0632	0.2458	2704	665	45973
4	13.8	11000	0.0427	0.191	2102	401	29001
3	10.6	11000	0.0259	0.1398	1538	215	16300
2	7.4	11000	0.0129	0.0925	1017	94	7527
1	4.2	11000	0.0043	0.0494	543	27	2283
				sum	**132928**	**111461**	**6050140**

Table Ex8.4.2.3 gives average frame ductility $\mu_F = 50.735/20 = 2.54$.

Plastic hinge length of wall is smaller out of:

$$L_p = 0.022 f_y d_b + 0.054 h_{inf} = 0.022 \times 500 \times 25/1000 + 0.054 \times 45 = 2.706 \text{ m.}$$

and

$$L_p = 0.2 L_W + 0.03 h_{inf} = 0.2 \times 10 + 0.03 \times 45 = 3.350 \text{ m}$$

That is, $L_p = 2.706$ m

Wall ductility is given by

$$\mu_W = 1 + \frac{1}{L_p \phi_{yW}}\left(\theta_d - \frac{\phi_{yW} h_{inf}}{2}\right) = 1 + \frac{1}{2.706 \times 0.0005}$$

$$\left(0.022 - \frac{0.0005 \times 45}{2}\right) = \mathbf{8.94.}$$

TABLE EX8.4.2.3
Computation of frame ductility in Example 8.4.2

Floor	h_i (m)	Δ_i	μ_{Fi}
20	65	1.261	2.483
19	61.8	1.191	2.816
18	58.6	1.12	2.816
17	55.4	1.05	2.816
16	52.2	0.98	2.816
15	49	0.909	2.816
14	45.8	0.839	2.816
13	42.6	0.768	2.815
12	39.4	0.698	2.804
11	36.2	0.629	2.778
10	33	0.56	2.738
9	29.8	0.493	2.684
8	26.6	0.428	2.614
7	23.4	0.365	2.531
6	20.2	0.304	2.432
5	17	0.246	2.319
4	13.8	0.191	2.192
3	10.6	0.14	2.05
2	7.4	0.092	1.893
1	4.2	0.049	1.506
0	0	sum	**50.735**

System ductility is given by

$$\mu_{ESDOF} = \frac{M_W \mu_W + M_{OTF} \mu_F}{M_W + M_{OTF}} = \frac{24.583 \times 8.94 + 44.083 \times 2.54}{24.583 + 44.083} = 4.83.$$

Trial effective time period $T_{e,trial} = \dfrac{n}{6}\sqrt{\mu_{ESDOF}} = \dfrac{20}{6}\sqrt{4.83} = 7.33$ s. Given, $r = 0.1$.

$$\xi_W = \frac{95}{1.3\pi}\left(1 - \mu_W^{-0.5} - 0.1r\mu_W\right)\left(1 + \left(T_{e,trial} + 0.85\right)^{-4}\right)$$

$$= \frac{95}{1.3\pi}\left(1 - 8.94^{-0.5} - 0.1 \times 0.1 \times 8.94\right)\left(1 + \left(7.33 + 0.85\right)^{-4}\right) = 13.41.$$

$$\xi_F = \frac{120}{1.3\pi}\left(1 - \sqrt{\mu_F} - 0.1r\mu_F\right)\left(1 + \left(T_{e,trial} + 0.85\right)^{-4}\right) = 10.19.$$

TABLE EX8.4.2.4
Distribution of base shear in Example 8.4.2

Floor	m_i (kg)	h_i (m)	$m_i h_i$	V_b	F_i (kN)
20	65	9000	585000		45764
19	61.8	11000	679800		21965
18	58.6	11000	644600		20828
17	55.4	11000	609400		19691
16	52.2	11000	574200		18553
15	49	11000	539000		17416
14	45.8	11000	503800		16278
13	42.6	11000	468600		15141
12	39.4	11000	433400		14004
11	36.2	11000	398200		12866
10	33	11000	363000	**268615**	11729
9	29.8	11000	327800		10592
8	26.6	11000	292600		9454
7	23.4	11000	257400		8317
6	20.2	11000	222200		7180
5	17	11000	187000		6042
4	13.8	11000	151800		4905
3	10.6	11000	116600		3768
2	7.4	11000	81400		2630
1	4.2	11000	46200		1493
		Sum	**7482000**		**268616**

$$\xi_{ESDOF} = \frac{M_W \xi_W + M_{OTF} \xi_F}{M_W + M_{OTF}} = \frac{24.583 \times 13.41 + 44.083 \times 10.19}{24.583 + 44.083} = 11.34\%.$$

For $\Delta_d = 0.839$ and $\xi_{ESDOF} = 11.34\%$, from EC-8 spectra at 0.65g level (Figure 8.7), $T_e = 4.42$ sec.

$$k_e = 4\pi^2 \frac{m_e}{T_e^2} = 4\pi^2 \frac{158528}{4.42^2} = 320348 \text{ kN/m.}$$

$$V_b = k_e \Delta_d = 268615 \text{ kN.}$$

The base shear is distributed over the floors as per Eq. (8.3.16).

$$\text{For } n > 10 F_i = V_t + 0.9 V_b \frac{m_i \Delta_i}{\sum_{i=1}^{n} m_i \Delta_i}$$

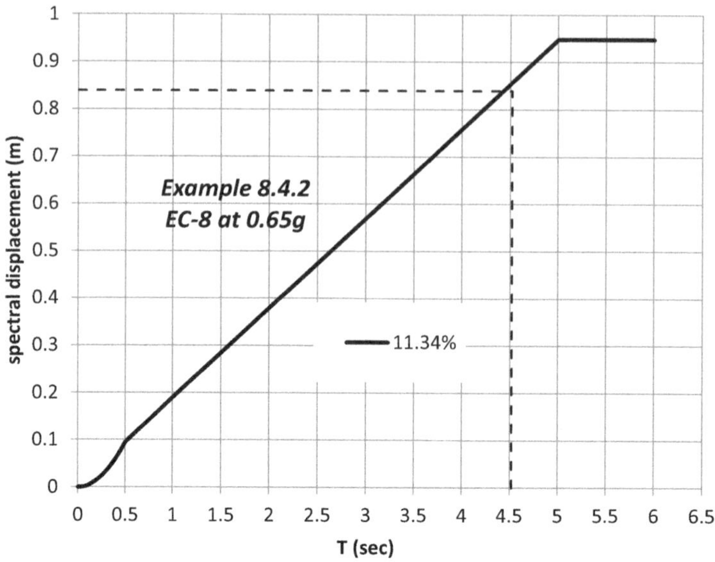

FIGURE 8.7   EC-8 spectrum for Example 8.4.2.

Here V_t is $0.1\,V_b = 0.1 \times 268615 = 26862$ kN, which is to be provided for roof level only. Rest of the 90% of the base shear is distributed over all floors including the roof.

Roof level lateral design force $= 26862 + 0.9 \times 268615 \times 585000/7482000 = 45764$ kN.

Lateral design force at 19th floor $= 0.9 \times 268615 \times 679800/7482000 = 21965$ kN. Similarly, calculations are done for other floors.

The floor lateral forces are shown in Table Ex8.4.2.4.

Example 8.4.3 *Find the lateral design forces in floors for a 15-storey dual system shown in Figure 8.8. The design drift is 3%. Take f_y as 550 MPa, diameter of rebar as 25 mm, interstorey height as 3.3 m and ground storey height as 4.4 m. The floor seismic weight is 5 kN/m² and that of roof is 4 kN/m². Assume that the frame takes 32% of the base shear. Use EC-8 design spectrum at the 0.7g level. Depths of beam along the short and long directions are 0.9 m and 0.7 m, respectively.*

Solution: The wall lengths are different in two directions of the plan. Seismic mass in each floor $= 36 \times 21 \times 5/9.81 = 385$ kN-s²/m, and seismic mass of roof $= 36 \times 21 \times 4/9.81 = 308$ kN-s²/m.

$$\varepsilon_y = 550/200000 = 0.00275.$$

Long direction of the plan

$$L_w = 12 \text{ m}, \quad l_b = 6 \text{ m}, \quad h_b = 0.7 \text{ m}.$$

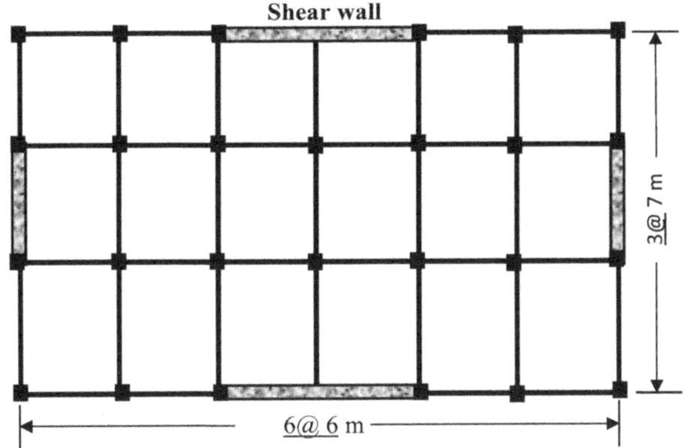

FIGURE 8.8 Plan of building in Example 8.4.3.

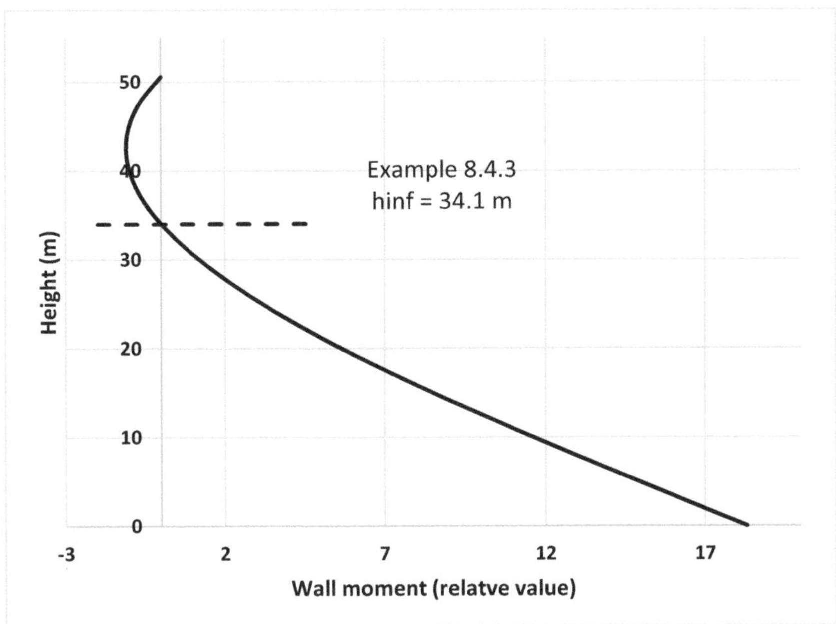

FIGURE 8.9 Inflection height of building in Example 8.4.3.

Table Ex8.4.3.1 is prepared. From this table, M_{OTF} = 34.5 kN-m (relative value), M_W = 18.305 kN-m (relative value). Also, this table gives (by interpolation) h_{inf} = 34.1 m. Also Figure 8.9 gives the same value of inflection height.

The ESDOF system properties are obtained from Table Ex8.4.3.2.

TABLE EX8.4.3.1
Computation of height of inflection for Example 8.4.3 (long direction)

Floor	h_i (m)	m_i (kN-s²/m)	$m_i h_i$	$m_i h_i$/sum $(m_i h_i)$	V_{ti} relative	M_{OTi} relative	V_F relative	V_w relative	M_w relative
15	50.6	308	15585	0.101	0.101	0	0.32	-0.219	0
14	47.3	385	18210	0.118	0.218	0.332	0.32	-0.102	-0.724
13	44	385	16940	0.1090	0.328	1.052	0.32	0.008	-1.06
12	40.7	385	15669	0.1010	0.429	2.133	0.32	0.109	-1.035
11	37.4	385	14399	0.0930	0.522	3.547	0.32	0.202	-0.677
10	34.1	385	13128	0.0850	0.6060	5.268	0.32	0.286	-0.012
9	30.8	385	11858	0.0770	0.683	7.269	0.32	0.363	0.933
8	27.5	385	10587	0.0680	0.751	9.523	0.32	0.431	2.131
7	24.2	385	9317	0.0600	0.811	12.002	0.32	0.491	3.554
6	20.9	385	8046	0.0520	0.863	14.679	0.32	0.543	5.175
5	17.6	385	6776	0.0440	0.907	17.528	0.32	0.587	6.968
4	14.3	385	5505	0.0360	0.943	20.522	0.32	0.623	8.906
3	11	385	4235	0.0270	0.970	23.632	0.32	0.65	10.96
2	7.7	385	2964	0.0190	0.989	26.833	0.32	0.669	13.105
1	4.4	385	1694	0.0110	1.000	30.097	0.32	0.68	15.313
	0					**34.500**			**18.305**
		sum	154916						

TABLE EX8.4.3.2
ESDOF system properties computation in Example 8.4.3 (long direction)

Floor	h_i	m_i	Δ_{iy}	Δ_i	$m_i \Delta_i$	$m_i \Delta_i^2$	$m_i h_i \Delta_i$
15	50.6	308	0.3063	1.4294	440	629	22277
14	47.3	385	0.2806	1.3304	512	681	24227
13	44	385	0.2548	1.2314	474	584	20860
12	40.7	385	0.2291	1.1324	436	494	17744
11	37.4	385	0.2033	1.0334	398	411	14880
10	34.1	385	0.1775	0.9344	360	336	12267
9	30.8	385	0.1519	0.8355	322	269	9907
8	27.5	385	0.1267	0.737	284	209	7803
7	24.2	385	0.1024	0.6395	246	157	5959
6	20.9	385	0.0796	0.5435	209	114	4373
5	17.6	385	0.0588	0.4494	173	78	3045
4	14.3	385	0.0403	0.3577	138	49	1969
3	11	385	0.0247	0.2689	104	28	1139
2	7.7	385	0.0126	0.1835	71	13	544
1	4.4	385	0.0042	0.1019	39	4	173
				sum	**4205**	**4056**	**147165**

TABLE EX8.4.3.3
Computation of frame ductility in Example 8.4.3
(long direction)

Floor	h_i (m)	Δ_i	μ_{Fi}
15	50.6	1.429	2.397
14	47.3	1.33	2.545
13	44	1.231	2.545
12	40.7	1.132	2.545
11	37.4	1.033	2.545
10	34.1	0.934	2.545
9	30.8	0.835	2.543
8	27.5	0.737	2.531
7	24.2	0.64	2.506
6	20.9	0.543	2.469
5	17.6	0.449	2.42
4	14.3	0.358	2.358
3	11	0.269	2.283
2	7.7	0.183	2.196
1	4.4	0.102	1.965
0	0	sum	**36.393**

$$\Delta_d = \frac{\sum_{i=1}^{n} m_i \Delta_i^2}{\sum_{i=1}^{n} m_i \Delta_i} = \frac{4056}{4206} = 0.964 \text{ m,}$$

$$m_e = \frac{\sum_{i=1}^{n} m_i \Delta_i}{\Delta_d} = \frac{4205}{0.964} = 4362 \text{ kg,}$$

$$H_e = \frac{\sum_{i=1}^{n} m_i \Delta_i h_i}{\sum_{i=1}^{n} m_i \Delta_i} = \frac{147165}{4205} = 34.998 \text{ m.}$$

$$\theta_{yF} = \frac{0.5 \varepsilon_y l_b}{h_b} = 0.5 \times 0.00275 \times 6/0.7 = 0.01179 \text{ radian.}$$

Table Ex8.4.3.3 gives frame ductility.

Frame ductility $\mu_F = 36.393/15 = 2.43$.

TABLE EX8.4.3.4
Distribution of base shear in Example 8.4.3 (long direction)

Floor	m_i (kg)	h_i (m)	$m_i h_i$	V_b (KN)	F_i (kN)
15	50.6	308	15585	7645	1457
14	47.3	385	18210		809
13	44	385	16940		752
12	40.7	385	15669		696
11	37.4	385	14399		639
10	34.1	385	13128		583
9	30.8	385	11858		527
8	27.5	385	10587		470
7	24.2	385	9317		414
6	20.9	385	8046		357
5	17.6	385	6776		301
4	14.3	385	5505		245
3	11	385	4235		188
2	7.7	385	2964		132
1	4.4	385	1694		75
		sum	154913		7645

Computation of base shear

Plastic hinge length of wall is smaller out of:

$$L_p = 0.022 f_y d_b + 0.054 h_{inf} = 2.412 \text{ m.}$$

and

$$L_p = 0.2 L_W + 0.03 h_{inf} = 3.422 \text{m}$$

Out of these two, the smaller value is $L_p = 2.412$ m.

$$\phi_{yW} = \frac{2\varepsilon_y}{L_W} = (2 \times 0.00275)/12 = 0.00046 \text{ per m.}$$

Wall ductility is given by,

$$\mu_W = 1 + \frac{1}{L_p \phi_{yW}} \left(\theta_d - \frac{\phi_{yW} h_{inf}}{2} \right) = \mathbf{23.61.}$$

System ductility is given by

$$\mu_{ESDOF} = \frac{M_W \mu_W + M_{OTF}\mu_F}{M_W + M_{OTF}} = \mathbf{9.77}.$$

Trial effective time period $T_{e,trial} = \dfrac{n}{6}\sqrt{\mu_{ESDOF}} = 7.81$ s. Given, $r = 0.1$.

$$\xi_W = \frac{95}{1.3\pi}\left(1 - \mu_W^{-0.5} - 0.1r\mu_W\right)\left(1 + \left(T_{e,trial} + 0.85\right)^{-4}\right) = 12.98.$$

$$\xi_F = \frac{120}{1.3\pi}\left(1 - \sqrt{\mu_F} - 0.1r\mu_F\right)\left(1 + \left(T_{e,trial} + 0.85\right)^{-4}\right) = 9.81.$$

$$\xi_{ESDOF} = \frac{M_W \xi_W + M_{OTF}\xi_F}{M_W + M_{OTF}} = 10.91\%.$$

For $\Delta_d = 0.965$ and $\xi_{ESDOF} = 10.91\%$, from EC-8 spectra at 0.7g level (Figure 8.10), $T_e = 4.66$ s.

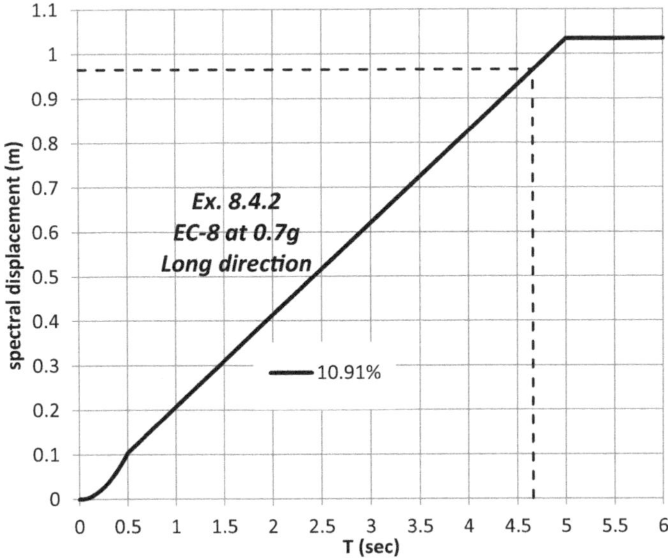

FIGURE 8.10   EC-8 spectrum for Example 8.4.3 (long direction).

TABLE EX8.4.3.5
ESDOF system properties computation in Example 8.4.3 (short direction)

Floor	h_i	m_i	Δ_{iv}	Δ_i	$m_i\Delta_i$	$m_i\Delta_i^2$	$m_ih_i\Delta_i$
15	50.6	308	0.5251	1.3661	421	575	21290
14	47.3	385	0.481	1.2671	488	618	23074
13	44	385	0.4368	1.1681	450	525	19787
12	40.7	385	0.3927	1.0691	412	440	16752
11	37.4	385	0.3485	0.9701	373	362	13968
10	34.1	385	0.3044	0.8711	335	292	11436
9	30.8	385	0.2603	0.7722	297	230	9157
8	27.5	385	0.2171	0.6742	260	175	7138
7	24.2	385	0.1756	0.5778	222	129	5383
6	20.9	385	0.1365	0.4838	186	90	3893
5	17.6	385	0.1007	0.3932	151	60	2665
4	14.3	385	0.0691	0.3068	118	36	1689
3	11	385	0.0424	0.2252	87	20	954
2	7.7	385	0.0215	0.1495	58	9	443
1	4.4	385	0.0073	0.0804	31	2	136
				sum	**3889**	**3563**	**137765**

$$k_e = 4\pi^2 \frac{m_e}{T_e^2} = 7925\,\text{kN/m.}$$

$$V_b = k_e\Delta_d = 7645 \text{ kN.}$$

The base shear is distributed as per Table Ex8.4.3.4.

Short direction of the plan

$$L_W = 7 \text{ m}, \ l_b = 7 \text{ m}, \ h_b = 0.9 \text{ m.}$$

The data for computation of the inflection height is same as that in the long direction. From Table Ex8.4.3.1, $M_{OTF} = 34.5$ kN-m (relative) and $M_W = 18.305$ kN-m (relative). $h_{inf} = 34.1$ m (this does not change with direction).

The ESDOF system properties differ in two directions because of the involvement of wall yield displacement, which contains ϕ_{yW}, which in turn uses the length of the wall. So, Table Ex8.3.4.5 is prepared.

From Table Ex8.4.3.5,

$$\Delta_d = \frac{\sum_{i=1}^{n} m_i\Delta_i^2}{\sum_{i=1}^{n} m_i\Delta_i} = \frac{3563}{3889} = 0.916 \text{ m,}$$

TABLE EX8.4.3.6
Computation of frame ductility in Example 8.4.3
(*short direction*)

Floor	h_i (m)	Δ_i	μ_{Fi}
15	50.6	1.366	2.524
14	47.3	1.267	2.805
13	44	1.168	2.805
12	40.7	1.069	2.805
11	37.4	0.97	2.805
10	34.1	0.871	2.805
9	30.8	0.772	2.801
8	27.5	0.674	2.778
7	24.2	0.578	2.732
6	20.9	0.484	2.661
5	17.6	0.393	2.568
4	14.3	0.307	2.45
3	11	0.225	2.31
2	7.7	0.15	2.146
1	4.4	0.08	1.709
0	0	sum	**38.704**

$$m_e = \frac{\sum_{i=1}^{n} m_i \Delta_i}{\Delta_d} = \frac{3889}{0.916} = 4246 \text{ kg},$$

$$H_e = \frac{\sum_{i=1}^{n} m_i \Delta_i h_i}{\sum_{i=1}^{n} m_i \Delta_i} = \frac{137765}{3889} = 35.424 \text{ m.}$$

$$\theta_{yF} = \frac{0.5\varepsilon_y l_b}{h_b} = 0.5 \times 0.00275 \times 7/0.9 = 0.01069 \text{ radian.}$$

Table Ex8.4.3.6 gives frame ductility.
Frame ductility μ_F = 38.704/15 = **2.58**.

Computation of the base shear

Plastic hinge length of wall is smaller out of:

$$L_p = 0.022 f_y d_b + 0.054 h_{inf} = 2.142 \text{ m.}$$

and, $L_p = 0.2 L_W + 0.03 h_{inf} = 2.422 \text{ m}$

TABLE EX8.4.3.7
Distribution of base shear in Example 8.4.3 (*short direction*)

Floor	h_i (m)	m_i (kN-s²/m)	$m_i h_i$	V_b (KN)	F_i (kN)
15	50.6	308	15585	7582	1445
14	47.3	385	18210		802
13	44	385	16940		746
12	40.7	385	15669		690
11	37.4	385	14399		634
10	34.1	385	13128		578
9	30.8	385	11858		522
8	27.5	385	10587		466
7	24.2	385	9317		410
6	20.9	385	8046		354
5	17.6	385	6776		298
4	14.3	385	5505		243
3	11	385	4235		187
2	7.7	385	2964		131
1	4.4	385	1694		75
		sum	154913		**7581**

That is, $L_p = 1.142$ m.

$$\phi_{yW} = \frac{2\varepsilon_y}{L_W} = (2 \times 0.00275)/7 = 0.00079 \text{ per m.}$$

Wall ductility is given by

$$\mu_W = 1 + \frac{1}{L_p \phi_{yW}} \left(\theta_d - \frac{\phi_{yW} h_{inf}}{2} \right) = \mathbf{10.88.}$$

System ductility is given by,

$$\mu_{ESDOF} = \frac{M_W \mu_W + M_{OTF} \mu_F}{M_W + M_{OTF}} = 5.46.$$

Trial effective time period $T_{e,trial} = \dfrac{n}{6} \sqrt{\mu_{ESDOF}} = 5.84$ sec. Given, $r = 0.1$.

$$\xi_W = \frac{95}{1.3\pi} \left(1 - \mu_W^{-0.5} - 0.1 r \mu_W \right) \left(1 + \left(T_{e,trial} + 0.85 \right)^{-4} \right) = 13.685$$

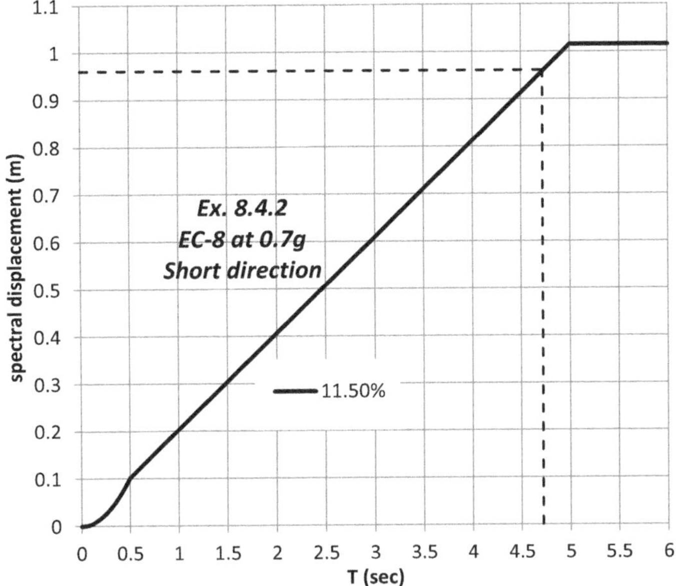

FIGURE 8.11   EC-8 spectrum for Example 8.4.3 (short direction).

$$\xi_F = \frac{120}{1.3\pi}\left(1 - \sqrt{\mu_F} - 0.1 r \mu_F\right)\left(1 + \left(T_{e,trial} + 0.85\right)^{-4}\right) = 10.34.$$

$$\xi_{ESDOF} = \frac{M_W \xi_W + M_{OTF} \xi_F}{M_W + M_{OTF}} = \mathbf{11.5\%}.$$

For $\Delta_d = 0.916$ and $\xi_{ESDOF} = 11.5\%$, from EC-8 spectra at 0.7g level (Figure 8.11), $T_e = 4.5$ s.

$$k_e = 4\pi^2 \frac{m_e}{T_e^2} = 8277 \text{ k N/m.}$$

$$V_b = k_e \Delta_d = 7582 \text{ kN.}$$

The base shear is distributed as shown in Table Ex8.4.3.7

It is worthwhile to compare the results of analyses in two directions. Table Ex8.4.3.8 shows the comparison.

TABLE EX8.4.3.8
Comparison of analyses results in short and long directions in Example 8.4.3

Item	Short direction	Long direction
Δ_d (m) (m)	0.916	0.965
m_e (kg)	4246	4359 (kN-s^2/m)
h_e (m)	35.424	34.997
$FFigh_{inf}$ (m)	34.06	34.06
μ_F	2.58	2.43
μ_W	10.88	23.61
μ_{ESDOF}	5.46	9.77
$T_{e,trial}$ (sec)	5.84	7.81
ξ_F	10.34	9.81
ξ_W	13.685	12.98
ξ_{ESDOF}	11.5	10.91
T_e (sec)	4.5	4.66
k_e (kN/m)	8277	7925
V_b kN	7582	7645

8.5 CLOSURE

In this chapter, the DDBD method for the dual system has been described. Examples are given for clarifying the application of the design philosophy. In dual system, most of the ductility arises out of shear wall. After the design, the actual performance of the building is to be evaluated by carrying out nonlinear analyses.

8.6 EXERCISES

Q 8.6.1 Write briefly the philosophy of DDBD for RC dual system.

Q 8.6.2 Write down the steps in DDBD for the RC dual system.

Q 8.6.3 Find the lateral design forces in floors for a 12-storey dual system building, which is to be designed for 2.0% drift. Take f_y as 500 MPa, diameter of longitudinal rebar as 20 mm, interstorey height as 3.3 m and ground storey height as 4.3 m. Floor mass is 10,000 kg in each floor and the roof mass is 6000 kg. Assume that the frame takes 30% of base shear. Length of wall is 6 m, length of beam is 6 m and depth of the beam is 0.8 m. Take post-yield stiffness ratio is 0.1. Use EC-8 spectrum for soil type B at 0.45g seismicity level.

Q 8.6.4 Find the lateral design forces in floors for a 20-storey dual system, which is to be designed for 2.4% drift. Take f_y as 500 MPa, diameter of the rebar as 25 mm, interstorey height as 3.3 m, ground storey height as 4.3 m. Floor mass is 11,000 kg and roof mass is 9000 kg. Assume that the frame takes 27% of base shear. Length of the wall is 10 m, length of the beam is 5 m and depth of the beam is 0.8 m. Take EC-8 design spectrum at 0.65g level.

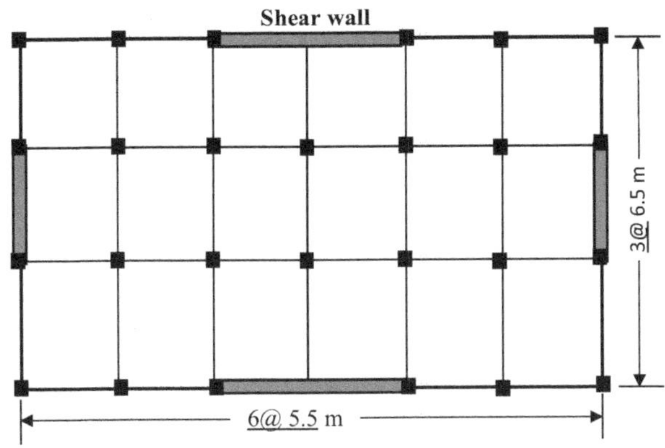

FIGURE 8.12   Plan of building in Question 8.6.5.

Q 8.6.5 Find the lateral design forces in floors for an 18-storey dual system shown in Figure 8.12. The design drift is 2.6%. Take f_y as 550 MPa, diameter of rebar as 25 mm, interstorey height as 3.4 m and ground storey height as 4.5 m. The floor seismic weight is 6 kN/m² and that of roof is 4 kN/m². Assume that the frame takes 25% of the base shear. Use EC-8 design spectrum at 0.65g level. Depths of beam along the short and long directions are 0.9 m and 0.7 m, respectively.

Q 8.6.6 What are the limitations of the DDBD method for the dual system?

FURTHER READINGS

Priestley, M.J.N., Calvi, G.M. and Kowalsky, M.J. (2007) *Displacement-Based Seismic Design of Structures*, IUSS Press. Italy.

Sullivan, T.J., Priestley, M.J.N. and Calvi, G.M. (2005) Development of an Innovative Seismic Design Procedure for Frame–Wall Structures, *Journal of Earthquake Engineering,* 9(2), 279–307.

Sullivan, T.J., Priestley, M.J.N. and Calvi, G.M. (2006a) Direct Displacement-Based Design of Frame-Wall Structures, *Journal of Earthquake Engineering,* 10 (Special Issue 1), 91–124.

Sullivan, T.J., Priestley, M.J.N. and Calvi, G.M. (2006b) Seismic Design of Frame-Wall Structures, Research Report No. ROSE-2006/02.

9 Unified Performance-Based Design of RC Frame Buildings

9.1 INTRODUCTION

In the preceding two chapters, we have described the DDBD method for frame building and dual system building. The DDBD methodology developed by Pettinga and Priestley (2005), as described in Chapter 7, is a handy design process for RC frame buildings. Similarly, the DDBD method for RC dual system (Sullivan et al., 2006a, 2006b), as described in Chapter 8, is an innovative design process for the dual system. However, these methods addressed only drift as the *single* target performance objective. The performance level (PL), which is another important design objective indicating the degree of damage, is not included in these methods. Furthermore, there is no guideline for taking the initial sizes of the members. As such, a large number of iterations may be required before arriving at the final member sizes. In each iteration, the whole design procedure, including the computation of ESDOF system properties, is to be repeated. These limitations have been overcome in unified performance-based design (UPBD) (Choudhury, 2007; Choudhury and Singh, 2013).

The term "unified performance-based design (UPBD)" was first coined by Choudhury (2007). The name signifies that both drift and performance level are addressed in the design method. The UPBD method for RC dual system was proposed by Choudhury (2007). Choudhury and Singh (2013) devised the UPBD method for RC frame buildings. This chapter describes the UPBD method for RC frame buildings.

In the UPBD method, the target design objectives for RC frame buildings are: (i) interstorey drift $\left(\theta_d\right)$ and, (ii) member performance level. The member performance level is designated in terms of plastic rotation in the members. As per capacity design (see Chapter 2), principle of "weak-beam strong-column" for buildings is applied so that the columns do not undergo inelastic deformation. It is the beam which has to form plastic hinges and dissipate energy through hysteresis. Thus, the member performance level for RC frame buildings can be expressed in terms of the plastic rotation of beams alone. In the following section, it is demonstrated how both these parameters, namely, drift and performance level, can be incorporated into the design formulation as the target design objectives. We shall also see how the member sizes are also given by the theoretical approach.

DOI: 10.1201/9781003441090-9

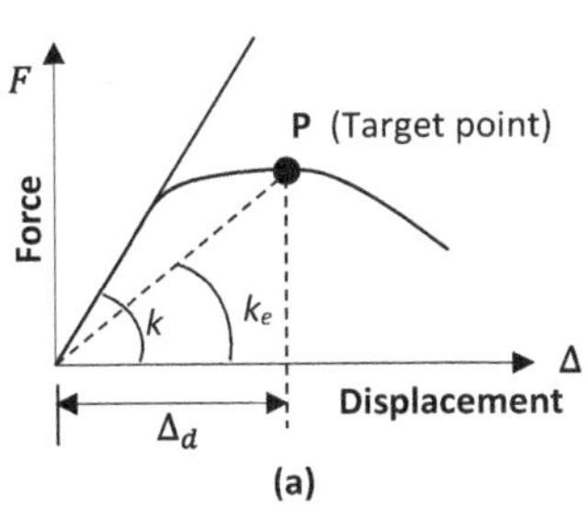

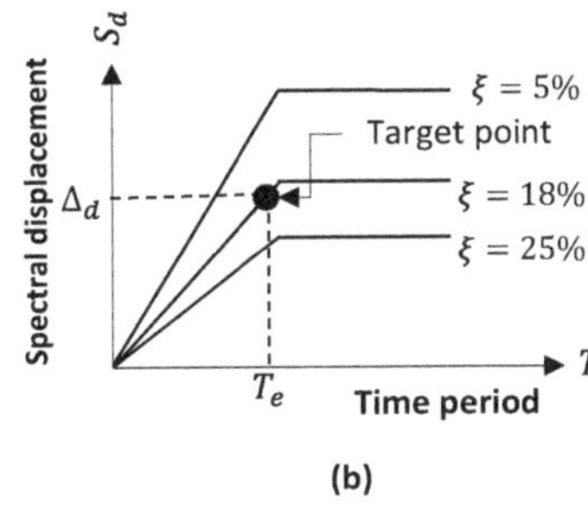

FIGURE 9.1 (a) Force–deformation curve and (b) displacement spectra.

Consider force–displacement relationship of a structure as shown Figure 9.1. For designating a structure corresponding to an accepted damage level (point P in Figure 9.1a), effective stiffness (k_e) or secant stiffness is to be used. The target displacement corresponding to the damage level is Δ_d. The time period of the structure is initially T (elastic time period), but this changes as the structure reaches a nonlinear state. The time period corresponding to any point on the nonlinear part of the force–deformation curve is designated as the effective time period (T_e). Instead of slope of tangent to point P, it is the slope of the line joining P from the origin. The time period relationship, in general, is given by Eq. (9.1.1).

$$T = 2\pi\sqrt{\frac{m}{k}} \tag{9.1.1}$$

Eq. (9.1.1) may be written as

$$k = 4\pi^2 \frac{m}{T^2} \tag{9.1.2}$$

For the target point (or any other point on the nonlinear part of the force–deformation curve) we can rewrite the Eq. (9.1.2) as Eq. (9.1.3):

$$k_e = 4\pi^2 \frac{m_e}{T_e^2} \tag{9.1.3}$$

In Eq. (9.1.3), k_e is effective stiffness, m_e is effective mass for equivalent single degree of freedom (ESDOF) system and T_e is effective time period corresponding to the target point.

From basic stiffness relationship, with V_b as base shear, we can write as Eq. (9.1.4).

$$V_b = k_e \Delta_d \tag{9.1.4}$$

In Figure 9.1(b), displacement spectra corresponding to the design spectrum is shown. As shown in this figure, corresponding to the design displacement and associated damping, the effective time period can be found out.

9.2 THEORETICAL DEVELOPMENT OF UPBD METHOD

Consider the 3D frame building shown in Figure 9.2(a). The building has masses at different floor levels and stiffnesses of members at various stories. From the structural dynamics theory, it is possible to convert the 3D building to an ESDOF system (Figure 9.2b). The ESDOF system shall have an equivalent single mass (m_e), equivalent height (H_e) and equivalent stiffness (k_e). The equivalent height corresponds to an equivalent overturning moment replicating the overturning moment of the real building. The ESDOF system shall undergo a displacement compatible to and arising out of the displacement pattern of the real building. The displacement pattern of real building depends upon its shape profile (generally, the first mode profile) and the amount of interstorey drift. The total angular displacement of the ESDOF system is the summation of the elastic angular displacement and plastic angular displacement. The total angular displacement (angular rotation) of the ESDOF system is the angular design drift (θ_d). Design drift means the interstorey drift value for which the building is to be designed. Thus, the design drift is the summation of the maximum elastic angular drift of the frame (θ_{yF}) and plastic rotation of the frame (θ_p). The above concept in reference to Figure 9.2b can be expressed by Eq. (9.2.1).

$$\theta_d = \theta_{yF} + \theta_p \tag{9.2.1}$$

As the entire plastic rotation of the ESDOF system arises out of the plastic rotation of beams (θ_{pb}) [columns to remain elastic as per capacity design], Eq. (9.2.1) can be written as Eq. (9.2.2)

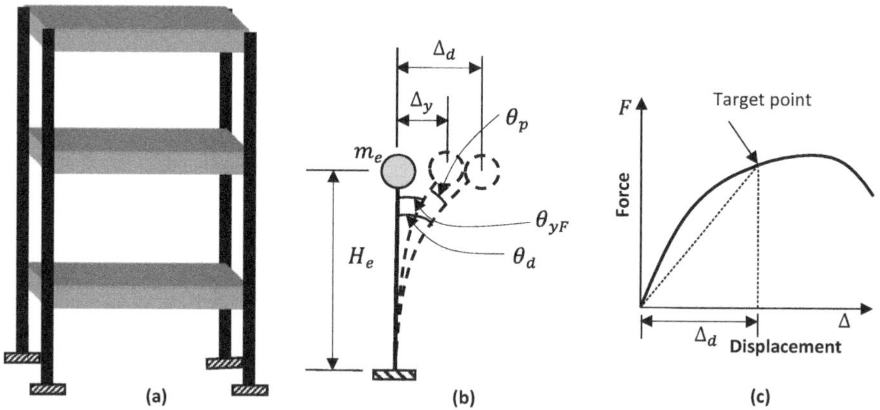

FIGURE 9.2 (a) MDOF building, (b) equivalent SDOF system and (c) force–displacement response.

$$\theta_d = \theta_{yF} + \theta_{pb} \tag{9.2.2}$$

Here, θ_{yF} is maximum elastic rotation of frame. As per Priestley (2003), the maximum frame elastic rotation is given by

$$\theta_{yF} = \frac{0.5\varepsilon_y l_b}{h_b} \tag{9.2.3}$$

In Eq. (9.2.3.), ε_y is the yield strain of rebar, l_b is the length of the beam along the direction of seismic action and h_b is overall depth of the beam.

Substituting Eq. (9.2.3) in (9.2.2), we get

$$\theta_d = \frac{0.5\varepsilon_y l_b}{h_b} + \theta_{pb}$$

$$\text{Or,} \quad \theta_d - \theta_{pb} = \frac{0.5\epsilon_y l_b}{h_b}$$

$$\text{Or,} \quad h_b = \frac{0.5\varepsilon_y l_b}{\theta_d - \theta_{pb}} \tag{9.2.4}$$

Eq. (9.2.4) gives a depth of beam that satisfies both interstorey drift and performance level. The performance level is expressed in terms of plastic rotation of members (here beams). The width of the beam is kept from half to two-thirds of beam depth as per practice. Thus, the beam size is known at the beginning of design.

9.3   INTERPRETATION OF EQ. (9.2.4)

We can rewrite Eq. (9.2.4) as Eq. (9.3.1):

$$\frac{h_b}{l_b} = \frac{0.5\varepsilon_y}{\theta_d - \theta_{pb}} \tag{9.3.1}$$

To plot this equation, we need to decide the range of drift that shall correspond to a particular performance level. Referring to Table C1-3 of FEMA-356, the drift for vertical members corresponding to different PLs are as follows:

IO: 1% transient drift with negligible permanent drift
LS: 2% transient drift with 1% permanent drift and
CP: 4% transient or permanent drift.

The permanent drift implies plastic rotation. As per the above guidelines, we shall assume the following ranges of drifts for different PLs:

IO: 1–1.5% (total drift)
LS: 1.5–3% (total drift),
CP: 3–4% (total drift).

Drift less than 1% may be assumed to be too stringent for practical design purposes. For any given PL, the maximum allowable value of θ_{pb} remains constant. For a given grade of steel ε_y is known and is fixed. So, as per Eq. (9.3.1), a graph can be drawn between h_b/l_b ratio and θ_d by varying θ_d. Let us take Fe415 steel (f_y = 415 MPa) and IO performance level. Here, ε_y = 415/(2 × 10⁵) = 0.002075. The average plastic rotation for beams for IO PL as per ASCE-SEI-41-17 (Table 10-7 of this document) for compliant buildings is 0.00625 radian. Such values are also available in FEMA-356. So, for IO PL, Eq. (9.3.1) takes the form (9.3.1a).

$$\frac{h_b}{l_b} = \frac{0.5 \times 0.002075}{\theta_d - 0.00625} \tag{9.3.1a}$$

The graph of corresponding to Eq. (9.3.1.a) is the curve AB as shown in Figure 9.3. So, AB corresponds to IO PL. The average plastic rotation of the beam for compliant buildings for LS PL is 0.02 radian. Putting this in Eq. (9.3.1), curve CD is obtained, which corresponds to LS PL. Similarly, curve EF is drawn, which corresponds to CP PL.

From the end of IO performance line, a horizontal line BM is drawn. Now for part CM of LS PL, the same h_b/l_b ratio can give IO PL as well. Naturally, one would go for IO instead of LS for this part of the curve, otherwise it will not be judicious or economical. Hence, this part of the curve total CD (i.e. CM) is discarded. From

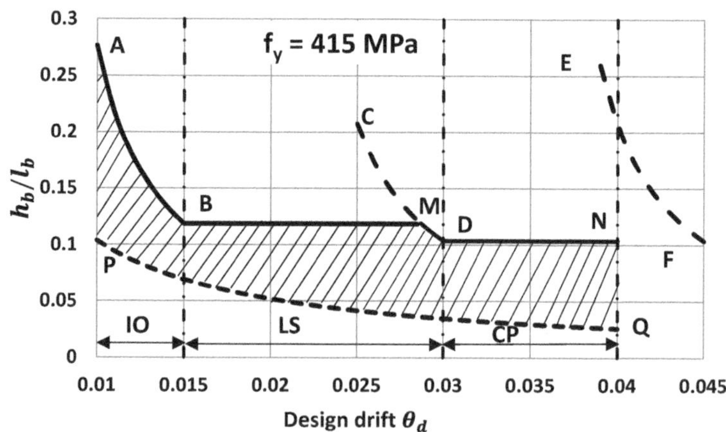

FIGURE 9.3 Interpretation of Eq. (9.3.1), for f_y = 415 MPa.

D, a horizontal line is drawn to meet 4% drift line at N. So, line ABMDN gives the upper bound of h_b/l_b ratio for different PLs. When no plastic rotation occurs (elastic system), θ_{pb} is zero in Eq. (9.3.1), which leads to curve PQ. So, PQ is the lower bound of h_b/l_b ratio.

The hatched zone between the upper bound and lower bound curves gives the viable zone of h_b/l_b ratios. In this viable zone, corresponding higher values of h_b/l_b ratio will correspond to a higher plastic rotation. The upper bound line gives plastic rotation corresponding to average values of plastic rotation of beams of compliant buildings as per ASCE-SEI-41-17. Similar diagrams can be readily constructed for other grades of steel (Figures 9.4 and 9.5).

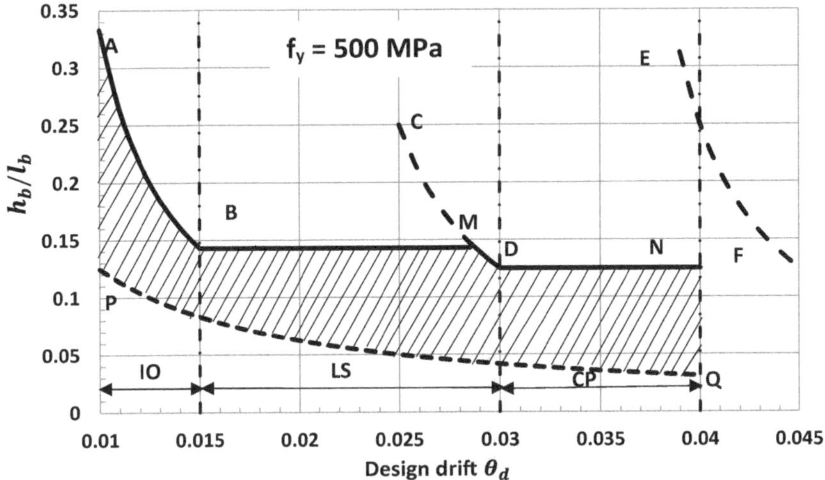

FIGURE 9.4 Interpretation of Eq. (9.3.1), for $f_y = 500$ MPa.

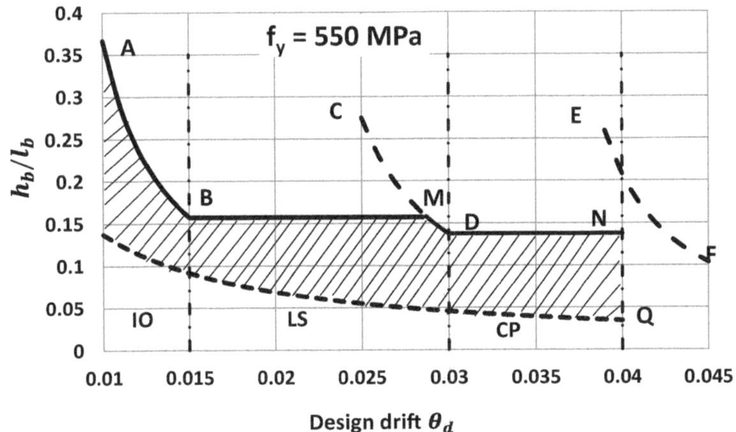

FIGURE 9.5 Interpretation of Eq. (9.3.1), for $f_y = 550$ MPa.

Example 9.3.1 *In a building plan the length of beam is 6 m. The building is to be designed for (a) IO performance level and 1.2% interstorey drift; (b) LS performance level and 2.5% interstorey drift; (c) CP performance level with 3.8% drift. Suggest a suitable beam size in each case for the building. The yield strength of steel rebar is 500 MPa.*

Solution: The yield strain rebar steel is $500/2 \times 10^5 = 0.0025$.

(a) The average plastic rotation for IO performance level for compliant buildings (from ASCE-SEI-41-17, Table 10-7) is 0.00625 radian. Referring to Eq. (9.2.4),

$$h_b = \frac{0.5\varepsilon_y l_b}{\theta_d - \theta_{pb}} = \frac{0.5 \times 0.0025 \times 6}{0.012 - 0.00625} = 1.3 \text{ m. The width of the beam is taken as 0.6 m.}$$

Suggested beam size is 0.6 m × 1.3 m.

(b) The average plastic rotation allowable for LS performance level for compliant buildings (from ASCE-SEI-41-17, Table 10-7) is 0.02 radian. Referring to Eq. (9.2.4), $h_b = \dfrac{0.5\epsilon_y l_b}{\theta_d - \theta_{pb}} = \dfrac{0.5 \times 0.0025 \times 6}{0.025 - 0.02} = 1.5$ m. Here, $h_b/l_b = 1.5/6 = 0.25$. Referring to Figure 9.4, this value lies above the line BM, and hence, discarded. The viable h_b/l_b ratio corresponding to line BM is 0.143. So, $h_b = 0.143 \times 6 = 0.858$ m, say, 0.85 m. The width of beam may be taken as 0.5 m. Suggested beam size is 0.5 m × 0.85 m.

(c) The average plastic rotation for CP performance level for compliant buildings (from ASCE-SEI-41-17, Table 10-7) is 0.035 radian. Referring to Eq. (9.2.4),

$$h_b = \frac{0.5\epsilon_y l_b}{\theta_d - \theta_{pb}} = \frac{0.5 \times 0.0025 \times 6}{0.038 - 0.035} = 2.5 \text{ m. Here, } h_b/l_b = 2.5/6 = 0.42. \text{ But}$$

from Figure 9.4, the maximum limit set by line DN is 0.125, Hence $h_b = 0.125 \times 6 = 0.75$ m, say, 0.7 m. The width of beam is taken as 0.4 m. Suggested beam size is 0.4 m × 0.7 m.

9.4   DESIGN STEPS IN UPBD METHOD FOR RC FRAME BUILDINGS

With the introduction of specific beam size satisfying target objectives as explained in Sections 9.2 and 9.3, there is no need for iteration to arrive at the beam size for satisfying target performance objectives. The design steps involved in UPBD method for RC frame buildings, including the computation of ESDOF system properties, are detailed below. The ESDOF system properties were described by Pettinga and Priestley (2005) and originated from the substitute structure concept (Shibata and Sozen, 1974).

The steps involved in the UPBD method for the design of RC frame buildings are given below.

Step 1: *Set the target objectives* for the building to be designed. Typical target objectives are interstorey drift (θ_d) and member performance level (designated by θ_{pb}). For example, a building may be designed for 2% drift and LS performance level.

Step 2: *Decide the hazard level* for which the building is to be designed. It involves deciding the design spectrum and its level. For example, a building may be designed as per EC-8 design spectrum for *type 1* structural condition and ground *type B*, at shaking level 0.6*g*.

Step 3: *Decide material properties* namely, characteristic strength of concrete (f_{ck}) and yield strength of rebar steel (f_y).

Step 4: *Compute beam depth* by Eq. (9.2.4) and by referring to Section 9.3. The plastic rotation of the beam is available from FEMA (Table 6-8 of FEMA-356) or SEI (Table 10-7 of ASCE-SEI-17), corresponding to the performance level desired. The width of the beam is kept from one-third to two-thirds of beam depth as per common practice. Take the preliminary sizes of columns from experience. One way to maintain the uniformity of buildings is to restrict the column steel to a percentage value (generally 3–4%). The column size calculation also has been discussed in Mayengbam and Choudhury (2014).

Step 5: *Take a shape profile* from the list as follows:

According to Priestley and Calvi (1997),

$$\text{For } n \leq 4: \qquad \phi_i = \frac{h_i}{H} \tag{9.4.1}$$

$$\text{For } 4 \leq n < 20: \quad \phi_i = \frac{h_i}{H}\left(\frac{16 - 0.5\dfrac{h_i}{H}(n-4)}{16 - 0.5(n-4)} \right) \tag{9.4.2}$$

$$\text{For } n \geq 20: \qquad \phi_i = \frac{2h_i}{H}\left(1 - 0.5\frac{h_i}{H} \right) \tag{9.4.3}$$

As per Pettinga and Priestley (2005),

$$\text{For} \leq 4: \quad \phi_i = \frac{h_i}{H} \tag{9.4.4}$$

$$\text{For} > 4: \quad \phi_i = \frac{4}{3}\frac{h_i}{H}\left(1 - 0.25\frac{h_i}{H} \right) \tag{9.4.5}$$

where h_i is height of *i*-th floor from base of building, H is total height of building from its base and ϕ_i is mode shape coefficient for *i*-th floor in first mode.

Find the critical storey displacement and other storey displacement:

$$\Delta_c = \theta_d h_{sc} \tag{9.4.6}$$

where h_{sc} is the critical storey height (critical storey is that which has largest mass or largest interstorey height).

$$\Delta_i = \phi_i \frac{\Delta_c}{\phi_c} \tag{9.4.7}$$

Here, Δ_i is storey sway in i-th floor level.

Step 6: *Compute the ESDOF system properties*: Δ_d, m_e, H_e as given by

$$\Delta_d = \frac{\sum_{i=1}^{n} m_i \Delta_i^2}{\sum_{i=1}^{n} m_i \Delta_i} \tag{9.4.8}$$

$$m_e = \frac{\sum_{i=1}^{n} m_i \Delta_i}{\Delta_d} \tag{9.4.9}$$

$$H_e = \frac{\sum_{i=1}^{n} m_i \Delta_i h_i}{\sum_{i=1}^{n} m_i \Delta_i} \tag{9.4.10}$$

where m_e is mass, Δ_d is design displacement and H_e is the effective height of ESDOF system, respectively.

Step 7: *Generate the displacement spectra* at various dampings corresponding to the design spectrum considered in step (2). (For the generation of displacement spectra, see Section 9.8).

Step 8: *Compute yield displacement.*

$$\Delta_y = \theta_{yF} H_e \tag{9.4.11}$$

$$\theta_{yF} = 0.5 \frac{\varepsilon_y l_b}{h_b} \tag{9.4.12}$$

Compute displacement ductility, $\mu = \Delta_d / \Delta_y$.

Compute equivalent damping in % from the relationship:

$$\xi_{eq} = 5 + 120 \left(\frac{1 - \dfrac{1}{\sqrt{\mu}}}{\pi} \right) \%$$

(9.4.13)

Step 9: *Compute effective time period* T_e [see Figure 9.1b] corresponding to Δ_d and damping ξ_{eq}.

Step 10: *Compute effective stiffness* k_e

$$k_e = 4\pi^2 \frac{m_e}{T_e}$$

(9.4.14)

Step 11: *Compute the base shear* from the relationship:

$$V_b = k_e \Delta_d$$

(9.4.15)

Step 12: *Distribute the base shear over the floors* as per equations,

$$\text{For } n \le 10 \quad F_i = V_b \frac{m_i \Delta_i}{\sum_{i=1}^{n} m_i \Delta_i}$$

(9.4.16)

$$\text{For } n > 10 \quad F_i = 0.9 V_b \frac{m_i \Delta_i}{\sum_{i=1}^{n} m_i \Delta_i}$$

(9.4.17)

$$F_{roof} = F_{i,roof} + 0.1 V_b$$

Step 13: Adopt the load combination and design the building.

The load combinations are:

$$DL + LL$$

$$DL + LL \pm F_x$$

$$DL + LL \pm F_y$$

where *DL* is dead load, *LL* is live load, F_x and F_y are lateral seismic forces (as per Eqs. (9.4.16) and (9.4.17)) in short and long directions of the plan of the building, respectively.

For design, take the expected strength of materials. From FEMA documents, the expected strength of materials are: for concrete $1.5\,f_{ck}$ and for rebar $1.25\,f_y$.

Step 14: *Evaluation* of the designed building.

The building designed has to be checked for its actual performance. The evaluation is done through nonlinear static analysis or pushover analysis (POA) and nonlinear time history analyses (NLTHA). The performance point is obtained from POA. The plastic hinges are noted at the end of NLTHA. The relative location of PP and PL (IO, LS, CP) gives the actual PL of the building. Drift is obtained from floor displacement time histories. If the target design objectives are not achieved, the design may be updated suitably.

9.5 DESIGN EXAMPLES

Example 9.5.1 *A building plan is shown in Figure 9.6. The design drift is 2.8% and target performance level is LS. Find the beam sizes as per UPBD method. Take f_y as 500 MPa for the rebar.*

Solution: The design drift θ_d = 0.028. The yield strain for rebar steel is $500/2 \times 10^5 = 0.0025$. The average plastic rotation allowable for LS performance level for compliant buildings (ASCE-SEI-41-17, Table 10-7) is 0.02 radian.

(a) *Short direction of the building*: Length of beam = 5 m.

$$\text{Depth of beam } h_b = \frac{0.5\epsilon_y l_b}{\theta_d - \theta_{pb}} = \frac{0.5 \times 0.0025 \times 5}{0.028 - 0.02} = 0.78 \text{ m. So, } h_b / l_b = 0.78/5 =$$

0.156. From Figure 9.4, the maximum value of h_b / l_b is 0.14. So, $h_b = 0.14$

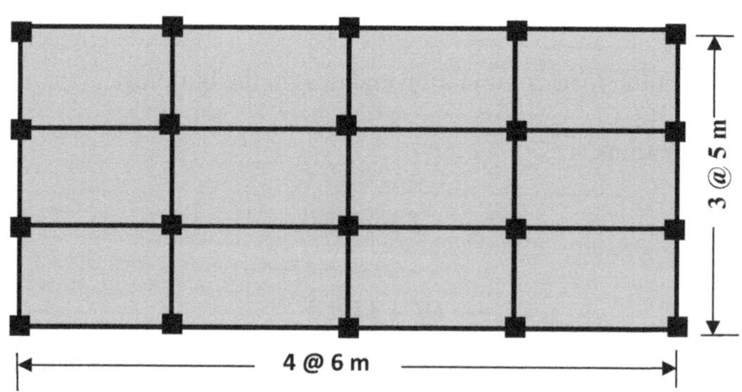

FIGURE 9.6 Building plan in Example 9.5.1.

× 5 = 0.7 m. Take the width of the beam as 0.3 m. The suggested beam size is 0.3 m × 0.7 m.

(b) *Long direction of the building*: Length of beam = 6 m.

Depth of beam $h_b = \dfrac{0.5\epsilon_y l_b}{\theta_d - \theta_{pb}} = \dfrac{0.5 \times 0.0025 \times 6}{0.028 - 0.02} = 0.936$ m. So, $h_b / l_b = $

0.936/6 = 0.156. From Figure 9.4, the maximum value of h_b / l_b is 0.14. So, $h_b = 0.14 \times 6 = 0.0.84$ m, say, 0.8 m. Take width of beam as 0.4 m. The suggested beam size is 0.4 m × 0.8 m.

Example 9.5.2 *A building plan is shown in Figure 9.7. Discuss how you shall apply UPBD method for this building. Consider 2.5% drift and LS performance level. Take $f_y = 500$ MPa.*

Solution: The building can be tackled in the usual way. The point to note is that we have to consider the dissimilar beam lengths in any particular direction. We need to calculate the yield rotations arising out of each beam length and take the average value. The frame yield rotation depends upon beam depth. So, beam depth is first found out by taking the highest beam length in a direction. In the UPBD method, you get a single beam size in one direction. The steel in each beam will differ based on the demand in each beam. This is also the practice for uniform shuttering in construction.

For LS PL, the plastic rotation allowed in beam is 0.02 radian (Table 10-7 of ASCE-SEI-41-7). ε_y = 500/200000= 0.0025. θ_d = 0.025.

(a) *Short direction*: Largest beam length is 5 m.

Find beam depth $h_b = \dfrac{0.5\varepsilon_y l_b}{\theta_d - \theta_{pb}} = \dfrac{0.5 \times 0.0025 \times 5}{0.025 - 0.02} = 1.25$ m. So, $h_b / l_b = $

1.25/ 5 = 0.25. From Figure 9.4, the maximum value of h_b / l_b is 0.12. So,

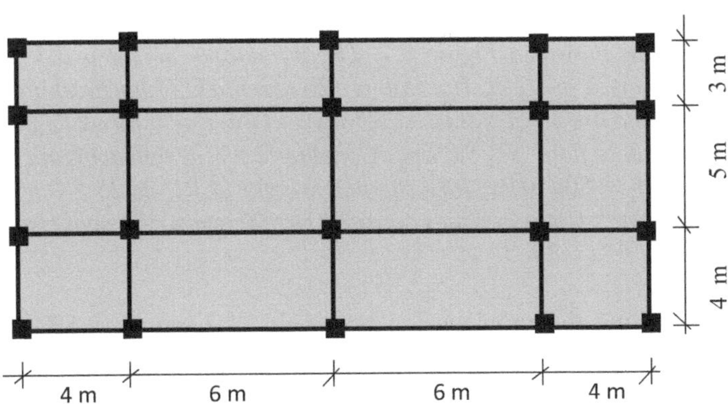

FIGURE 9.7 Building plan in Example 9.5.2.

$h_b = 0.12 \times 5 = 0.6$ m. Take width of the beam as 0.3 m. The suggested beam size is 0.3 m × 0.6 m.

For beam of length 4 m: $\theta_{yF} = \dfrac{0.5\varepsilon_y l_b}{h_b} = \dfrac{0.5 \times 0.0025 \times 4}{0.6} = 0.00833$ radian.

For beam of length 5 m: $\theta_{yF} = \dfrac{0.5\varepsilon_y l_b}{h_b} = \dfrac{0.5 \times 0.0025 \times 5}{0.6} = 0.0104$ radian.

For beam of length 3 m: $\theta_{yF} = \dfrac{0.5\varepsilon_y l_b}{h_b} = \dfrac{0.5 \times 0.0025 \times 3}{0.6} = 0.00625$ radian.

The average value of $\theta_{yF} = (0.00833 + 0.0104 + 0.00625)/3 = 0.00831$ radian.

This value is to be used for the computation of yield displacement in the short direction.

(b) *Long direction*: Largest beam length is 6 m.

Find beam depth $h_b = \dfrac{0.5\varepsilon_y l_b}{\theta_d - \theta_{pb}} = \dfrac{0.5 \times 0.0025 \times 6}{0.025 - 0.02} = 1.5$ m. So, $h_b / l_b =$

1.5/6 = 0.25. From Figure 9.4, the maximum value of h_b / l_b is 0.12. So, $h_b = 0.12 \times 6 = 0.72$ m, say 0.7 m. Take the width of the beam as 0.35 m. The suggested beam size is 0.35 m × 0.7 m.

For beam of length 4 m: $\theta_{yF} = \dfrac{0.5\varepsilon_y l_b}{h_b} = \dfrac{0.5 \times 0.0025 \times 4}{0.7} = 0.00714$ radian.

For beam of length 6 m: $\theta_{yF} = \dfrac{0.5\varepsilon_y l_b}{h_b} = \dfrac{0.5 \times 0.0025 \times 6}{0.7} = 0.0107$ radian.

The rest of the two beams are of similar lengths.

The average value of $\theta_{yF} = (0.00714 + 0.0107)/2 = 0.00893$ radian. This value is to be used for the computation of yield displacement in the long direction.

Example 9.5.3 *Carry out the entire procedural steps for UPBD method for a 7-storey building with plan shown in Figure 9.7. The interstorey height is 3.3 m, except in ground storey where it is 4.2 m. The slab is 120 mm thick. Floor finishing is 1 kN/m². Exterior infill walls are 1 brick thick (0.25 m) and interior infill walls are half brick (0.125 m) thick. LL in floor is 2.5 kN/m². Consider 2.5% design drift and LS performance level. Take f_y = 500 MPa. Take the unit weight of RC as 25 kN/m³ and that of masonry as 20 kN/m³. Column size is constant as 500 mm × 500 mm. Use EC-8 spectrum for soil type B at 0.35g level.*

Solution: In solving the problem, we take help of Example 9.5.2. Plan size is 20 m × 12 m.

For LS PL, the plastic rotation allowed in beam is 0.02 radian (Table 10-7 of ASCE-SEI-41-7). $\varepsilon_y = 500/200000 = 0.0025$. $\theta_d = 0.025$.

As shown in Example 9.5.2, the short directional beam size is 0.3 m × 0.6 m and long directional beam size is 0.35 m × 0.7 m. In the computer model, the weights are generally captured by the software from the member sizes and given loadings. Here, we shall see how it is computed manually. It may be noted that the seismic weight is contributed by the floor mass, a fraction of LL, beams, mass of half of column height at top and bottom of floor and mass of half of the infill weight at the top and bottom of the floor.

Floor weight

DL due to slab = 0.12 × 25 × 20 × 12 = 720 kN.
DL due to floor finish = 1 × 20 × 12 = 240 kN.
Fraction of LL = 0.25 × 2.5 × 2 0 × 12 = 150 kN. [As LL intensity is less than 3 kN/m².]

The weights of beams, columns and infill are tabulated in Table Ex9.5.3.1.

Total floor weight = 720 + 240 + 150 + 2443 = 3553 kN.

Mass of floor = 3553 × 1000/9.81 kg = 362,181 kg.

Mass of the first floor will be slightly more than this due to extra height of column. But this part has been ignored here. Weight of roof = 720 + 2443/2 = 1941.5 kN. So, the roof mass is 197,910 kg.

The steps as in Section 9.4 are followed.

Step 1: *Target objectives:* θ_d = 2.5%, PL is LS (θ_{pb} = 0.02 radian).

Step 2: *Hazard level*: Demand spectrum of EC-8 spectrum for soil type B at 0.45g level is considered.

Step 3: *Material:* Yield strength of rebar f_y = 500 MPa. The strength of the concrete will not be required for the question.

Step 4: *Compute beam depth*: The beam sizes have been computed above and these are: 0.3 m × 0.6 m in the short direction and 0.35 m × 0.7 m in the long direction.

TABLE EX9.5.3.1

Weights of column, beam and infill in a floor for Example 9.5.3

Item	L (m)	B (m)	H (m)	No	Unit wt (kN/m³)	Weight (kN)
Column	2.65	0.5	0.5	20	25	331
Beam in short dir	12	0.3	0.6	5	25	270
Beam in long dir	20	0.35	0.7	4	25	490
Exterior infill	64	0.25	2.65	1	20	848
Interior infill	76	0.125	2.65	1	20	504
					Sum	2443

Note: The height of column = 3.3 – average beam depth = 3.3 – 0.65 = 2.65 m. Similarly, with infill height.

Step 5: *Take a shape profile and displacement profile*:

As per Pettinga and Priestley (2005),

$$\text{For } n > 4 : \quad \phi_i = \frac{4}{3}\frac{h_i}{H}\left(1 - 0.25\frac{h_i}{H}\right)$$

$$\Delta_c = \theta_d h_{sc} = 0.025 \times 4.2 = 0.105 \text{ m [ground storey is critical storey]}.$$

$$\Delta_i = \phi_i \frac{\Delta_c}{\phi_c}$$

The shape and displacement profiles are given in Table Ex9.5.3.2.

Step 6: *ESDOF system properties*:

Eqs. (9.4.8) to (9.4.10) are used and the values are tabulated in Table Ex9.5.3.2.

$$\Delta_d = \frac{\sum_{i=1}^{n} m_i \Delta_i^2}{\sum_{i=1}^{n} m_i \Delta_i} = 233720/689925 = 0.339 \text{ m}.$$

$$m_e = \frac{\sum_{i=1}^{n} m_i \Delta_i}{\Delta_d} = 689925/0.339 = 2036610 \text{ kg}.$$

$$H_e = \frac{\sum_{i=1}^{n} m_i \Delta_i h_i}{\sum_{i=1}^{n} m_i \Delta_i} = 11001733/689925 = 15.946 \text{ m}.$$

TABLE EX9.5.3.2
Computation for ESDOF system properties for Example 9.5.3

i from bottom	m_i	h_i	ϕ_i	Δ_i	$m_i \Delta_i$	$m_i \Delta_i^2$	$m_i \Delta_i h_i$
1	362181	**4.2**	0.2231	0.105	38029	3993	159722
2	362181	7.5	0.3841	0.181	65468	11834	491008
3	362181	10.8	0.5325	0.251	90758	22743	980190
4	362181	14.1	0.6683	0.314	113901	35820	1605998
5	362181	17.4	0.7915	0.372	134895	50242	2347167
6	362181	20.7	0.9020	0.424	153741	65261	3182429
7	197910	24	1	0.471	93134	43828	2235219
				Sum	**689925**	**233720**	**11001733**

Step 7: *Displacement spectra:* The displacement spectra corresponding to the design spectrum is shown in Figure 9.8.

Step 8: *Yield displacement and ductility*

Short direction

As calculated in Ex. 5.4.2, average value of $\theta_{yF} = 0.00831$ radian.

$$\Delta_y = \theta_{yF} H_e = 0.00831 \times 15.946 = 0.133 \text{ m}.$$

$$\mu = \Delta_d / \Delta_y = 0.339/0.133 = 2.55.$$

Long direction

As calculated in Ex. 5.4.2, average value of $\theta_{yF} = 0.00893$ radian.

$$\Delta_y = \theta_{yF} H_e = 0.00893 \times 15.946 = 0.143 \text{ m}.$$

$$\mu = \Delta_d / \Delta_y = 0.339/0.143 = 2.37.$$

Step 9: *Effective damping and effective time period*

Short direction

$$\xi_{eq} = 5 + 120 \left(\frac{1 - \dfrac{1}{\sqrt{\mu}}}{\pi} \right)\% = 5 + 120 \left(\frac{1 - \dfrac{1}{\sqrt{2.55}}}{\pi} \right) = 19.3\%.$$

T_e corresponding to $\Delta_d (= S_d = 0.339 \text{ m})$ and damping ξ_{eq} (19.3%) from Figure 9.8 is 4.1 s.

Long direction

$$\xi_{eq} = 5 + 120 \left(\frac{1 - \dfrac{1}{\sqrt{\mu}}}{\pi} \right)\% = 5 + 120 \left(\frac{1 - \dfrac{1}{\sqrt{2.37}}}{\pi} \right) = 18.4\%.$$

T_e corresponding to $\Delta_d (= S_d = 0.339 \text{ m})$ and damping ξ_{eq} (18.4%) from Figure 9.9 is 3.95 s.

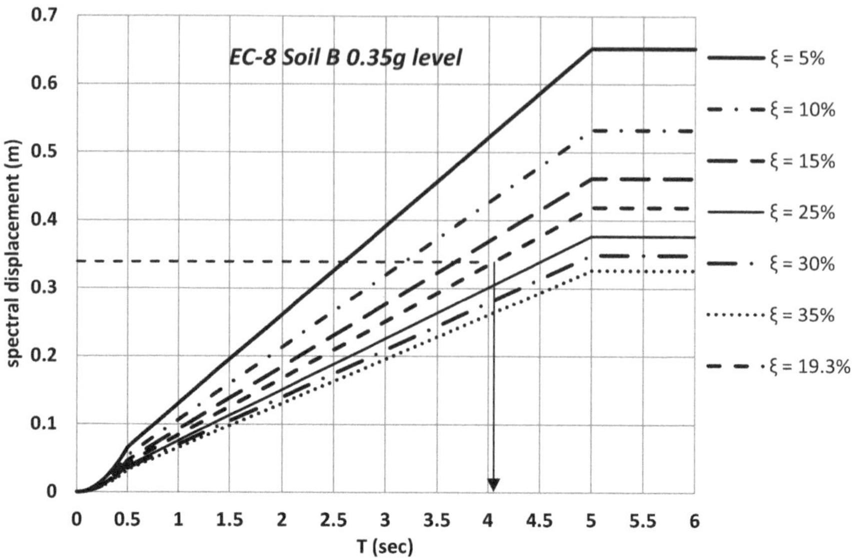

FIGURE 9.8   Displacement spectra for EC-8 spectrum for soil type B at 0.35g level (for short direction).

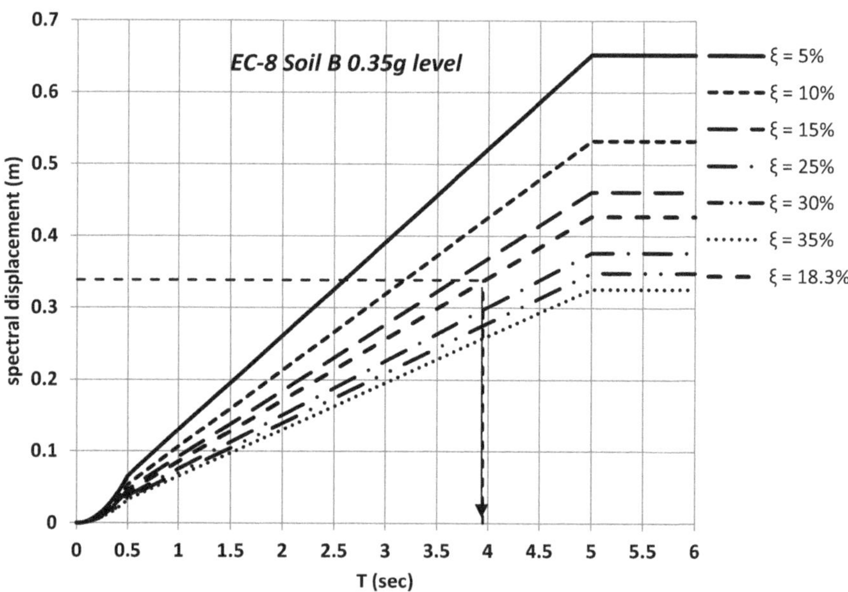

FIGURE 9.9   Displacement spectra for EC-8 spectrum for soil type B at 0.35g level (for long direction).

TABLE EX9.5.3.3
Computation lateral floor forces in Example 9.5.3

		Short direction		Long direction	
	$m_i \Delta_i$	V_b	F_i (kN)	V_b	F_i (kN)
	38029	1621	89.4	1746	96.2
	65468		153.8		165.7
	90758		213.2		229.7
	113901		267.6		288.2
	134895		316.9		341.4
	153741		361.2		389.1
	93134		218.8		235.7
Sum	689925		1621.0		1746.0

Step 10: *Effective stiffness*

$$\text{Short direction: } k_e = 4\pi^2 \frac{m_e}{T_e^2} = 4\pi^2 \frac{2036610}{4.1^2} \text{ N/m} = 4783 \text{ kN/m.}$$

$$\text{Long direction: } k_e = 4\pi^2 \frac{m_e}{T_e^2} = 4\pi^2 \frac{2036610}{3.95^2} \text{ N/m} = 5153 \text{ kN/m.}$$

Step 11: *Compute base shear*:

Short direction: $V_b = k_e \Delta_d = 4783 \times 0.339 = 1621$ kN.
Long direction: $V_b = k_e \Delta_d = 5153 \times 0.339 = 1746$ kN.

Step 12: *Distribute the base shear over the floors*:

$$\text{For } n \leq 10 \, F_i = V_b \frac{m_i \Delta_i}{\sum_{i=1}^{n} m_i \Delta_i}.$$

The floor lateral forces are tabulated in Table Ex9.5.3.3.

Example 9.5.4 *Design the 10-storey RC frame building as shown in Figure 9.10 for 2.8% design drift and LS performance level. The interstorey height is 3.3. The slab is 125 mm thick. Floor finishing is 2 kN/m². Exterior infill walls are 1 brick thick (0.25 m) and interior infill walls are half brick (0.125 m) thick. Assume 20% opening in infill walls. LL in floor is 2.5 kN/m². Take f_y =500 MPa. Take the unit weight of RC as 25 kN/m³ and that of masonry as 20 kN/m³. Column size is constant as 600 mm × 600 mm up to 4th storey and for remaining stories it is 500 mm × 500 mm. Use EC-8*

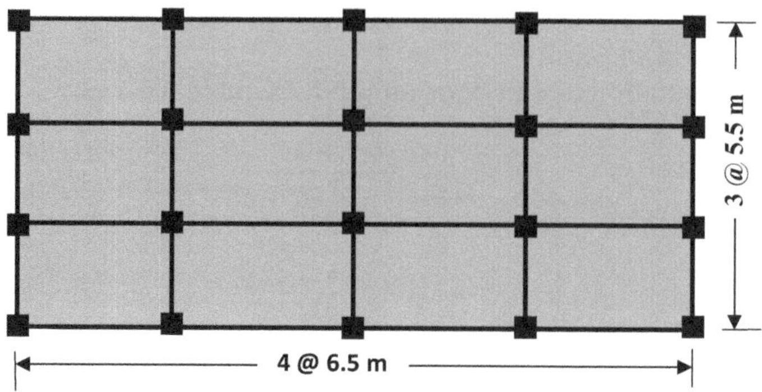

FIGURE 9.10 Plan of building in Example 9.5.4.

spectrum for soil type B at 0.5g level. Evaluate the building performance and show the achievement of target criteria. Take concrete characteristic strength as 30 MPa.

Solution: Plan size is 26 m × 16.5 m (429 m²).

As per the discussion in Section 4.11, plastic rotation for primary members corresponding to LS is 0.01875 radian. $\varepsilon_y = 500/200000 = 0.0025$; $\theta_d = 0.028$.

Beam size: *Short direction*:

$$h_b = \frac{0.5\varepsilon_y l_b}{\theta_d - \theta_{pb}} = \frac{0.5 \times 0.0025 \times 5.5}{0.028 - 0.01875} = 0.743 \text{ m. So, } h_b/l_b = 0.743/5.5 = 0.135,$$

which is within the upper bound limit of h_b/l_b (Figure 9.4); hence assume beam depth as 0.75 m and bam width as 0.4 m. Take beam size as 0.4 m × 0.75 m.

Beam size: *Long direction*:

$$h_b = \frac{0.5\varepsilon_y l_b}{\theta_d - \theta_{pb}} = \frac{0.5 \times 0.0025 \times 6.5}{0.028 - 0.01875} = 0.878 \text{ m. So, } h_b/l_b = 0.878/6.5 = 0.135,$$

which is within the upper bound limit (Figure 9.4). Take beam size as 0.4 m × 0.85 m.

Floor weight

DL intensity in slab = 0.125 × 25 + 2 (including finish) = 5.125 kN/m².

   Floor DL = 5.125 × 429 = 2199 kN.

   Fraction of LL intensity in floor = 0.25 × 2.5 =0.625 kN/m². [As LL intensity is within 3 kN/m², 25% LL contributes to seismic load.]

   LL weight on floor = 0.625 × 429 = 268 kN.

   Total seismic weight in each floor = 2199 + 268 = 2467 kN.

   Considering roof treatment weight in roof as 2 kN/m², roof load = 2199 kN.

   The weights of beams, columns and infill are tabulated in Table Ex9.5.4.1.

   Total floor weight in each floor = 2467 + 3648 = 6115 kN.

   Mass of floor = 6115 × 1000/9.81 kg = 623,343 kg.

TABLE EX9.5.4.1
Weights of column, beam and infill in a floor for Example 9.5.4

Item	L (m)	B (m)	H (m)	No	Unit wt (kN/m³)	Weight (kN)
Column to 4th level	2.45	0.6	0.6	20	25	441
Column above 4th level	2.45	0.5	0.5	20	25	306
Beam in short direction	5.5	0.4	0.75	15	25	619
Beam in long direction	6.5	0.4	0.85	16	25	884
Exterior Infill (20% opening)	85	0.25	2.575	1	20	876
Interior infill (20% opening)	101.5	0.125	2.575	1	20	523
					Sum	3648

Note: The height of column = 3.4 − average beam depth = 3.3 − 0.85 = 2.45 m. Similarly, with infill height.

Weight of roof = 2199 + 3648/2 = 4023 kN. So, roof mass is 410,091 kg.

Shape profile and displacement profile:
As per Pettinga and Priestley (2005),

$$\text{For } n > 4: \quad \phi_i = \frac{4}{3}\frac{h_i}{H}\left(1 - 0.25\frac{h_i}{H}\right)$$

$$\Delta_c = \theta_d h_{sc} = 0.028 \times 3.3 = 0.0924\,\text{m} \left[\text{all storeys of equal height.}\right].$$

$$\Delta_i = \phi_i \frac{\Delta_c}{\phi_c}$$

The ESDOF system properties computation is shown in Table Ex9.5.4.2.

$$\Delta_d = \frac{\sum_{i=1}^{n} m_i \Delta_i^2}{\sum_{i=1}^{n} m_i \Delta_i} = 1289526/2528898 = 0.510\,\text{m}.$$

$$m_e = \frac{\sum_{i=1}^{n} m_i \Delta_i}{\Delta_d} = 2528898/0.510 = 4959438\,\text{kg}.$$

$$H_e = \frac{\sum_{i=1}^{n} m_i \Delta_i h_i}{\sum_{i=1}^{n} m_i \Delta_i} = 55308685/2528898 = 21.871\,\text{m}.$$

TABLE EX9.5.4.2
Computation for ESDOF system properties for Example Ex9.5.4

i from bottom	m_i	h_i	ϕ_i	Δ_i	$m_i\Delta_i$	$m_i\Delta_i^2$	$m_i\Delta_i h_i$
1	623343	**3.3**	0.130	0.092	57597	5322	190070
2	623343	6.6	0.253	0.180	112240	20210	740785
3	623343	9.9	0.370	0.263	163930	43111	1622903
4	623343	13.2	0.480	0.341	212665	72555	2807184
5	623343	16.5	0.583	0.415	258448	107156	4264385
6	623343	19.8	0.680	0.483	301276	145614	5965266
7	623343	23.1	0.770	0.547	341151	186709	7880584
8	623343	26.4	0.853	0.607	378072	229309	9981099
9	623343	29.7	0.930	0.661	412039	272364	12237568
10	410091	33	1.000	0.711	291480	207175	9618842
				Sum	**2528898**	**1289526**	**55308685**

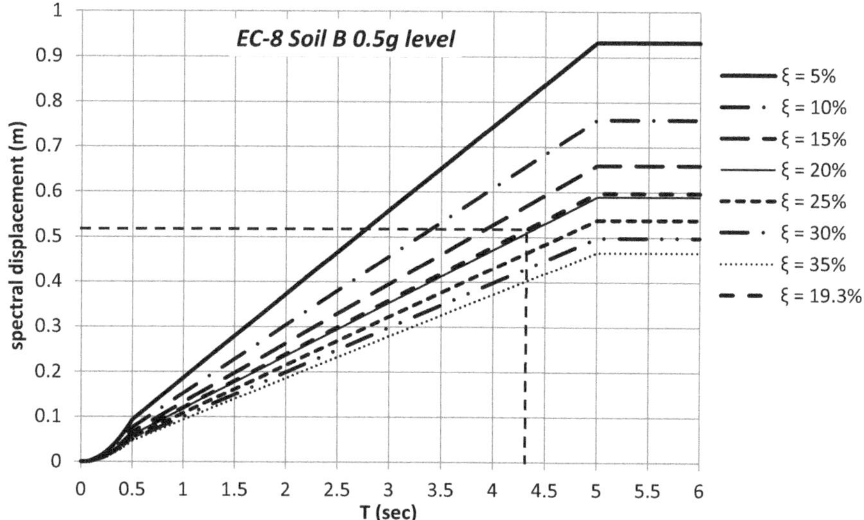

FIGURE 9.11 Displacement spectra in Ex9.5.4.2 (for short direction).

Analysis for Short direction (*beam size 0.4 m × 0.75 m, beam length 5.5 m*)

$$\theta_{yF} = \frac{0.5\varepsilon_y l_b}{h_b} = \frac{0.5 \times 0.0025 \times 5.5}{0.75} = 0.00917 \text{ rad.}$$

$$\Delta_y = \theta_{yF} H_e = 0.00917 \times 21.869 = 0.200 \text{ m.}$$

TABLE EX9.5.4.3
Computation lateral floor forces in the short direction

Level	$m_i\,\Delta_i$	Short direction	
		V_b	F_i (kN)
1	57955	5433	124
2	112938		241
3	164948		352
4	213987		457
5	260054		555
6	303148		647
7	343271		733
8	380422		812
9	414600		885
10	292858		625
Sum	2544181		5433

$$\mu = \Delta_d / \Delta_y = 0.510/0.200 = 2.55.$$

$$\xi_{eq} = 5 + 120\left(\frac{1 - \dfrac{1}{\sqrt{\mu}}}{\pi}\right)\% = 5 + 120\left(\frac{1 - \dfrac{1}{\sqrt{2.55}}}{\pi}\right) = \mathbf{19.3\%}.$$

T_e corresponding to $\Delta_d\,(= S_d = 0.510$ m) and damping ξ_{eq} (19.3%) from Fig 9.11 is **4.3** sec.

$$k_e = 4\pi^2 \frac{m_e}{T_e^2} = 4\pi^2 \frac{4989745}{4.3^2}\text{ N/m} = 10654\text{ kN/m}.$$

$$V_b = k_e \Delta_d = 10654 \times 0.510 = 5433\text{ kN}.$$

$$\text{For } n \le 10 \quad F_i = V_b \frac{m_i \Delta_i}{\sum_{i=1}^{n} m_i \Delta_i}.$$

The floor lateral forces are tabulated in Table Ex9.5.4.3.

Analysis for Long direction (beam size 0.45 m × 0.9 m, beam length 6.5 m)

$$\theta_{yF} = \frac{0.5\varepsilon_y l_b}{h_b} = \frac{0.5 \times 0.0025 \times 6.5}{0.85} = 0.00956\text{ rad.}$$

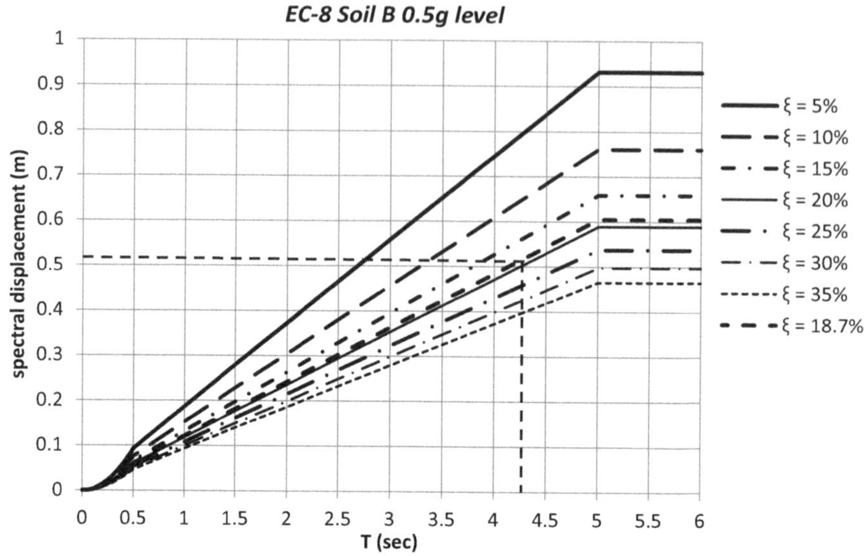

FIGURE 9.12   Displacement spectra for Example 9.5.4 (for long direction).

$$\Delta_y = \theta_{yF} H_e = 0.00956 \times 21.869 = 0.2091 \text{ m.}$$

$$\mu = \Delta_d / \Delta_y = 0.510/0.2091 = 2.439.$$

$$\xi_{eq} = 5 + 120\left(\frac{1 - \dfrac{1}{\sqrt{\mu}}}{\pi}\right)\% = 5 + 120\left(\frac{1 - \dfrac{1}{\sqrt{2.439}}}{\pi}\right) = \mathbf{18.7\%.}$$

T_e corresponding to Δ_d $(= S_d = 0.510 \text{ m})$ and damping ξ_{eq} (18.7%) from Figure 9.12 is 4.25 s.

$$k_e = 4\pi^2 \frac{m_e}{T_e^2} = 4\pi^2 \frac{4989745}{4.25^2} \text{ N/m} = 10906 \text{ kN/m.}$$

$$V_b = k_e \Delta_d = 10906 \times 0.510 = 5562 \text{ kN.}$$

$$\text{For } n \leq 10 \quad F_i = V_b \frac{m_i \Delta_i}{\sum_{i=1}^{n} m_i \Delta_i}.$$

The floor lateral forces are tabulated in Table Ex9.5.4.4.

TABLE EX9.5.4.4
Computation lateral floor forces in the short direction for Example 9.5.4

Level	$m_i \Delta_i$	Long direction	
		V_b	F_i (kN)
1	57955	5562	126.7
2	112938		246.9
3	164948		360.6
4	213987		467.8
5	260054		568.5
6	303148		662.7
7	343271		750.4
8	380422		831.7
9	414600		906.4
10	292858		640.2
Sum	2544181		5562

9.5.1 Modelling Example

The building is modelled in SAP2000 (CSI Inc.) software. The analysis is done along both long and short directions. The pushover curves are shown in Figure 9.13 in the short direction and in Figure 9.14 along the long direction. The performance point in both the curves lies beyond IO and within LS PL. This indicates that the building satisfies LS PL, which is the target PL.

Inter-storey drift diagram (IDR)
For drawing IDR diagram, NLTHA are to be carried out under SCGMs. In Figure 9.15, a typical drift diagram for a ground motion in short direction of the building is shown. As per FEMA-356, five ground motions may be taken when the highest value of output parameter is to be reported. If ten or more earthquakes are taken, the average value output parameter may be reported. As per ASCE-SEI-17 minimum of 11 ground motions to be used for evaluation purpose and the average value is to be reported. It is found that the achieved drift is within the target value, hence, satisfactory. The drift diagram in the long direction is not shown here.

9.6 DUCTILITY VS. DAMPING

Pettinga and Priestley (2005) defined the relation between ductility and damping for frame building as

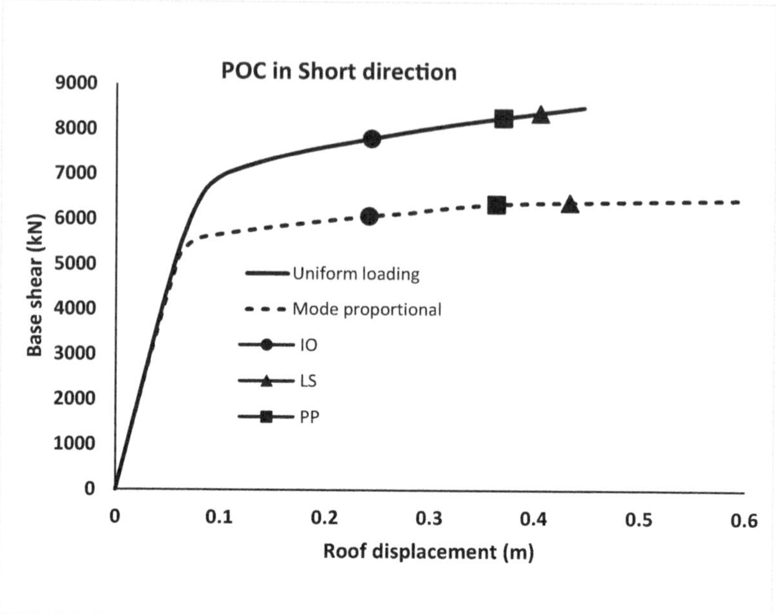

FIGURE 9.13   Pushover curves for Example 9.5.4 (for short direction).

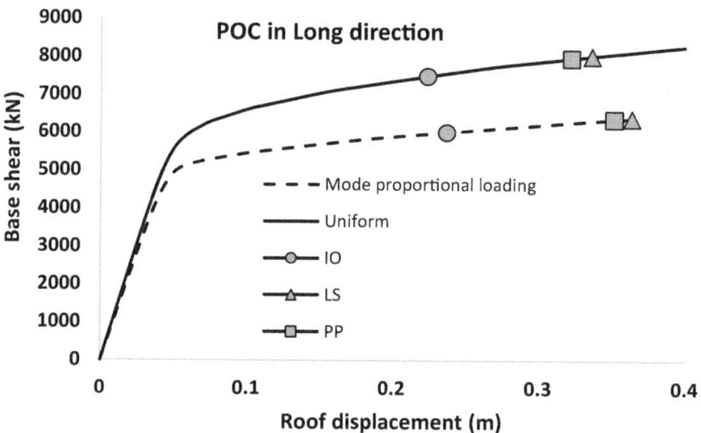

FIGURE 9.14   Pushover curves for Example 9.5.4 (for long direction).

$$\xi_{eq} = 5 + 120 \left(\frac{1 - \dfrac{1}{\sqrt{\mu}}}{\pi} \right) \%.$$

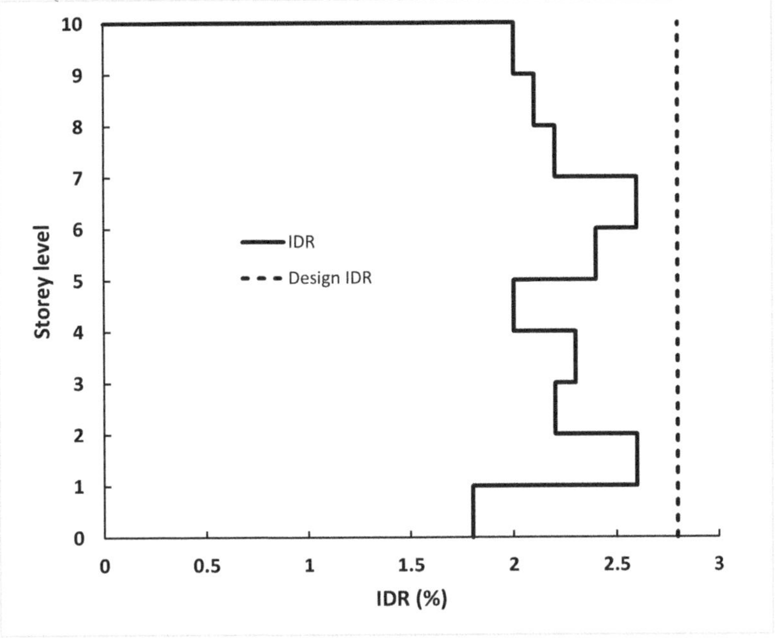

FIGURE 9.15   IDR diagram for Example 9.5.4 (for short direction).

On the other hand, Priestley et al. (2007) defines the relationship as

$$\xi_{eq} = 5 + 56.5\left(\frac{\mu - 1}{\mu\pi}\right)\%.$$

These two relations are plotted in Figure 9.16. It is found from this figure that Pettinga and Priestley (2005) gives lower values of effective damping.

9.7   CRITICAL CASES

In some cases, the given drift may be less than the plastic rotation corresponding to the design PL. In such cases, the Eq. (9.2.4) does not work. However, the case is not impractical altogether. We discuss this aspect below.

Let us consider the equation for yield rotation of frame:

$$\theta_{yF} = \frac{0.5\varepsilon_y l_b}{h_b}$$

A graph can be drawn between l_b/h_b ratio and θ_{yF} for various grades of steel (Figure 9.17). The beam depth is generally restricted to 1/10th to 1/12th of beam length. Here, the plot has been shown for l_b/h_b ratio up to 17.

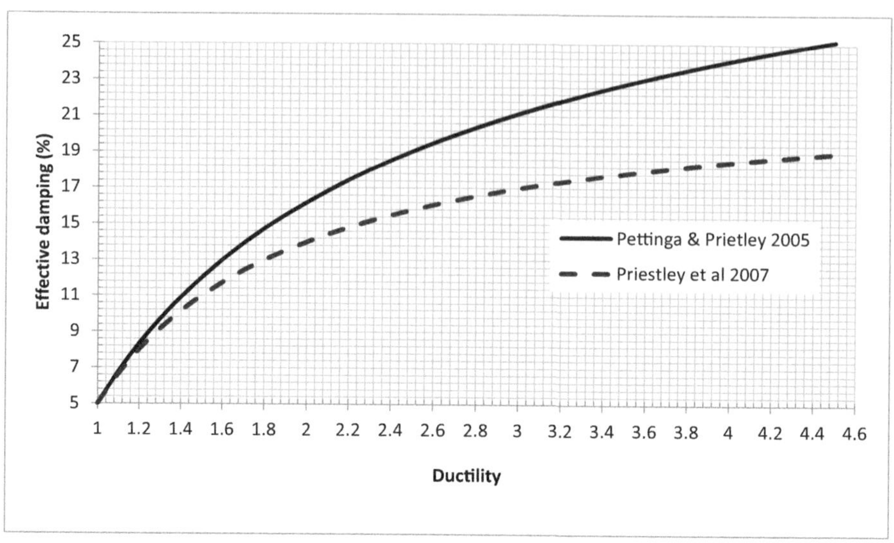

FIGURE 9.16   Displacement ductility vs. effective damping.

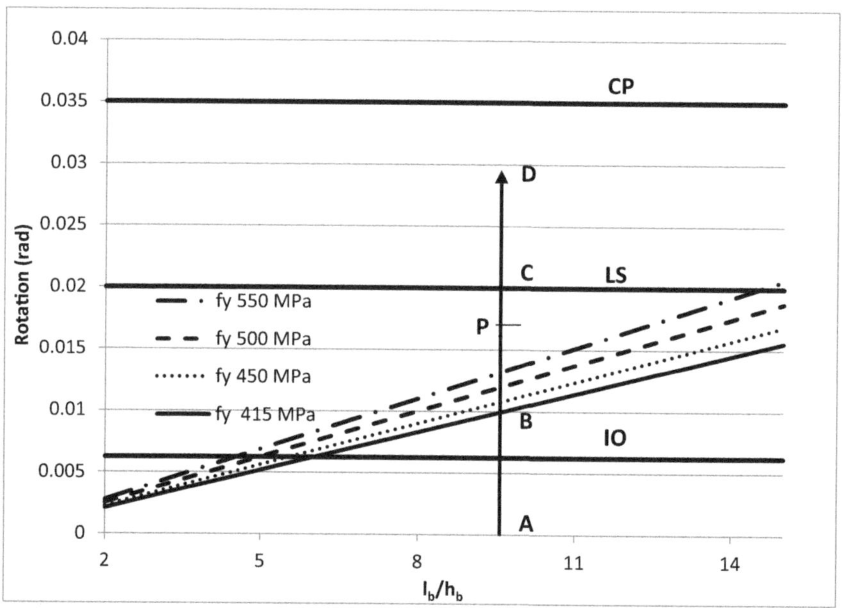

FIGURE 9.17   Interpretation of yield rotation and plastic rotation.

On the same figure, we show the limits of average plastic rotations for IO, LS and CP as horizontal lines. If we take f_y as 415 MPa and l_b/h_b ratio at point A, then AB is the amount of yield rotation. If we design for LS performance level, the plastic rotation allowable is AC. Total rotation allowable is AB + AC. In the figure AB +

$AC = 0.01 + 0.02 = 0.03$ radian ($= AD$). So, for this combination of l_b/h_b ratio and steel grade, a system can be designed for 3% drift and LS PL.

Now, take up the case when $\theta_d = \theta_{pb}$ is negative. Here, θ_d is less than θ_{pb}. It indicates that we are below the horizontal line corresponding to the PL considered. P is such a point for LS PL and Fe415 steel. So, here, AP is total rotation (θ_d) AB is yield rotation and BP is plastic rotation. Here, the members of the structure will not reach the last limit of LS performance level, but the PL designator for the structure is still LS, as point P is above IO line and below LS line.

9.8   GENERATION OF DISPLACEMENT SPECTRA

This topic was discussed in Chapter 3. Here it is further explained with examples.

Typically, the design codes give the acceleration response spectrum as the design spectrum. For displacement-based design, one needs to convert the acceleration spectra to displacement spectra. The basic equation for the conversion is given as Eq. (9.8.1).

$$S_a = \omega^2 S_d \tag{9.8.1}$$

where S_a is spectral acceleration, ω is natural frequency of SDOF system and S_d is spectral displacement.

Noting that $\omega = \dfrac{2\pi}{T}$ (T is time period of vibration of SDOF system). Eq. (9.8.1) can be written as Eq. (9.8.2).

$$S_a = \left(\frac{2\pi}{T}\right)^2 S_d$$

$$\text{Or,} \quad S_d = \frac{T^2}{4\pi^2} S_a \tag{9.8.2a}$$

$$\text{Or,} \quad S_d = \frac{T^2}{4\pi^2}\frac{S_a}{g} g \tag{9.8.2b}$$

The design spectrum is given in terms of time period along the x-axis vs. S_a/g ratio along the y-axis. Every point i in the design spectrum is given as $\left(T_i, \left(S_a/g\right)_i\right)$. From Eq. (9.8.2b),

$$S_{d,i} = \frac{T_i^2}{4\pi^2}\left(\frac{S_a}{g}\right)_i g \tag{9.8.3}$$

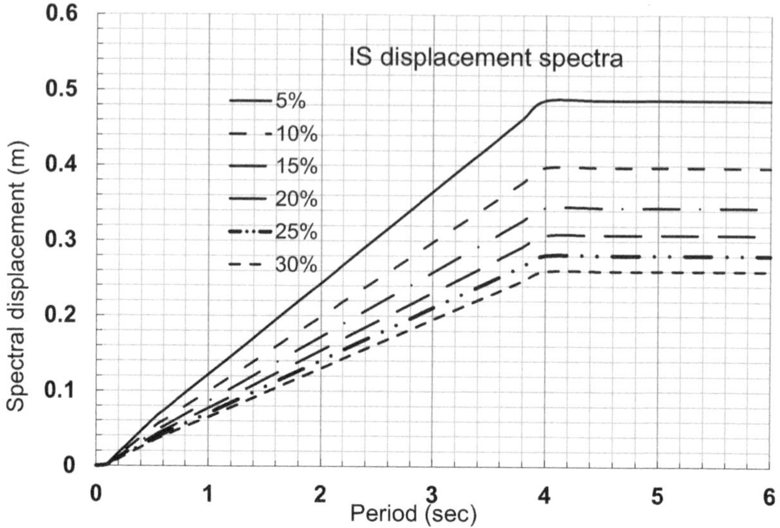

FIGURE 9.18 Displacement spectra corresponding to IS 1893 for medium soil at MCE (0.36g level).

Using Eq. (9.8.3), all points $\left(T_i, \left(S_a/g\right)_i\right)$ can be converted to $\left(T_i, S_{d,i}\right)$. Now the vertical axis will indicate spectral displacement.

The design spectrum is given at 5% damping at elastic condition. The displacement spectra for other damping may be found from Eq. (9.8.4), available in literature.

$$S_{d,i,\xi\%} = \frac{T_i^2}{4\pi^2}\left(\frac{S_a}{g}\right)_{i,5\%} g \times \left(\frac{10}{5+\xi\%}\right)^{0.5} \tag{9.8.4}$$

Using Eq. (9.8.4), the displacement spectra at various dampings for (i) Indian design spectrum (IS 1893-206) for medium soil and a MCE level (0.36g seismicity level) and (ii) EC-8 design spectrum for soil type B at 0.45g seismicity level is shown in Figures 9.18 and 9.19, respectively.

9.9 CLOSURE

The UPBD method for RC frame buildings has been presented in this chapter. The theoretical background of the method was discussed. The interpretation of the basic equation was elaborated. The step-wise procedure for design was described. Design examples were provided to discuss various aspects.

The advantage of UPBD method is that it be can used to design the building for target drift and target PL. It also gives the member size at the beginning, so that the

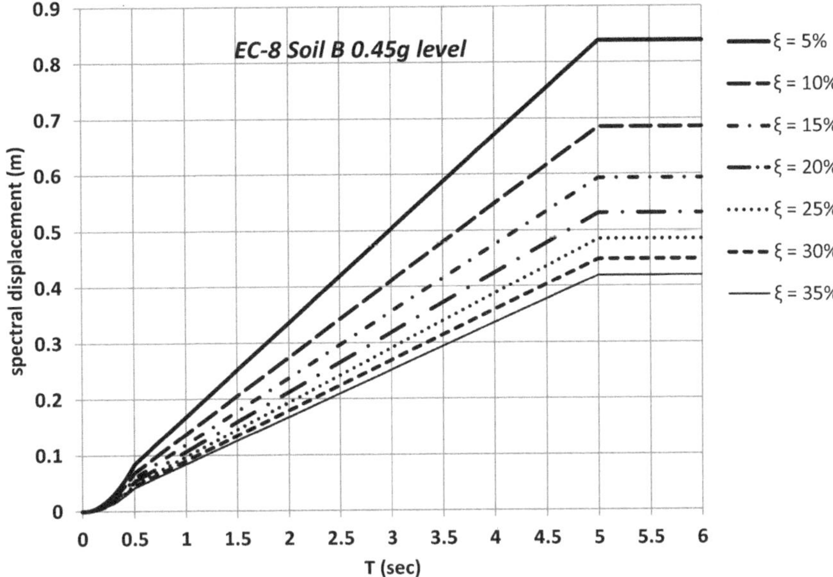

FIGURE 9.19 Displacement spectra corresponding to EC-8 for B soil at 0.45g level.

iteration for member size is not required. The method has been found to give consistent results. Experience showed that in all cases the performance level is achieved. In some cases, drift may be a little away from the target value. This happens due to the effect of column on drift. A variation of achieved drift to within ± 20% of target value may be considered satisfactory.

9.10 EXERCISES

Q 9.10.1 What are the limitations of DDBD method of design? How these are overcome in UPBD method?

Q 9.10.2 Give an outline of the UPBD method for RC frame buildings.

Q 9.10.3 A building plan is shown in Figure 9.20. The design drift is 2.2% and target performance level is LS. Find the beam sizes as per UPBD method. Take Fe550 grade rebar.

Q 9.10.4 Design an eight-storey RC frame building for IO PL and 1.4% drift. The plan is shown in Figure 9.21. The interstorey height is 3.4 m, except in ground storey where it is 4.0 m. The floor slab is 120 mm thick. Floor finishing is 2 kN/m². Exterior infill walls are 1 brick thick (0.25 m) and interior infill walls are half brick (0.125 m) thick. LL in floor is 3 kN/m² and that in roof is 1.5 kN/m². Take f_y =500 MPa. Take unit weight of RC as 25 kN/m³ and that of masonry as 20 kN/m³. Column size is constant as 500 mm×500 mm for bottom three stories

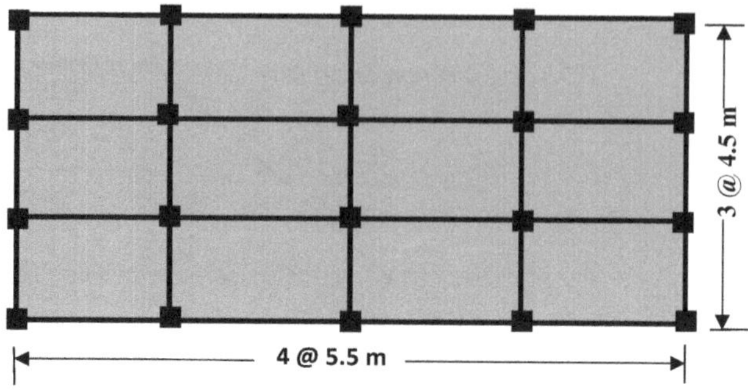

FIGURE 9.20 Building plan in Exercise 9.3.

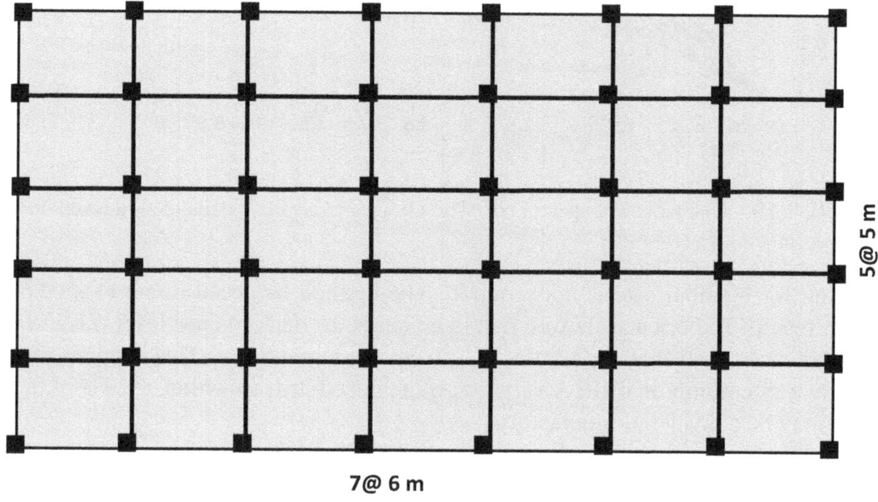

FIGURE 9.21 Building plan in Exercise 9.4.

and 450 mm×450 mm for other stories. Use EC-8 spectrum for soil type B at 0.45g level.

Q 9.10.5 For Q9.10.4, carry out modelling in software and check if targets are reached or not.

FURTHER READINGS

Choudhury, S. (2007) Performance-Based Seismic Design of Hospitals, Ph.D. Thesis, Indian Institute of Technology, Roorkee, India, 2008.

Choudhury, S. and Singh, Sunil M. (2013) A Unified Approach to Performance-Based Design of RC Frame Buildings, *Journal of Institution of Engineers (I)-Series-A*, 94(2), 73–82, DOI 10.1007/s40030-013-0037-8.

Das, S. and Choudhury, S. (2019a) Evaluation of Effective Stiffness of RC Column Sections by Support Vector Regression Approach, *Neural Computing & Applications*, April 2019, DOI 10:1007/s00521-019-04190-0.

Das, S. and Choudhury, S. (2019b) Influence of effective stiffness on the performance of RC frame buildings designed using displacement-based method and evaluation of column effective stiffness using ANN, June 2019, *Engineering Structures* https://doi.org/10.1016 /engstruct.2019.109354

Das, S., Mansouri, I., Choudhury, S., Gandomi, A. H. and Wan, H.J. (2021) A Prediction Model for the Calculation of Effective Stiffness Ratios of Reinforced Concrete Columns, *Materials Journal* 14 (7). https://doi.org/10.3390/ma14071792.

Mayengbam, S.S. and Choudhury, S. (2014) Determination of column size for displacement-based design of reinforced concrete frame buildings, *Journal of Earthquake Engineering and Structural Dynamics*, 43(8), 1149–1172. Article first published online: 21 Nov 2013 I DOI: 10.1002/eqe.2391.

Pettinga, J.D. and Priestley, M.J.N. (2005) Dynamic Behaviour of Reinforced Concrete Frames Designed with Direct Displacement-Based Design, *Journal of Earthquake Engineering*, 9(Special Issue 2), 309–330.

Priestley, M.J.N. (2003) Myths and Fallacies in Earthquake Engineering, Revisited, European School for Advanced Studies in Reduction of Seismic Risk, 9th Mallet-Milne Lecture.

Priestley, M.J.N. and Calvi, G.M. (1997) Concepts and Procedures for Direct Displacement-based Design and Assessment, Proc. of International Workshop on Seismic Design Methodologies for Next Generation of Codes, Bled, Slovenia.

Shibata, A. and Sozen, M.A. (1974) Substitute-Structure Method to Determine Design Forces in Earthquake-resistant Reinforced Concrete Frames. Engineering Studies, L Research Series No. 412, University of Illinois.

Sullivan, T.J., Priestley, M.J.N. and Calvi, G.M. (2006a) "Direct Displacement-Based Design of Frame-Wall Structures", *Journal of Earthquake Engineering, V. 10 Special Issue 1*, pp. 91–124.

Sullivan, T.J., Priestley, M.J.N. and Calvi, G.M. (2006b) "Seismic Design of Frame-Wall Structures", Research *Report No. ROSE-2006/02*.

10 Unified Performance-Based Design of RC Dual System

10.1 INTRODUCTION

In Chapter 8, we studied the DDBD method for the dual system. As was stated in that chapter, a dual system consists of frame elements and shear walls. Shear wall is a planar element in the building, which takes the lateral shear arising out of earthquake or wind, in its own plane. We have also studied the frame–shear wall interaction. In this chapter, we shall establish the UPBD method for the design of RC dual system. The term unified performance-based design (UPBD) was first coined by Choudhury (2007). The UPBD method for RC frame buildings has been already introduced in Chapter 9. The UPBD method for frame–shear wall buildings was developed by Choudhury (2007). It is an extension of the DDBD methodology developed by Sullivan et al. (2006). In the DDBD method developed by Sullivan et al. (2006a, 2006b), only drift was taken as the target design criterion. In the UPBD method, both drift and performance levels are incorporated in the design philosophy. Apart from satisfying drift and PL, the UPBD method also gives the size of the wall in the beginning of the design, thus avoiding iteration. The sizes of beams are determined as per by Choudhury and Singh (2013), as discussed in Chapter 9. The size of columns can be found out from Mayengbam and Choudhury (2014).

The primary design target objectives for RC frame–shear wall buildings are: (i) interstorey drift $\left(\theta_d\right)$ and (ii) member performance level. The member performance level is designated in terms of plastic rotation in the members. As per capacity design (see Chapter 6) and the principle of weak-beam strong-column for buildings, the columns do not undergo inelastic deformation. It is the beam that has to form plastic hinges and dissipate energy. Shear wall is a cantilever deep beam. The lion's share of the hysteretic energy in a dual system is dissipated through the shear wall. This is because of the enormous length of the wall. Thus, the member performance level for RC frame–shear wall buildings can be expressed in terms of the plastic rotation of the shear wall. In the following section, it is shown how both these parameters, namely, drift and performance level, can be incorporated in the design philosophy as the target design objectives. We shall also see how the member sizes are also conceptualized by theoretical formulation.

DOI: 10.1201/9781003441090-10

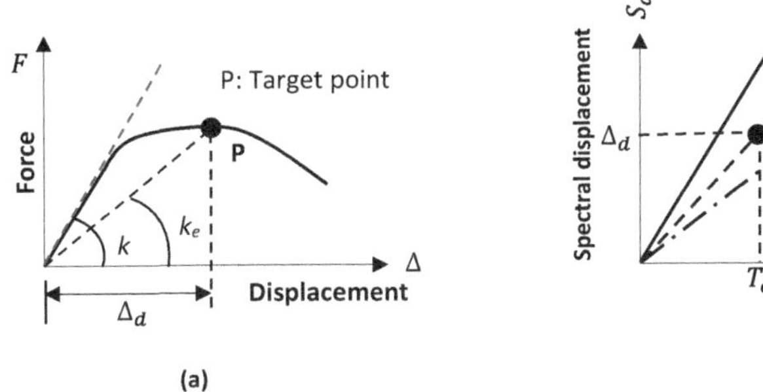

FIGURE 10.1   (a) Force–deformation curve and (b) displacement spectra.

For designating a structure corresponding to an accepted damage state (point P in Figure 10.1(a), effective stiffness (k_e) (also called secant stiffness) is to be used. The target displacement (also called design displacement) that corresponds to the damage level is Δ_d. The time period of the structure is initially T (elastic time period), but this keeps changing as the structure reaches the nonlinear state. The time period corresponding to any point on the nonlinear part of the force–deformation curve is designated as the effective time period (T_e). The time period relationship in general is given by Eq. (10.1.1), in which m_e is effective mass of the equivalent single degree of freedom (ESDOF) system.

$$k_e = 4\pi^2 \frac{m_e}{T_e^2} \tag{10.1.1}$$

From basic stiffness–displacement relation, with V_b as the base shear, we can write Eq. (10.1.2):

$$V_b = k_e \Delta_d \tag{10.1.2}$$

10.2   THEORETICAL DEVELOPMENT OF UPBD METHOD FOR THE DUAL SYSTEM

A 3D dual system was shown in Figure 8.1 in Chapter 8. Consider the 2D dual system building in Figure 10.2(a). From structural dynamics theory, it is possible to convert the 2D/3D building to an equivalent single degree of freedom (ESDOF) system, as has been discussed in earlier chapters. The ESDOF system has an equivalent single mass (m_e) and equivalent height (H_e). The equivalent height corresponds to an equivalent overturning moment replicating the overturning moment of the real building. The ESDOF system undergoes a displacement compatible to and in tune with the displacement pattern of the real building.

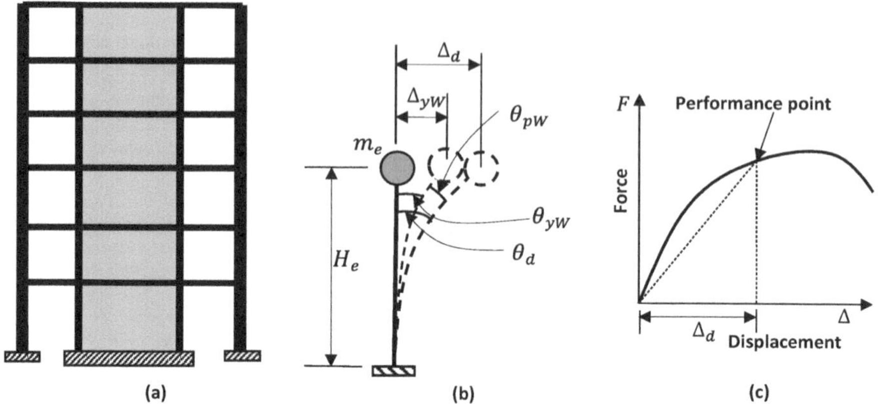

FIGURE 10.2 (a) MDOF dual system, (b) equivalent SDOF system and (c) force–displacement response.

The displacement pattern of a real building depends upon its shape profile. The shape profile of a dual system is given by the yield displacement of the wall at different floor levels.

We use the following symbols:

θ_d = target design drift
θ_{yW} = maximum elastic rotation of wall
θ_{pW} = plastic rotation of wall
m_e = effective mass of ESDOF system
H_e = effective height of ESDOF system
L_W = horizontal length of wall
h_{inf} = height of inflection for wall
ϕ_{yW} = yield curvature of wall
Δ_{yW} = yield displacement of wall = $\theta_{yW} H_e$
ε_y = yield strain of rebar

The total angular displacement of the ESDOF system is the summation of the elastic angular displacement and plastic angular displacement. The total angular displacement of the ESDOF system is the angular design drift (θ_d). Though the frame and wall are integrally built, the deformation of the wall is predominant in a dual system. Design drift means the interstorey drift value for which the building is designed. The design drift is the summation of the maximum elastic drift of the wall (θ_{yW}) and plastic rotation of the wall (θ_{pW}). The above concept is in reference to Figure 10.2(b) and can be expressed as

$$\theta_d = \theta_{yW} + \theta_{pW} \tag{10.2.1}$$

The yield rotation of wall is given by Eq. (10.2.2) (Priestley 2003):

$$\theta_{yW} = \frac{\phi_{yW} h_{inf}}{2} \tag{10.2.2}$$

The yield curvature of wall is given by Eq. (10.2.3):

$$\phi_{yW} = \frac{2\varepsilon_y}{L_W} \tag{10.2.3}$$

Putting Eqs. (10.2.2) and (10.2.3) in Eq. (10.2.1), we get,

$$\theta_d = \frac{2\varepsilon_y h_{inf}}{2L_W} + \theta_{pW}$$

$$\text{Or,} \quad \theta_d = \frac{\varepsilon_y h_{inf}}{L_W} + \theta_{pW}$$

$$\text{Or,} \quad L_W = \frac{\varepsilon_y h_{inf}}{\theta_d - \theta_{pW}} \tag{10.2.4}$$

Eq. (10.2.4) gives the length of the wall that satisfies the given drift and performance level (quantified by the plastic rotation of wall). Depending on the target PL, the corresponding plastic rotation of the wall (available in FEMA documents and SEI-41) can be fed into the equation.

10.2.1 Thickness of the Wall

Based on the plan arrangement, we can select the number of walls in any particular direction of the plan of the building. In a dual system, shear wall takes a major part of the design base shear. However, the frame is designed to take about 25–40% of the base shear. Let V_t denote the total design base shear (which we shall calculate soon), V_W denote the base shear taken by the wall and V_F denotes the base shear taken by the frame system. We can write Eq. (10.2.5) as follows:

$$V_t = V_W + V_F \tag{10.2.5a}$$

$$V_W = V_t - V_F \tag{10.2.5b}$$

If there are N_w is the number of walls in a particular direction of the plan and t_w is the thickness of each wall, then shear stress in a wall is given by

$$\tau = \frac{V_w / N_w}{L_w t_w} \qquad (10.2.6a)$$

Some codes (like, IS 1893 2016) suggest that 80% of the length of the wall is effective in taking shear. So, Eq. (10.2.6a) is modified to Eq. (10.2.6b):

$$\tau = \frac{V_w / N_w}{0.8 L_w t_w} \qquad (10.2.6b)$$

The maximum shear stress permissible in concrete is given by $\tau_{c,\,max}$. For safe design, Eq. (10.2.6c) has to be satisfied.

$$\frac{V_w / N_w}{0.8 L_w t_w} \leq \tau_{c,\,max} \qquad (10.2.6c)$$

E. (10.2.6c) gives the safe value of t_w.

In actual design, a part of wall shear will be carried by concrete and the balance shear by horizontal shear steel. Accordingly, the horizontal steel is obtained.

10.3   INTERPRETATION OF THE EQUATION: EQ. (10.2.4)

We can rewrite Eq. (10.2.4) as Eq. (10.3.1):

$$\frac{L_w}{h_{inf}} = \frac{\varepsilon_y}{\theta_d - \theta_{pW}} \qquad (10.3.1)$$

As discussed in Chapter 9, to plot Eq. (10.3.1), we need to decide the range of drift that corresponds to a performance level. Referring to Table C1-3 of FEMA-356, the drift for vertical members corresponding to different PLs are: IO: 1% transient drift with negligible permanent drift, LS: 2% transient drift with 1% permanent drift and, CP: 4% transient or permanent drift. The permanent drift implies plastic rotation. As per these guidelines, we shall assume the following ranges of drifts for different PLs:

IO: 1–1.5% drift
LS: 1.5–3% drift
CP: 3–4% drift.

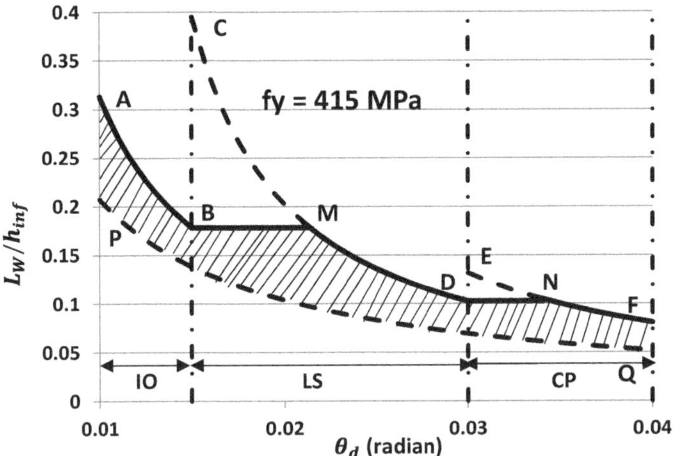

FIGURE 10.3 Interpretation of Eq. (10.3.1), for f_y = 415 MPa.

Drift less than 1% may be assumed as too stringent for practical design purposes. For any given PL, the maximum allowable value of θ_{pW} remains constant. For a given grade of steel, ε_y is known and is fixed. So, a graph can be drawn between L_W/h_{inf} ratio and θ_d by varying θ_d. Let us take Fe415 steel (f_y = 415 MPa) and IO performance level. Here, ε_y = 415/2 × 10^5 = 0.002075. The average plastic rotation for RC walls for IO level as per ASCE-SEI-41-17 (Table 10-19 of this document) for compliant buildings is 0.003375 radian. So, Eq. (10.3.1) takes the form (10.3.1a):

$$\frac{L_W}{h_{inf}} = \frac{0.002075}{\theta_d - 0.003375}$$

(10.3.1a)

The graph of corresponding to Eq. (10.3.1.a) is the curve AB as shown in Figure 10.3. So, AB corresponds to the IO performance level. The average plastic rotation of wall for compliant buildings for LS PL is 0.00975 radian. Putting this in Eq. (10.3.1), curve CD is obtained, which corresponds to LS performance level. Similarly, curve EF corresponds to CP performance level (plastic rotation of wall is 0.01425 radian). From the end of IO performance horizontal line, BM is drawn. Now, for part CM of LS performance, the same L_W/h_{inf} can give IO performance as well. Hence, this part of curve CD is discarded. From point D, horizontal line is drawn to meet 4% drift line at N. So, line ABMDN gives the upper bound of L_W/h_{inf} ratio for different performances. When no plastic rotation occurs (elastic system), θ_{pW} is zero in Eq. (10.3.1), which leads to curve PQ. So, PQ is lower bounds of L_W/h_{inf} ratio.

The hatched zone between the upper bound and lower bound curves give the viable zone of L_W/h_{inf} ratios. In the viable zone, the corresponding higher values of L_W/h_{inf} ratio will give more plastic rotation. The upper bound line gives plastic rotation corresponding to average values of compliant building walls as per ASCE-SEI-41-17. Similar diagrams can be readily constructed for other grades of steel (Figures 10.4 and 10.5).

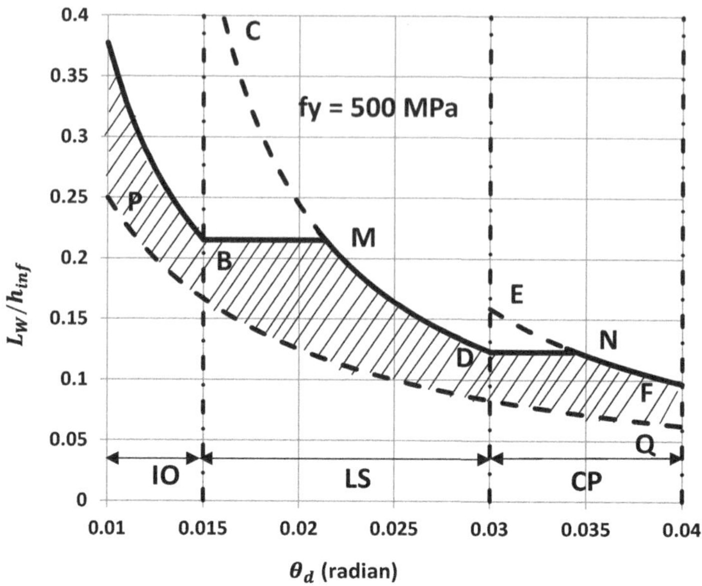

FIGURE 10.4   Interpretation of Eq. (10.2.4), for f_y = 500 MPa.

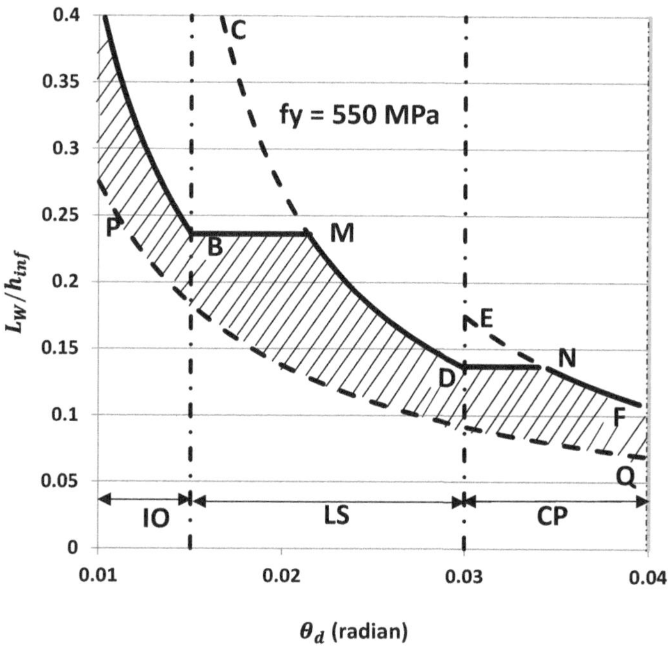

FIGURE 10.5   Interpretation of Eq. (10.2.4), for f_y = 550 MPa.

Example 10.3.1: *In a dual system building plan the length of beam is 6 m. The building is to be designed for (a) IO performance level and 1.3% interstorey drift, (b) LS performance level and 2.8% interstorey drift and (c) CP performance level with 3.6% drift. Suggest suitable wall length in each case for the building. The grade of steel is Fe500. Take height of inflection as 20 m.*

Solution: The yield strain for Fe500 grade steel = 500/2 × 10^5 = 0.0025.
(a) The average plastic rotation for IO performance level wall for compliant buildings (from ASCE-SEI-41-17, Table 10-19) is 0.003375 radian. Referring to Eq. (10.2.4),

$$L_W = \frac{\varepsilon_y h_{inf}}{\theta_d - \theta_{pW}} = \frac{0.0025 \times 20}{0.013 - 0.003375} = 5.2 \text{m}.$$

Ratio L_W / h_{inf} = 5.2/20 = 0.26. From Figure 10.4, corresponding to θ_d = 1.3%, maximum allowable value of L_W / h_{inf} = 0.24. Hence, L_W = 0.24 × 20 = 4.8 m.

(b) The average plastic rotation allowable for LS performance level wall for compliant dual system buildings (from ASCE-SEI-41-17, Table 10-19) is 0.00975 radian. Referring to Eq. (10.2.4),

$$L_W = \frac{\varepsilon_y h_{inf}}{\theta_d - \theta_{pW}} = \frac{0.0025 \times 20}{0.028 - 0.00975} = 2.7 \text{ m}.$$ Referring to Figure 10.4, corresponding to θ_d = 2.8%, maximum allowable value of L_W / h_{inf} = 0.13. Hence, L_W = 0.13 × 20 = 2.6 m.

(c) The average plastic rotation allowable for LS performance level for wall for compliant dual system buildings (from ASCE-SEI-41-17, Table 10-19) is 0.01425 radian. Referring to Eq. (10.2.4),

$$L_W = \frac{\varepsilon_y h_{inf}}{\theta_d - \theta_{pW}} = \frac{0.0025 \times 20}{0.036 - 0.01425} = 2.7 \text{ m}.$$ Referring to Figure 10.4, corresponding to θ_d = 2.8%, maximum allowable value of L_W / h_{inf} = 0.13. Hence, L_W = 0.12 × 20 = 2.4 m.

10.4 DESIGN STEPS IN UPBD METHOD FOR RC DUAL SYSTEM BUILDINGS

The design steps involved in UPBD method for RC dual system buildings, including the computation of ESDOF system properties, are detailed below.

Step 1: *Set the target objectives* for the building to be designed. The typical target objectives are interstorey drift (θ_d) and member performance level (designated by θ_{pb} for beam and θ_{pW} for wall).

Step 2: *Decide the hazard level* for which the building is to be designed. It involves deciding the design spectrum and its level. For example, a building may be designed as per EC-8 design spectrum for *type 1* structural condition and ground *type B*, at shaking level 0.6*g*. As dual system is taller than normal buildings, a higher level of seismicity may be needed to meet the displacement demand.

Step 3: *Decide material properties* namely characteristic strength of concrete (f_{ck}) and yield strength of rebar steel (f_y). The concrete strength will be required when you model it in the computer.

Step 4: *Compute length of wall* from Eq. (10.2.4) and Section 10.3.

Step 5: *Compute beam depth* as discussed in Chapter 9.

Step 6: *Compute* the *inflection height of wall.*

See Chapter 8 for this computation. This gives relative overturning moment in frame M_{OTF}, and wall base moment M_W.

The following expressions are helpful in preparing the table for inflection height.

$$M_{OTi} = M_{OT,i+1} + V_{t,i+1}\left(h_{i+1} - h_i\right) \tag{8.4.1}$$

$$M_{Wi} = M_{W,i+1} + V_{W,i+1}\left(h_{i+1} - h_i\right) \tag{8.4.2}$$

Step 7: *Compute shape profile*

The displacement profile of the wall is obtained from the yield displacement of walls in different floors as expressed by Eqs. (10.4.1 to 10.4.2). The final deflections at floor levels is given by Eq. (10.4.3).

$$\Delta_{iy} = \frac{\phi_{yW} h_{inf} h_i}{2} - \frac{\phi_{yW} h_{inf}^2}{6} \quad \text{when } h_1 \geq h_{inf} \tag{10.4.1}$$

$$\Delta_{iy} = \frac{\phi_{yW} h_i^2}{2} - \frac{\phi_{yW} h_i^3}{6h_{inf}} \quad \text{when } h_i < h_{inf} \tag{10.4.2}$$

$$\Delta_i = \Delta_{iy} + \left(\theta_d - \frac{\phi_{yW} h_{inf}}{2}\right) h_i \tag{10.4.3}$$

$$\phi_{yW} = \frac{2\varepsilon_y}{L_W} \tag{10.4.4}$$

where Δ_{iy} is the yield displacement of wall, h_i is height of i-th floor from base, h_{inf} is inflection height of the wall, θ_{yF} is the yield rotation of the frame, θ_d is the design drift, Δ_i is lateral deflection of i-th floor. Yield rotation of wall is ϕ_{yW}.

Step 8: *Compute the ESDOF system properties*: Δ_d, m_e, H_e

$$\Delta_d = \frac{\sum_{i=1}^{n} m_i \Delta_i^2}{\sum_{i=1}^{n} m_i \Delta_i} \tag{10.4.5}$$

$$m_e = \frac{\sum_{i=1}^{n} m_i \Delta_i}{\Delta_d} \tag{10.4.6}$$

$$H_e = \frac{\sum_{i=1}^{n} m_i \Delta_i h_i}{\sum_{i=1}^{n} m_i \Delta_i} \tag{10.4.7}$$

where m_e is the effective mass, Δ_d is the design displacement and H_e is the effective height of ESDOF system, respectively.

Step 9: *Generate the displacement spectra* at various dampings for the design spectrum considered in step (2). (For generation, see Chapter 3).

Step 10: *Compute frame ductility*:

$$\mu_{F,i} = \left(\frac{\Delta_i - \Delta_{i-1}}{h_i - h_{i-1}} \right) \frac{1}{\theta_{yF}} \tag{10.4.8}$$

$$\theta_{yF} = \frac{0.5 \varepsilon_y l_b}{h_b} \tag{10.4.8a}$$

where $\mu_{F,i}$ is the frame ductility at floor i. This will require a table to be prepared. If the beam depth is uniform over height of building, as is the case in UPBD method, the average value of $\mu_{F,i}$ can be taken as μ_F.

Step 11: *Compute wall ductility*

Wall ductility (μ_W) is given by Eq. (10.4.9), in which L_p is the lower value out of Eqs. (10.4.10a and 10.4.10b).

$$\mu_W = 1 + \frac{1}{L_p \phi_{yW}} \left(\theta_d - \frac{\phi_{yW} h_{inf}}{2} \right) \tag{10.4.9}$$

$$L_p = 0.022 f_y d_b + 0.054 h_{inf} \tag{10.4.10a}$$

$$L_p = 0.2 L_W + 0.03 h_{inf} \tag{10.4.10b}$$

Smaller out of two values of L_p is to be considered.

In Eq. (10.4.10a) f_y is in MPa and d_b and hinf are is in m (dimensionally not homogenous).

Step 12: *Compute system ductility trial effective time period*

System ductility μ_{sys} is given by Eq. (10.4.11). Trial effective time period $T_{e,trial}$ is given by Eq. (10.4.12).

$$\mu_{sys} = \frac{M_W \mu_W + M_{OTF} \mu_F}{M_W + M_{OTF}} \tag{10.4.11}$$

$$T_{e,trial} = \frac{n}{6} \sqrt{\mu_{sys}} \tag{10.4.12}$$

Step 13: *Compute system damping*

Wall damping is given by Eq. (10.4.13), Frame damping is given by Eq. (10.4.14), and system damping is given by Eq. (10.4.15)).

$$\xi_W = \frac{95}{1.3\pi} \left(1 - \mu_W^{-0.5} - 0.1 r \mu_W \right) \left(1 + \left(T_{e,trial} + 0.85 \right)^{-4} \right) \tag{10.4.13}$$

$$\xi_F = \frac{120}{1.3\pi} \left(1 - \mu_F^{-0.5} - 0.1 r \mu_F \right) \left(1 + \left(T_{e,trial} + 0.85 \right)^{-4} \right) \tag{10.4.14}$$

$$\xi_{sys} = \frac{M_W \xi_W + M_{OTF} \xi_F}{M_W + M_{OTF}} \tag{10.4.15}$$

Step 14: *Compute design base shear*

Corresponding to design displacement Δ_d and system damping ξ_{ESDOF}, from displacement spectra, we get the effective time period, T_e. Effective stiffness is given by Eq. (10.4.16). Base shear is given by Eq. (10.4.17):

$$k_e = 4\pi^2 \frac{m_e}{T_e^2} \qquad (10.4.16)$$

$$V_b = k_e \Delta_d \qquad (10.4.17)$$

Step 15: *Distribute the base shear over the floors*

The lateral forces F_i are obtained by using Eq. (10.4.18). Here, n is number of storeys.

$$\text{For } n \leq 10 \quad F_i = V_b \frac{m_i \Delta_i}{\sum_{i=1}^{n} m_i \Delta_i} \qquad (10.4.18a)$$

$$\text{For } n > 10 \quad F_i = 0.9 V_b \frac{m_i \Delta_i}{\sum_{i=1}^{n} m_i \Delta_i} \qquad (10.4.18b)$$

At roof, add $0.1 V_b$.

Step 16: *Apply load combination and design the building.*

The load combinations are:

$$DL + LL$$

$$DL + LL \pm F_x$$

$$DL + LL \pm F_y$$

where DL is dead load, LL is live load, F_x and F_y are lateral seismic forces (as per Eqs. (10.4.18)) in the short and the long directions of the plan of the building, respectively.

For design, take the expected strength of materials. From FEMA documents, the expected strength of concrete is $1.5 f_{ck}$ and that of steel is $1.25 f_y$.

Step 17: *Evaluation of the designed building*

The building designed has to be checked for its actual performance. The evaluation is done through nonlinear static analysis (POA) and nonlinear dynamic analyses (NLTHA). The performance point is obtained from POA. The plastic hinges are noted at the end of NLTHA. This along with PP will give PL of the building. Drift is obtained from floor displacement time histories. If the target design objectives are not achieved, design may be updated suitably.

10.5 DESIGN EXAMPLES

Example 10.5.1 *A 15-storey dual system building plan is 30 m × 18 m in size. The design drift is 2.5% and target performance level is LS. The floor thickness is 120 mm, superimposed DL in floor is 1 kN/m² in floor. Exterior infill walls are 250 mm thick and interior walls are 125 mm thick. Column size is 600 mm × 600 m all through. Find the lateral seismic design forces. Take f_y as 500 MPa for rebar, and concrete strength as 30 MPa. Use EC-8 spectrum for soil type B at 0.55g level. Interstorey height is 3.4 m and ground storey height is 4.4 m. Consider 20% opening in infill walls. Assume that frame carries 30% of base shear. The diameter of the longitudinal bar in wall is 20 mm.*

Solution:

Initial computations:

The design drift θ_d = 0.025 radian. The yield strain rebar steel is $500/2 \times 10^5$ = 0.0025. The average plastic rotation allowable for LS performance level for compliant buildings (ASCE-SEI-41-17) is 0.02 radian for beams and 0.00975 radian for walls. Building height is 52 m. Using Eq. (10.6.1) $h_{inf} = (1 - 0.3) \times 52$ = 36.4 m.

$$L_W = \frac{\varepsilon_y h_{inf}}{\theta_d - \theta_{pW}} = \frac{0.0025 \times 36.4}{0.025 - 0.00975} = 5.96 \text{ m}.$$ Let us take wall and beam length as

6 m. Let us take preliminary wall thickness as 180 mm. We adopt wall positions in the given plot plan as shown in Figure 10.6.

$$h_b = \frac{0.5\varepsilon_y l_b}{\theta_d - \theta_{pb}} = \frac{0.5 \times 0.0025 \times 6}{0.025 - 0.02} = 1.5 \text{ m}.$$ But from Figure 9.4 (Chapter 9), at

2.5% drift h_b/l_b = 0.14. which gives, l_b = 0.14 × 6 = 0.84 m. Let us take beam size 400 mm × 800 mm. $\phi_{yW} = 2\varepsilon_y / L_w$ = 0.00083 per m.

The floor weights are tabulated in Table Ex10.5.1.1. Column height = 3.4 − 0.8 = 2.6 m. Floor weight is 6356 kN; floor mass is 6356/9.81 = 648 kN-s²/m.

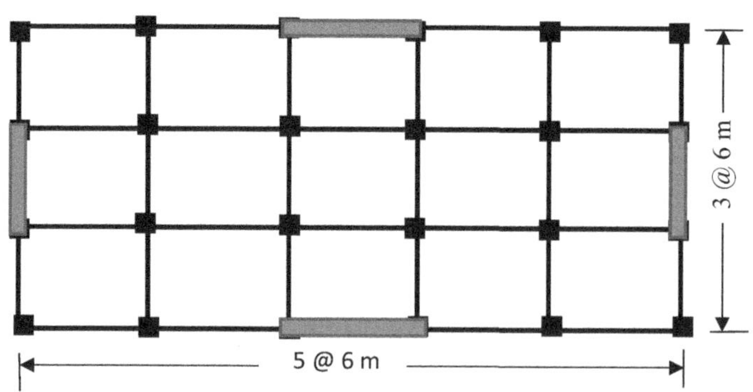

FIGURE 10.6 Wall disposition for Example 10.5.1.

TABLE EX10.5.1.1
Floor weight in Example 10.5.1

Item	L (m)	B (m)	H (m)	N	Unit wt (kN/m³)	Weight (kN)
Column	2.6	0.6	0.6	16	25	374
Beam short dir	18	0.4	0.8	6	25	864
Beam long dir	30	0.4	0.8	4	25	960
Exterior infill	96	0.25	2.65	1	20	1018
Interior infill	132	0.125	2.65	1	20	700
Wall in short	6	2.6	0.18	2	25	140
Wall in long dir	6	2.6	0.18	2	25	140
Slab DL	30	18	0.12	1	25	1620
Floor SDL	30	18	1	1		540
	80% opening in infill wall considered				Sum	6356

Roof weight: Slab and beam weights are same as that in floor. Half load of column, wall and infill will be added. Superimposed DL will not be added. Roof weight = 864 + 960 + 1620 + 0.5(374 +140 + 140 + 1018 + 700) kN = 4630 kN. Roof mass = 4630/9.81 = 472 kN-s²/m.

Compute inflection height of the wall
Table Ex10.2 is prepared.

From the Table by interpolation, or from Figure 10.7 h_{inf} = 35.7 m. Remember that we used this value as 36.4 m. The difference is not high. L_W/h_{inf} = 6/35.7 = 0.16. Figure 10.4 shows that this value is within the upper and lower bound curves; hence, admissible. The table gives M_{OTF} = 35.294 kN-m (relative value) and M_W = 19.694 kN-m (relative value).

Compute shape profile and ESDOF system properties
Using Eqs. (10.4.1) to (10.4.7), Tables Ex10.5.1.2 and Ex10.5.1.3 are prepared.

$$\Delta_d = \frac{\sum_{i=1}^{n} m_i \Delta_i^2}{\sum_{i=1}^{n} m_i \Delta_i} = 3825/5145 = 0.743 \text{ m}$$

$$m_e = \frac{\sum_{i=1}^{n} m_i \Delta_i}{\Delta_d} = 5145/0.743 = 6924 \text{ kN-s}^2/\text{m}$$

$$H_e = \frac{\sum_{i=1}^{n} m_i \Delta_i h_i}{\sum_{i=1}^{n} m_i \Delta_i} = 188550/5145 = 36.647 \text{ m}$$

TABLE EX10.5.1.2

Computation of height of inflection for Example 10.5.1

Floor	h_i (m)	m_i (ton-mass)	$m_i h_i$	$m_i h_i/$ sum $(m_i h_i)$	V_{ti} relative	M_{OTi} relative	V_F relative	V_w relative	M_w relative
15	52	472	24544	0.093	0.093	0.000	0.30	−0.207	0.0000
14	48.6	648	31493	0.119	0.211	0.315	0.30	−0.089	−0.7050
13	45.2	648	29290	0.111	0.322	1.034	0.30	0.022	−1.0059
12	41.8	648	27086	0.102	0.424	2.129	0.30	0.124	−0.9310
11	38.4	648	24883	0.094	0.518	3.572	0.30	0.218	−0.5085
10	35	648	22680	0.086	0.604	5.333	0.30	0.304	0.2334
9	31.6	648	20477	0.077	0.681	7.386	0.30	0.381	1.2663
8	28.2	648	18274	0.069	0.75	9.702	0.30	0.450	2.5620
7	24.8	648	16070	0.061	0.811	12.252	0.30	0.511	4.0921
6	21.4	648	13867	0.052	0.863	15.008	0.30	0.563	5.8285
5	18	648	11664	0.044	0.907	17.943	0.30	0.607	7.7428
4	14.6	648	9461	0.036	0.943	21.027	0.30	0.643	9.8068
3	11.2	648	7258	0.027	0.97	24.232	0.30	0.670	11.9923
2	7.8	648	5054	0.019	0.989	27.531	0.30	0.689	14.2708
1	4.4	648	2851	0.011	1	30.894	0.30	0.700	16.6142
	0					**35.294**			**19.6942**
		sum	264952						

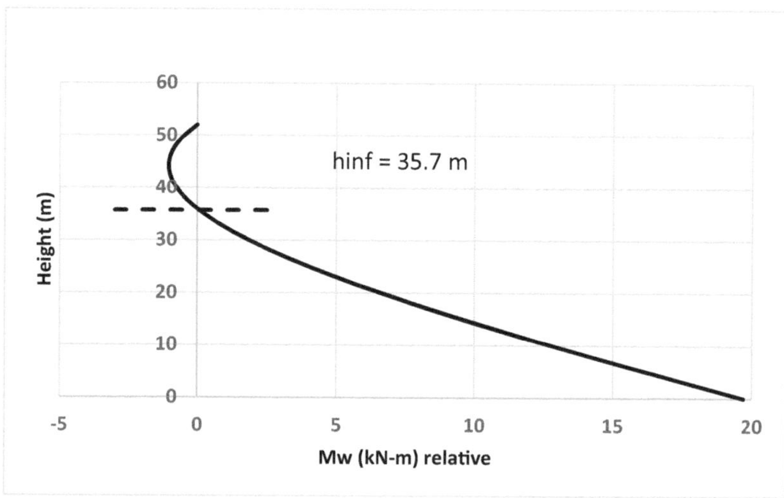

FIGURE 10.7　Inflection height.

TABLE EX10.5.1.3
ESDOF system properties computation in Example 10.5.1

Floor	h_i	m_i	Δ_{iy}	Δ_i	$m_i\Delta_i$	$m_i\Delta_i^2$	$m_i\Delta_i h_i$
15	52	472	0.5965	1.1230	530.1	595.2	27563
14	48.6	648	0.5376	1.0296	667.2	687.0	32426
13	45.2	648	0.4920	0.9497	615.4	584.4	27815
12	41.8	648	0.4439	0.8671	561.9	487.2	23487
11	38.4	648	0.3941	0.7829	507.3	397.2	19481
10	35	648	0.3436	0.6980	452.3	315.7	15830
9	31.6	648	0.2933	0.6133	397.4	243.7	12558
8	28.2	648	0.2441	0.5296	343.2	181.8	9678
7	24.8	648	0.1969	0.4480	290.3	130.1	7200
6	21.4	648	0.1527	0.3694	239.3	88.4	5122
5	18	648	0.1123	0.2946	190.9	56.2	3436
4	14.6	648	0.0767	0.2245	145.5	32.7	2124
3	11.2	648	0.0468	0.1602	103.8	16.6	1163
2	7.8	648	0.0235	0.1025	66.4	6.8	518
1	4.4	648	0.0077	0.0523	33.9	1.8	149
	0			Sum	**5145**	**3825**	**188550**

Compute frame ductility

Prepare Table 10.4 using Eqs. (10.4.8) and (10.4.8a).

$$\mu_F = 35.91/15 = 2.39.$$

Computation of wall ductility and system ductility

$$L_p = 0.022 f_y d_b + 0.054 h_{inf} = 0.022 \times 500 \times 0.02 + 0.05 \times 35.7 = 2.148 \text{ m.}$$

$$L_p = 0.2 L_W + 0.03 h_{inf} = 0.2 \times 6 + 0.03 \times 35.7 = 2.271 \text{ m.}$$

Lower value is 2.148 m.

$$\mu_W = 1 + \frac{1}{L_p \phi_{yW}} \left(\theta_d - \frac{\phi_{yW} h_{inf}}{2} \right) = 6.66$$

$$\mu_{ESDOF} = \frac{M_W \mu_W + M_{OTF} \mu_F}{M_W + M_{OTF}} = 3.92.$$

TABLE EX10.5.1.4
Computation of frame ductility in Example 10.5.1

Floor	h_i (m)	Δ_i	Δ_{Fi}
15	50.5	1.1230	3.02
14	47.2	1.0296	2.59
13	43.9	0.9497	2.67
12	40.6	0.8671	2.72
11	37.3	0.7829	2.74
10	34	0.6980	2.74
9	30.7	0.6133	2.70
8	27.4	0.5296	2.64
7	24.1	0.4480	2.54
6	20.8	0.3694	2.42
5	17.5	0.2946	2.26
4	14.2	0.2245	2.08
3	10.9	0.1602	1.87
2	7.6	0.1025	1.62
1	4.3	0.0523	1.30
	0	0.0000	
		Sum	**35.91**

Computation of system damping

$$T_{e,trial} = \frac{n}{6}\sqrt{\mu_{ESDOF}} = 4.95 \text{ s.}$$

$$\xi_W = \frac{95}{1.3\pi}\left(1 - \mu_W^{-0.5} - 0.1r\mu_W\right)\left(1 + \left(T_{e,trial} + 0.85\right)^{-4}\right) = 12.7\%.$$

$$\xi_F = \frac{120}{1.3\pi}\left(1 - \mu_F^{-0.5} - 0.1r\mu_F\right)\left(1 + \left(T_{e,trial} + 0.85\right)^{-4}\right) = 9.70\%.$$

$$\xi_{ESDOF} = \frac{M_W \xi_W + M_{OTF} \xi_F}{M_W + M_{OTF}} = \mathbf{10.8\%.}$$

The displacement spectra corresponding to EC-8 design spectrum for type B soil at 0.55g level is shown in Figure 10.8. The effective time period from this figure is 4.51 s.

This differs only slightly from the trial time period taken. So, the result is accepted.

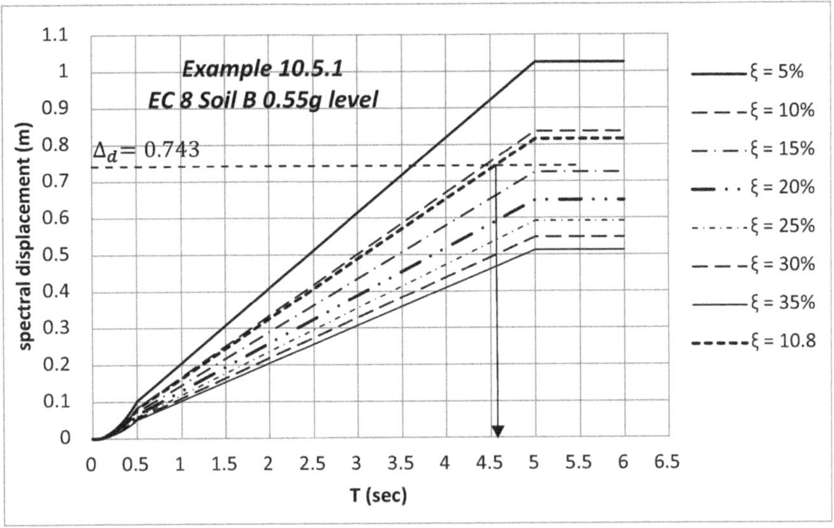

FIGURE 10.8   Displacement spectra for 10.8% damping.

Computation of design base shear

$$k_e = 4\pi^2 \frac{m_e}{T_e^2} = 4\pi^2 \frac{6924}{4.51^2} = 13439 \text{ kN/m.}$$

$$V_b = k_e \Delta_d = 13439 \times 0.743 = 9985 \text{ kN.}$$

Check for wall thickness

Here, $V_W = 0.7 V_b = 5687$ kN. $N_W = 2$ in any one direction of plan. $\tau_{c,max}$ is 3.5 MPa

for M30 concrete. Consider wall thickness as 180 mm.

$$\tau = \frac{V_W / N_W}{0.8 L_W t_W} = \frac{5687 \times 1000/2}{0.8 \times 6000 \times 180} = 3.29 \text{ MPa} < 3.5 \text{ MPa; hence, safe.}$$

Wall thickness of 180 mm is adopted.

Distribution of base shear over the floors
Table Ex10.5.1.4 is prepared.

Example 10.5.2 *A 20-storey dual system building plan is 40 m × 35 m in size, as given in Figure 10.9. The design drift is 2.6% and target performance level is LS. The seismic load is 15 kN/m² in floor and 10 kN/m² at roof. Take as f_y as 500 MPa for*

TABLE EX10.5.1.5
Distribution of base shear over floors in Example 10.5.1

Floor	h_i (m)	m_i (kN-s²/)	$m_i h_i$	V_b (kN)	F_i (kN)
15	52	472	24544	9985	925
14	48.6	648	31493		1187
13	45.2	648	29290		1104
12	41.8	648	27086		1021
11	38.4	648	24883		938
10	35	648	22680		855
9	31.6	648	20477		772
8	28.2	648	18274		689
7	24.8	648	16070		606
6	21.4	648	13867		523
5	18	648	11664		440
4	14.6	648	9461		357
3	11.2	648	7258		274
2	7.8	648	5054		190
1	4.4	648	2851		107
		Sum	**264952**		**9985**

rebar, and concrete strength as 30 MPa. Use EC-8 spectrum for soil type B at 0.7g level. Interstorey height is 3.3 m and ground storey height is 4.3 m. Assume that the frame carries 27% of base shear. Find lateral seismic design forces along both the directions of the building. Beams in the short and long directions are 4 m and 3.5 m respectively. The rebar diameter of vertical steel in wall is 20 mm.

Solution:

Initial computations:

Number of storeys, $N = 20$, $H = 19 \times 3.3 + 4.3 = 67$ m. Design drift $\theta_d = 0.026$. The yield strain rebar steel is $500/2 \times 10^5 = 0.0025$. The average plastic rotation allowable for LS performance level for compliant buildings (ASCE-SEI-41-17) is 0.02 radian for beams and, 0.00975 radian for walls. 27% of base shear is to carried by the frame. Using Eq. (10.6.1),

$$h_{inf} = \left(1 - v_f\right)H = (1 - 0.27) \times 67 = 48.91 \text{ m (preliminary assessment).}$$

$$L_W = \frac{\varepsilon_y h_{inf}}{\theta_d - \theta_{pW}} = \frac{0.0025 \times 48.91}{0.025 - 0.00975} = 8.01 \text{ m.}$$

Let us take wall length as 8 m in the long direction of plan. We have considered 6 walls in each direction. Let us take preliminary wall thickness as 180 mm. We adopt wall positions in the given plan as shown in Figure 10.9 with (wall lengths different in different directions).

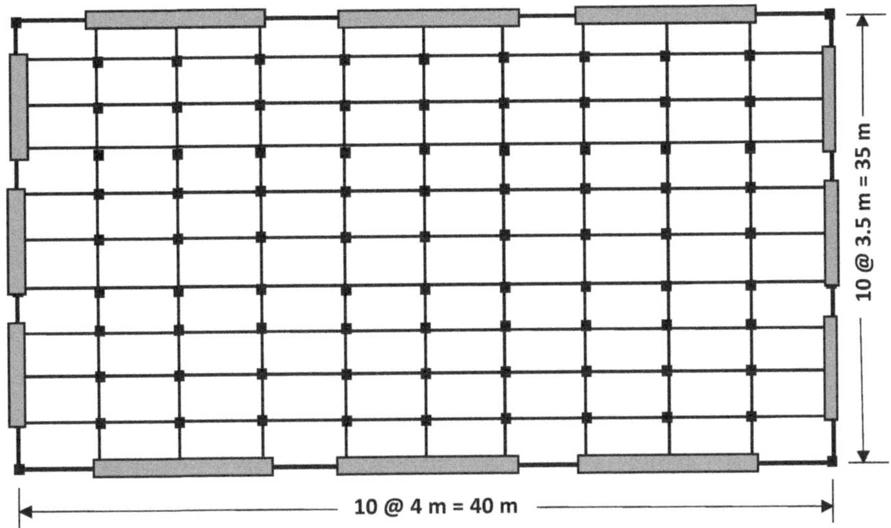

FIGURE 10.9   Building plan for Example 10.5.2.

The short directional wall is taken as 7 m in length. Approximate value of is L_w/h_{inf} = 7/48.91 = 0.143. From Figure 10.4, it is found to be within feasible zone.

Long direction
Wall length = 8 m
Beam length is taken as 4 m.

$$h_b = \frac{0.5\varepsilon_y l_b}{\theta_d - \theta_{pb}} = \frac{0.5 \times 0.0025 \times 4}{0.026 - 0.02} = 0.83 \text{ m}.$$

Let us take beam depth 0.8 m (h_b/l_b = 0.1). From Figure 9.4 (Chapter 9), at 2.6% drift it is a valid ratio. Let us take beam size as 400 mm × 800 mm.

Short direction
Let us take wall length as 7 m and beam length as 3.5 m.

$$h_b = \frac{0.5\varepsilon_y l_b}{\theta_d - \theta_{pb}} = \frac{0.5 \times 0.0025 \times 3.5}{0.026 - 0.02} = 0.726 \text{ m}.$$

Let us take beam depth 0.7 m (h_b/l_b = 0.1). From Figure 9.4 (Chapter 9), at 2.6% drift, this is a valid ratio. Let us take beam size as 350 mm × 700 mm.

$$\phi_{yW} = 2\varepsilon_y/L_w.$$

TABLE EX10.5.2.1
Some data for Example 10.5.2

Item	Short direction	Long direction	Remarks
Wall length	7 m	8 m	4 walls in each direction
Wall thickness	180 mm	180 mm	
Beam depth	700 mm	800 mm	
Beam size	350 mm × 700 mm	400 mm × 800 mm	
ϕ_{yW}	0.000714 per m	0.000625 per m	

We tabulate the initial values in Table Ex10.5.2.1.

The floor weight is $40 \times 35 \times 15 = 21000$ kN. Roof weight is 14,000 kN.

   Floor mass = 21000/9.81 = 2141 ton-mass ($kN\text{-}s^2/m$).

   Roof mass = 14000/9.81 = 1427 ton-mass ($kN\text{-}s^2/m$).

Computation of inflection height of the wall

We use the following two relations (discussed elsewhere):

$$M_{OTi} = M_{OT,i+1} + V_{t,i+1}\left(h_{i+1} - h_i\right)$$

$$M_{Wi} = M_{W,i+1} + V_{W,i+1}\left(h_{i+1} - h_i\right)$$

Table Ex10.5.2.2 is prepared.

From Table 10.7, $M_{OTF} = 45.12$ kN-m (relative value), $M_W = 27.03$ kN-m (relative value). From the same Table or, from Figure 10.10, $h_{inf} = 48.3$ m.

Computation for *short* direction of the building

$L_w = 7$ m, $l_b = 3.5$ m, $h_b = 700$ mm, the $\phi_{yW} = 0.000714$ per m.

Computation of displacement profile and ESDOF system properties (short direction)

Using Eqs. (10.4.1 to 10.4.7) Table Ex10.5.2.3 is prepared.

   From Table 10.8,

$$\Delta_d = \frac{\sum_{i=1}^{n} m_i \Delta_i^2}{\sum_{i=1}^{n} m_i \Delta_i} = 27331/28484 = 0.960 \text{ m}$$

$$m_e = \frac{\sum_{i=1}^{n} m_i \Delta_i}{\Delta_d} = 28484/0.96 = 29671 \text{ kN-s}^2/m$$

TABLE EX10.5.2.2
Computation of height of inflection for Example 10.5.2

Floor	h_i (m)	m_i (ton-mass)	$m_i h_i$	$m_i h_i /$ sum $(m_i h_i)$	V_{ti} relative	M_{OTi} relative	V_F relative	V_w relative	M_w relative
20	67	1427	95609	0.065	0.065	0	0.27	−0.205	0
19	63.7	2141	136382	0.092	0.157	0.213	0.27	−0.113	−0.6776
18	60.4	2141	129316	0.087	0.244	0.731	0.27	−0.026	−1.0509
17	57.1	2141	122251	0.083	0.327	1.537	0.27	0.057	−1.1356
16	53.8	2141	115186	0.078	0.405	2.617	0.27	0.135	−0.9474
15	50.5	2141	108121	0.073	0.478	3.953	0.27	0.208	−0.5022
14	47.2	2141	101055	0.068	0.546	5.530	0.27	0.276	0.1843
13	43.9	2141	93990	0.064	0.610	7.333	0.27	0.340	1.0963
12	40.6	2141	86925	0.059	0.669	9.346	0.27	0.399	2.2181
11	37.3	2141	79859	0.054	0.723	11.553	0.27	0.453	3.5339
10	34	2141	72794	0.049	0.772	13.938	0.27	0.502	5.0279
9	30.7	2141	65729	0.044	0.816	16.485	0.27	0.546	6.6844
8	27.4	2141	58663	0.040	0.856	19.180	0.27	0.586	8.4875
7	24.1	2141	51598	0.035	0.891	22.005	0.27	0.621	10.4216
6	20.8	2141	44533	0.030	0.921	24.945	0.27	0.651	12.4708
5	17.5	2141	37468	0.025	0.946	27.984	0.27	0.676	14.6194
4	14.2	2141	30402	0.021	0.967	31.108	0.27	0.697	16.8516
3	10.9	2141	23337	0.016	0.983	34.299	0.27	0.713	19.1516
2	7.6	2141	16272	0.011	0.994	37.542	0.27	0.724	21.5038
1	4.3	2141	9206	0.006	1.000	40.821	0.27	0.730	23.8922
0	0					**45.121**			**27.031**

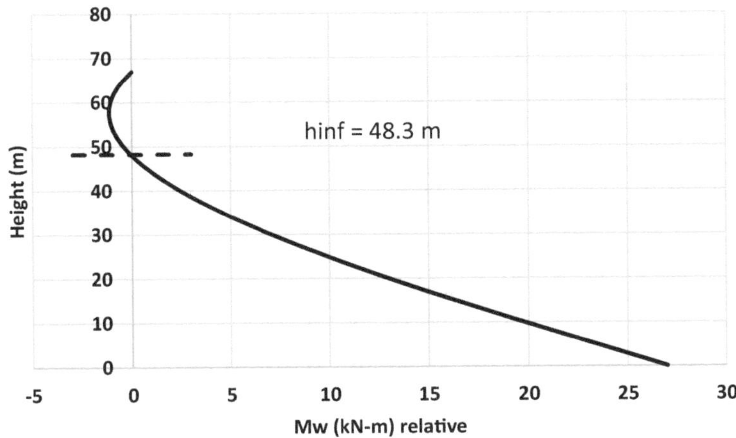

FIGURE 10.10 Inflection height in Example 10.5.2.

TABLE EX10.5.2.3
ESDOF system properties computation in Example 10.5.2 (short direction)

Floor	h_i	m_i	Δ_{iy}	Δ_i	$m_i\Delta_i$	$m_i\Delta_i^2$	$m_i\Delta_i h_i$
20	67	1427	0.8780	1.4643	2089.5	3059.6	139998
19	63.7	2141	0.8211	1.3785	2951.3	4068.3	187999
18	60.4	2141	0.7642	1.2927	2767.6	3577.6	167164
17	57.1	2141	0.7072	1.2069	2583.9	3118.5	147542
16	53.8	2141	0.6503	1.1211	2400.2	2690.8	129132
15	50.5	2141	0.5934	1.0353	2216.5	2294.7	111934
14	47.2	2141	0.5365	0.9495	2032.8	1930.1	95950
13	43.9	2141	0.4798	0.8639	1849.6	1597.8	81196
12	40.6	2141	0.4238	0.7790	1667.8	1299.2	67714
11	37.3	2141	0.3690	0.6954	1488.8	1035.2	55531
10	34	2141	0.3160	0.6135	1313.5	805.8	44658
9	30.7	2141	0.2653	0.5339	1143.1	610.3	35093
8	27.4	2141	0.2174	0.4572	978.8	447.5	26820
7	24.1	2141	0.1729	0.3838	821.7	315.4	19804
6	20.8	2141	0.1323	0.3143	673.0	211.5	13998
5	17.5	2141	0.0962	0.2493	533.7	133.1	9340
4	14.2	2141	0.0650	0.1892	405.1	76.6	5752
3	10.9	2141	0.0392	0.1346	288.2	38.8	3142
2	7.6	2141	0.0195	0.0860	184.2	15.9	1400
1	4.3	2141	0.0064	0.0440	94.3	4.2	405
				Sum	**28484**	**27331**	**1344572**

$$H_e = \frac{\sum_{i=1}^{n} m_i\Delta_i h_i}{\sum_{i=1}^{n} m_i\Delta_i} = 1344572/28484 = 47.204 \text{ m.}$$

Computation of frame ductility (short direction)
Prepare Table Ex10.5.2.4 using Eqs. (10.4.8) and (10.4.8a).

$$\theta_{yF} = \frac{0.5\varepsilon_y l_b}{h_b} = 0.00625$$

$$\mu_{F,i} = \left(\frac{\Delta_i - \Delta_{i-1}}{h_i - h_{i-1}}\right)\frac{1}{\theta_{yF}}$$

$\mu_F = 70.5/20 = \textbf{3.53}.$

TABLE EX10.5.2.4
Computation of frame ductility in Example 10.5.2 (Short direction)

Floor	h_i (m)	Δ_i	μ_{Fi}
20	67	1.4643	4.16
19	63.7	1.3785	4.16
18	60.4	1.2927	4.16
17	57.1	1.2069	4.16
16	53.8	1.1211	4.16
15	50.5	1.0353	4.16
14	47.2	0.9495	4.15
13	43.9	0.8639	4.12
12	40.6	0.7790	4.06
11	37.3	0.6954	3.97
10	34	0.6135	3.86
9	30.7	0.5339	3.72
8	27.4	0.4572	3.56
7	24.1	0.3838	3.37
6	20.8	0.3143	3.15
5	17.5	0.2493	2.91
4	14.2	0.1892	2.65
3	10.9	0.1346	2.35
2	7.6	0.0860	2.04
1	4.3	0.0440	1.64
	0	0.0000	
		Sum	70.50

Computation wall ductility and system ductility (short direction)

$$L_p = 0.022 f_y d_b + 0.054 h_{inf} = 0.022 \times 500 \times 0.02 + 0.054 \times 48.3 \text{ m} = 2.828 \text{ m}.$$

$$L_p = 0.2 L_W + 0.03 h_{inf} = 0.2 \times 7 + 0.03 \times 48.3 = 2.849 \text{ m}.$$

So, $L_p = 2.828$ m.

$$\mu_W = 1 + \frac{1}{L_p \phi_{yW}} \left(\theta_d - \frac{\phi_{yW} h_{inf}}{2} \right)$$

$$= 1 + \frac{1}{2.828 \times 0.000714} \left(0.026 - \frac{0.000714 \times 48.3}{2} \right) = 5.33$$

$$\mu_{ESDOF} = \frac{M_W \mu_W + M_{OTF} \mu_F}{M_W + M_{OTF}} = \frac{27.03 \times 5.33 + 45.12 \times 3.53}{27.03 + 45.12} = 4.202.$$

Computation of system damping (short direction)

$$T_{e,trial} = \frac{n}{6}\sqrt{\mu_{ESDOF}} = \frac{20}{6}\sqrt{4.202} = \mathbf{6.833}\,\text{sec}.$$

$$\xi_W = \frac{95}{1.3\pi}\left(1 - \mu_W^{-0.5} - 0.1r\mu_W\right)\left(1 + \left(T_{e,trial} + 0.85\right)^{-4}\right) = 11.95\%.$$

$$\xi_F = \frac{120}{1.3\pi}\left(1 - \mu_F^{-0.5} - 0.1r\mu_F\right)\left(1 + \left(T_{e,trial} + 0.85\right)^{-4}\right) = 12.7\%.$$

$$\xi_{ESDOF} = \frac{M_W\xi_W + M_{OTF}\xi_F}{M_W + M_{OTF}} = \mathbf{12.42\%}.$$

Computation of design base shear (short direction)

The displacement spectra corresponding to EC-8 design spectrum for B type soil at 0.7g level is shown in Figure 10.11.

From Figure 10.11, $T_e = 4.51$ s. This differs from $T_{e,trial}$ (6.833 s). So, put 4.51 s as $T_{e,trial}$ in Equations of ξ_W and ξ_F.

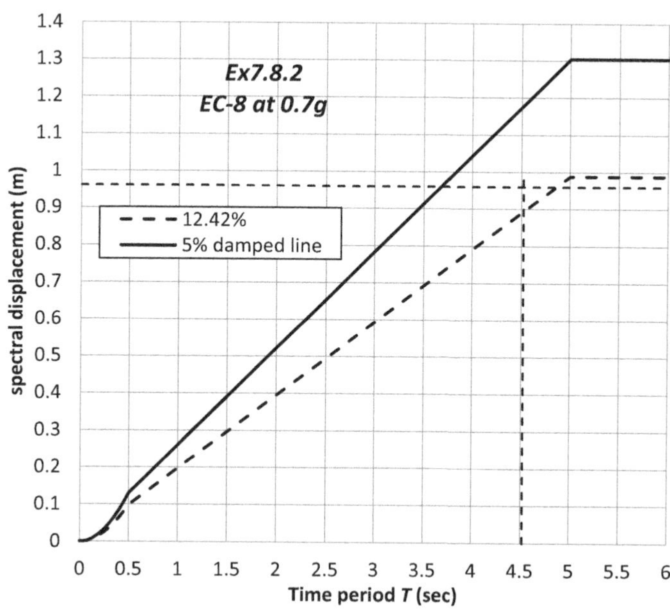

FIGURE 10.11 Displacement spectra in Example 10.5.2 (short direction) (first trial).

New values of parameters with this time period are given below.

$$\xi_W = 12.79\%$$

$$\xi_F = 15.11\%$$

$$\xi_{ESDOF} = 14.24\%$$

The new displacement spectrum is shown in Figure 10.12, from which $T_e = \mathbf{5}$ s.

It is noted that the target design displacement line is slightly above the 14.23% damped spectrum. This will not lead to any serious deviation of the results.

$$k_e = 4\pi^2 \frac{m_e}{T_e^2} = 4\pi^2 \frac{29671}{5^2} = 46854\,\text{kN/m}.$$

$$V_b = k_e \Delta_d = 46854 \times 0.96 = 44,980\,\text{kN}.$$

Check for wall thickness (short direction)

Here, $V_W = (1 - 0.27) \times V_b = 32836$ kN. $N_W = 6$ in any one direction of plan. For M30 concrete, $\tau_{c,max}$ is 3.5 MPa (IS 456). Shear stress in concrete,

$$\tau = \frac{V_W / N_W}{0.8 L_W t_W} = \frac{32836 \times 1000/6}{0.8 \times 7000 \times 180} = 5.43 \text{ MPa} > 3.5 \text{ MPa; hence, not safe.}$$

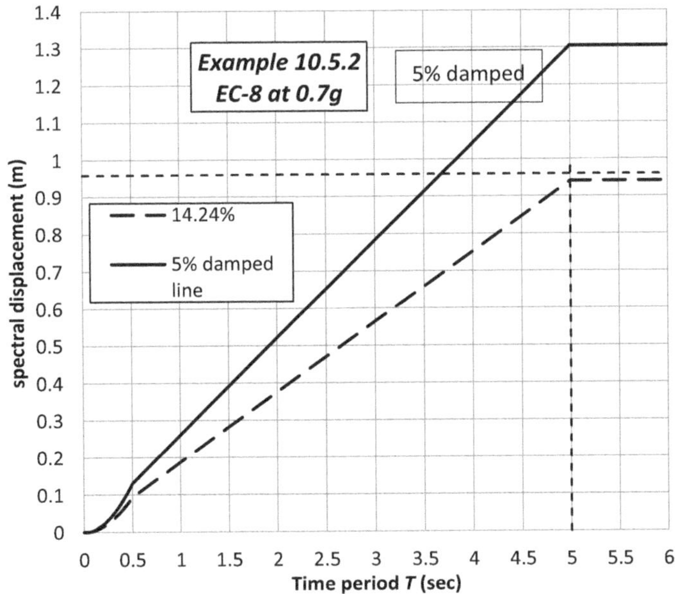

FIGURE 10.12   Displacement spectra updated in Example 10.5.2 (short direction) (final).

TABLE EX10.5.2.5
Distribution of base shear over floors in Example 10.5.2 (short direction)

Floor	h_i (m)	m_i (kN-s²/m)	m_ih_i	V_b (kN)	F_i (kN)
20	67	1427	95609	44980	2908
19	63.7	2141	136382		4149
18	60.4	2141	129316		3934
17	57.1	2141	122251		3719
16	53.8	2141	115186		3504
15	50.5	2141	108121		3289
14	47.2	2141	101055		3074
13	43.9	2141	93990		2859
12	40.6	2141	86925		2644
11	37.3	2141	79859		2429
10	34	2141	72794		2214
9	30.7	2141	65729		1999
8	27.4	2141	58663		1784
7	24.1	2141	51598		1570
6	20.8	2141	44533		1355
5	17.5	2141	37468		1140
4	14.2	2141	30402		925
3	10.9	2141	23337		710
2	7.6	2141	16272		495
1	4.3	2141	9206		280
	Sum		**1478695**		**44980**

Try wall thickness 300 mm.

$$\tau = \frac{32835 \times 1000/6}{0.8 \times 7000 \times 300} = 3.25 \text{ MPa} < 3.5 \text{ MPa; hence } safe.$$

Wall thickness of 300 mm is adopted. In the beginning, we had taken wall thickness as 180 mm.

Distribution of the base shear over the floors (Short direction)
    Table Ex10.5.2.5 is prepared by using Eq. (10.4.18).

Computation for *long* direction of the building

$$L_w = 8 \text{ m}, \ l_b = 4 \text{ m}, \ h_b = 800 \text{ mm}, \ \phi_{yW} = 0.000625 \text{ per m}.$$

Computation of displacement profile and ESDOF system properties (long direction)
Using Eqs. (10.4.1 to 10.4.7) Table Ex10.5.2.6 is prepared.

TABLE EX10.5.2.6
ESDOF system properties computation in Example 10.5.2 (long direction)

Floor	h_i	m_i	Δ_{iy}	Δ_i	$m_i\Delta_i$	$m_i\Delta_i^2$	$m_i\Delta_i h_i$
20	67	1427	0.7683	1.3545	1932.9	2618.2	129504
19	63.7	2141	0.7185	1.2758	2731.6	3485.0	174001
18	60.4	2141	0.6687	1.1972	2563.1	3068.4	154812
17	57.1	2141	0.6188	1.1185	2394.6	2678.3	136734
16	53.8	2141	0.5690	1.0398	2226.2	2314.7	119768
15	50.5	2141	0.5192	0.9611	2057.7	1977.7	103915
14	47.2	2141	0.4694	0.8824	1889.3	1667.1	89173
13	43.9	2141	0.4198	0.8039	1721.2	1383.7	75560
12	40.6	2141	0.3708	0.7260	1554.4	1128.6	63110
11	37.3	2141	0.3229	0.6492	1390.0	902.4	51847
10	34	2141	0.2765	0.5740	1228.9	705.4	41783
9	30.7	2141	0.2321	0.5008	1072.1	536.9	32914
8	27.4	2141	0.1902	0.4300	920.6	395.9	25225
7	24.1	2141	0.1513	0.3622	775.4	280.9	18688
6	20.8	2141	0.1158	0.2978	637.6	189.9	13262
5	17.5	2141	0.0841	0.2373	508.0	120.5	8890
4	14.2	2141	0.0568	0.1811	387.7	70.2	5505
3	10.9	2141	0.0343	0.1297	277.7	36.0	3027
2	7.6	2141	0.0171	0.0836	179.0	15.0	1360
1	4.3	2141	0.0056	0.0432	92.6	4.0	398
				Sum	**26541**	**23579**	**1249476**

From Table Ex10.5.2.6,

$$\Delta_d = \frac{\sum_{i=1}^{n} m_i\Delta_i^2}{\sum_{i=1}^{n} m_i\Delta_i} = 23579/26541 = 0.888 \text{ m}$$

$$m_e = \frac{\sum_{i=1}^{n} m_i\Delta_i}{\Delta_d} = 26541/0.888 = 29875 \text{ kN-s}^2/\text{m}$$

$$H_e = \frac{\sum_{i=1}^{n} m_i\Delta_i h_i}{\sum_{i=1}^{n} m_i\Delta_i} = 1249476/26541 = 47.078 \text{ m.}$$

Note that h_{inf} (= 48.3 m) does not change with direction.

Computation of frame ductility (long direction)

$$\theta_{yF} = \frac{0.5\varepsilon_y l_b}{h_b} = 0.00625$$

Here, l_b/h_b ratio is same for long and short directions (= 7/0.7 and 8/0.8 = 10). Hence, the Table 10.9 remains valid for long direction too.

$$\text{So, } \mu_F = 3.53.$$

Computation wall ductility and system ductility (long direction)

$$L_p = 0.022 f_y d_b + 0.054 h_{inf} = 0.022 \times 500 \times 0.02 + 0.054 \times 48.3 \text{ m} = 2.828 \text{ m.}$$

$$L_p = 0.2 L_W + 0.03 h_{inf} = 0.2 \times 8 + 0.03 \times 48.3 = 3.049 \text{ m.}$$

So, $L_p = 2.828$ m.

$$\phi_{yW} = \frac{2\varepsilon_y}{L_w} = \frac{2 \times 500/200000}{8} = 0.000625 \text{ per m.}$$

$$\mu_W = 1 + \frac{1}{L_p \phi_{yW}}\left(\theta_d - \frac{\phi_{yW} h_{inf}}{2}\right) = 1 + \frac{1}{2.828 \times 0.000625}\left(0.026 - \frac{0.000625 \times 48.3}{2}\right)$$

$$= 7.17$$

$$\mu_{ESDOF} = \frac{M_W \mu_W + M_{OTF}\mu_F}{M_W + M_{OTF}} = \frac{27.03 \times 7.17 + 45.12 \times 3.53}{27.03 + 45.12} = \mathbf{4.89.}$$

Computation of system damping (long direction)

$$T_{e,trial} = \frac{n}{6}\sqrt{\mu_{ESDOF}} = \frac{20}{6}\sqrt{4.89} = 7.37 \text{ s.}$$

$$\xi_W = \frac{95}{1.3\pi}\left(1 - \mu_W^{-0.5} - 0.1 r \mu_W\right)\left(1 + \left(T_{e,trial} + 0.85\right)^{-4}\right) = 12.9\%.$$

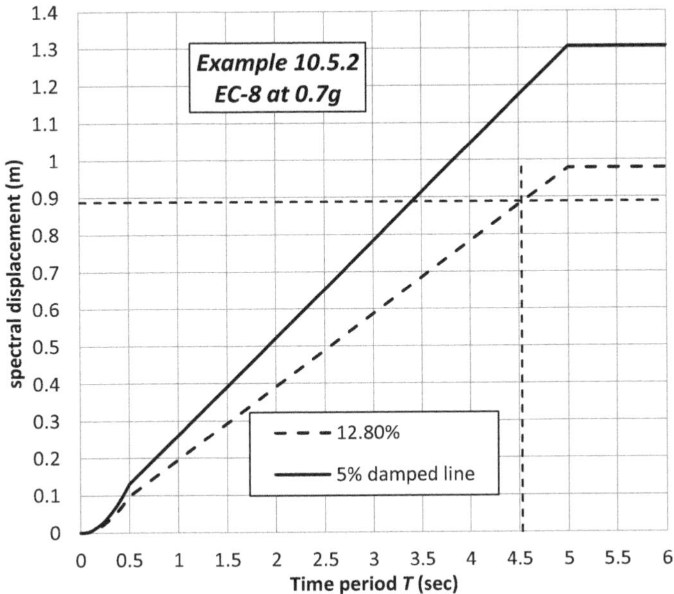

FIGURE 10.13 Displacement spectra in Example 10.5.2 (long direction) (first trial).

$$\xi_F = \frac{120}{1.3\pi}\left(1 - \mu_F^{-0.5} - 0.1r\mu_F\right)\left(1 + \left(T_{e,trial} + 0.85\right)^{-4}\right) = 12.7\%.$$

$$\xi_{ESDOF} = \frac{M_W\xi_W + M_{OTF}\xi_F}{M_W + M_{OTF}} = 12.8\%.$$

Computation of design base shear (long direction)

The displacement spectra corresponding to EC-8 design spectrum for B type soil at 0.7g level is shown in Figure 10.13.

From Figure 10.13, $T_e = 4.51$ s. This differs from $T_{e,trial}$ (7.37 s). So, put 4.51 sec as $T_{e,trial}$ in equations of ξ_W and ξ_F.

New values of parameters with this time period is given below.

$$\xi_W = 12.99\%$$

$$\xi_F = 16.3\%$$

$$\xi_{ESDOF} = 15.1\%$$

The new displacement spectrum is shown in Figure 10.14, from which, $T_e = 4.56$ sec.

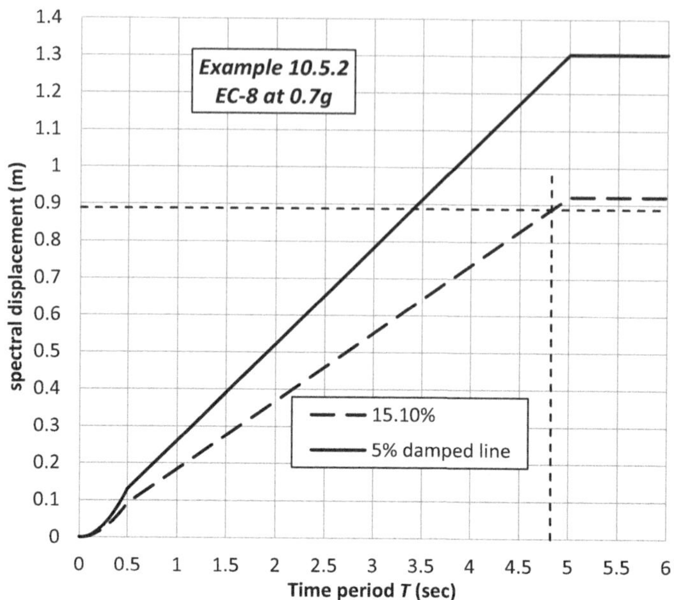

FIGURE 10.14 Displacement spectra updated in Example 10.5.2 (long direction) (final).

This small difference between the two time periods (4.51 and 4.56 s) is negligible. Hence no more iteration is done.

$$k_e = 4\pi^2 \frac{m_e}{T_e^2} = 4\pi^2 \frac{29875}{4.56^2} = 56720 \text{ kN/m.}$$

$$V_b = k_e \Delta_d = 56720 \times 0.888 = 50367 \text{ kN.}$$

Check for wall thickness (Long direction)

Here, $V_W = (1 - 0.27) \times V_b = 36768$ kN. $N_W = 6$ in any one direction of plan. For M30 concrete, $\tau_{c,max}$ is 3.5 MPa (IS 456).

$$\tau = \frac{V_W/N_W}{0.8 L_W t_W} = \frac{36768 \times 1000/6}{0.8 \times 8000 \times 180} = 5.31 \text{ MPa} > 3.5 \text{ MPa; hence, } not \ safe.$$

Try wall thickness 300 mm.

$$\tau = \frac{36768 \times 1000/6}{0.8 \times 8000 \times 300} = 3.19 \text{ MPa} < 3.5 \text{ MPa; hence } safe.$$

Wall thickness of 300 mm is adopted in the long direction. In the beginning, we had taken wall thickness as 180 mm.

TABLE EX10.5.2.7
Distribution of base shear over floors in Example 10.5.2 (Long direction)

Floor	h_i (m)	m_i (kN-s^2/m)	$m_i h_i$	V_b (kN)	F_i (kN))
20	67	1427	95609	50367	3257
19	63.7	2141	136382		4645
18	60.4	2141	129316		4405
17	57.1	2141	122251		4164
16	53.8	2141	115186		3923
15	50.5	2141	108121		3683
14	47.2	2141	101055		3442
13	43.9	2141	93990		3201
12	40.6	2141	86925		2961
11	37.3	2141	79859		2720
10	34	2141	72794		2479
9	30.7	2141	65729		2239
8	27.4	2141	58663		1998
7	24.1	2141	51598		1758
6	20.8	2141	44533		1517
5	17.5	2141	37468		1276
4	14.2	2141	30402		1036
3	10.9	2141	23337		795
2	7.6	2141	16272		554
1	4.3	2141	9206		314
	Sum		**1478695**		**50367**

Distribution of the base shear over the floors (long direction)
Table Ex10.5.2.7 is prepared by using Eq. (10.4.18).

10.6   RELATIONSHIP BETWEEN h_{inf} AND H

The inflection height of wall does not depend on wall size (length and thickness). It depends on height of building, interstorey height and, may be slightly on floor masses. So, it should be possible to find an empirical relationship between total height of building and inflection height. Computer program was written in C++ and the variation of inflection height was studied by varying heights and other parameters. By analysing the large number of building data, an empirical relationship has been established as given in Eq. (10.6.1).

$$h_{inf} = \left(1 - v_f\right) H \tag{10.6.1}$$

where v_f is the fraction of shear taken by frame (in ratio) and H is total height of the building.

TABLE 10.6.1
Verification of expression in (10.6.1)

Sl	Example	% Frame shear	Height of building H (m)	h_{inf} (m) Eq. (10.6.1)	Actual
1	Ex8.4.1	35	35	22.75	21.6
2	Ex8.4.2	30	65	45.5	45.0
3	Ex8.4.3	32	50.6	34.4	34.1
4	Ex10.5.1	30	52	36.4	35.7
5	Ex10.5.2	27	67	48.91	48.3

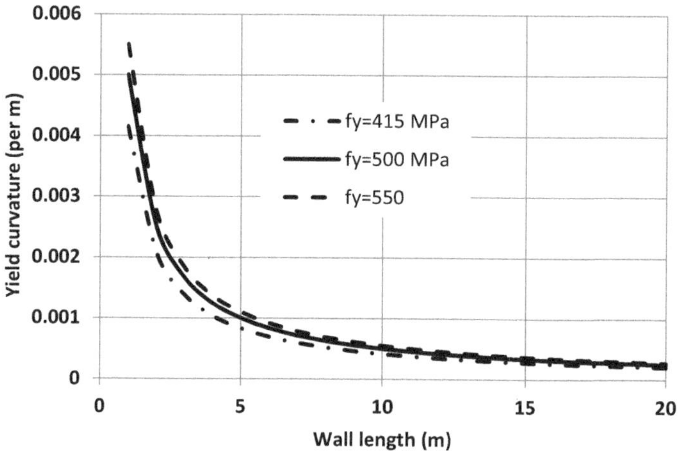

FIGURE 10.15 Variation of yield curvature of wall with length of the wall.

Verification of the empirical formula Eq. (10.6.1)
The verification is done here by utilizing the results obtained in various dual system buildings in Chapter 9 and 10. Table 10.6.1 is prepared by utilizing the data. It is observed that the expression (10.6.1) works well.

10.7 EFFECT OF LENGTH OF THE WALL

First, the yield curvature of wall is related to its length. Using Eq. (10.4.4), Figure 10.15 is obtained, which shows the variation of yield curvature (ϕ_{yW}) with wall length and yield strength of steel. The figure shows that yield curvature value is high at small wall length and yield curvature value is low at large wall length. From Sullivan et al. (2006), the ultimate curvature of wall is given by Eq. (10.7.1).

$$\phi_{uW} = \frac{0.072}{L_W} \qquad (10.7.1)$$

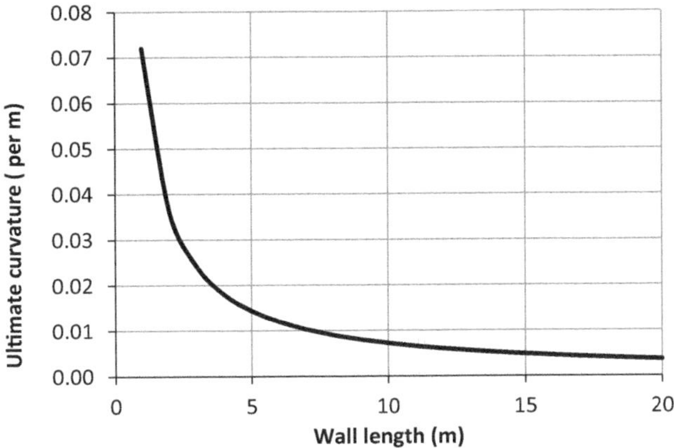

FIGURE 10.16 Variation of ultimate curvature of wall with length of wall.

Ultimate curvature is independent of grade of steel. The variation of ultimate curvature of wall with its length is shown in Figure 10.16. The pattern of variation is similar to yield curvature. The curvature ductility ($\mu_{\phi W}$) is given by Eq. (10.7.2). Eq. (10.7.2) shows that the curvature ductility of wall depends only on the grade of steel.

$$\mu_{\phi W} = \frac{\phi_{uW}}{\phi_{yW}} = \frac{0.072/L_W}{L_W/2\varepsilon_y} = \frac{0.036}{\varepsilon_y} \tag{10.7.2}$$

$$\begin{cases} \mu_{\phi W} = 28.8, & \text{for } f_y = 250 \text{ MPa} \\ \mu_{\phi W} = 17.3, & \text{for } f_y = 415 \text{ MPa} \\ \mu_{\phi W} = 14.4, & \text{for } f_y = 500 \text{ MPa} \\ \mu_{\phi W} = 13.1, & \text{for } f_y = 550 \text{ MPa} \end{cases} \tag{10.7.3}$$

Eqs. (10.7.2) and (10.7.3) show that maximum curvature ductility of wall depends only on the grade of steel.

For studying the variation of parameters like $\mu_F, \mu_W, \mu_{sys} \xi_F, \xi_W, \xi_{sys}, \Delta_d, m_e, k_e,$ V_b, a 10-storey frame-wall building of height 36 m is considered. The wall length is varied from 3 to 20 m. The percentage shear carried by frame is taken as 35%. Beam length is taken as 5 m and beam depth is 0.9 m. EC-8 spectra for soil type at 0.45g level was considered. The floor mass is taken as 800 kN-s²/m and the roof mass is taken as 700 kN-s²/m.

From analysis of the data, Figure 10.17 is drawn. It is found that the wall ductility always increases with wall length. The frame ductility is more or less constant because no variation in frame parameters was involved in the study. The system ductility also increases with wall length, but the increase is much lower than wall

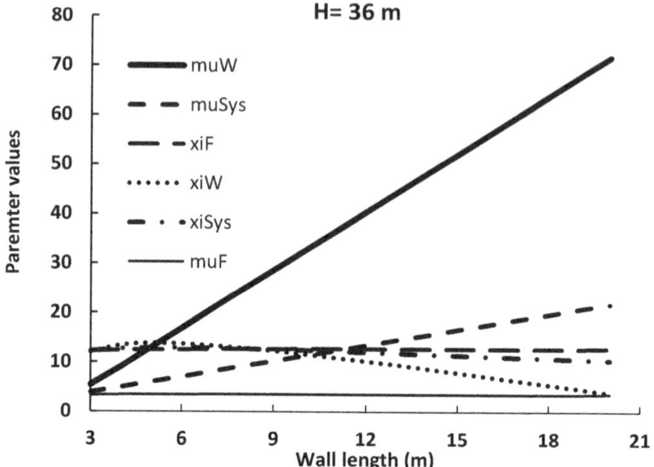

FIGURE 10.17   Effect of wall length on various ESDOF system parameters.

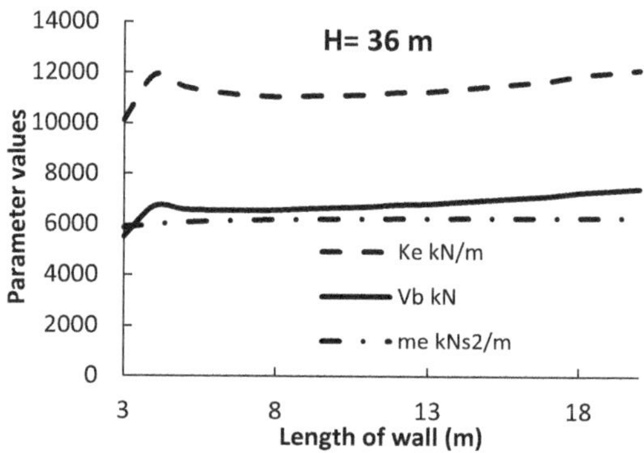

FIGURE 10.18   Effect of wall length on k_e, m_e and V_b.

ductility. The damping of wall reduces with wall length. The system damping does not vary much with wall length. The design displacement (not shown in the figure) was found to vary from 0.54 to 0.61 m.

Figure 10.18 shows that the variation of V_b, k_e and m_e with wall length is negligible.

10.8   EFFECT OF HEIGHT OF THE BUILDING

Example 10.8.1 *To study the effect of height of buildings on various design parameters, we vary the building height keeping wall length, beam size and column size constant. This is a bit hypothetical, because the wall length will increase with height of the building. We actually increase the number of storeys of the building. To judge approximately the effect of height of building, we consider storey height as 3.4 m, ground storey height as 4.4 m, floor mass as 800 kN-s²/m, roof mass as 700 kN-s²/*

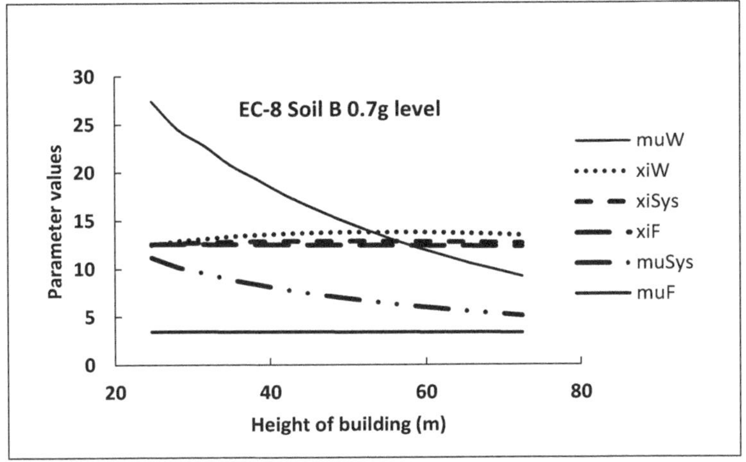

FIGURE 10.19 Variation of design parameters with building height.

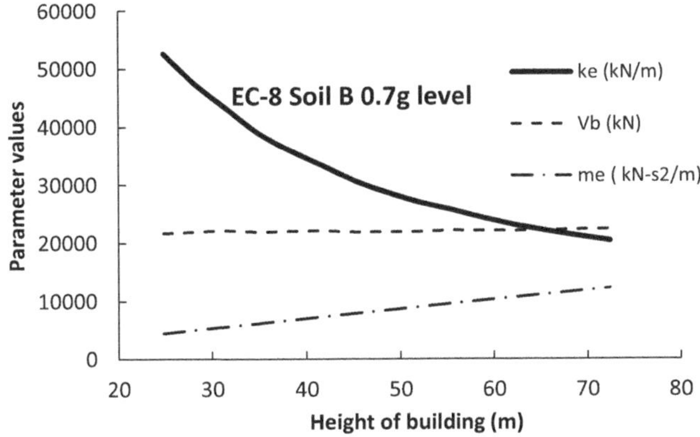

FIGURE 10.20 Variation of k_e, V_b and m_e with building height.

m, length of wall as 8 m, length of beam as 5m, depth of beam as 0.9 m, diameter of bar as 20 mm, yield strength of rebar as 500 MPa, post yield stiffness ratio as 01 and frame carries 30% of base shear. As the displacement demand increases with height, we consider EC-8 spectrum for soil type B at seismicity level 0.8g. We calculate all system properties and plot them.

Figure 10.19 shows that the wall ductility falls with building height when wall length is kept constant. The frame ductility is almost invariant. The system ductility decreases with building height. The damping in wall, frame and system are almost identical in magnitude and their variation with building height is negligible.

Figure 10.20 shows that the effective stiffness falls with building height when wall length is kept constant. This is natural as taller the building is, more flexible it becomes. The design base shear is almost invariant with height. The effective mass

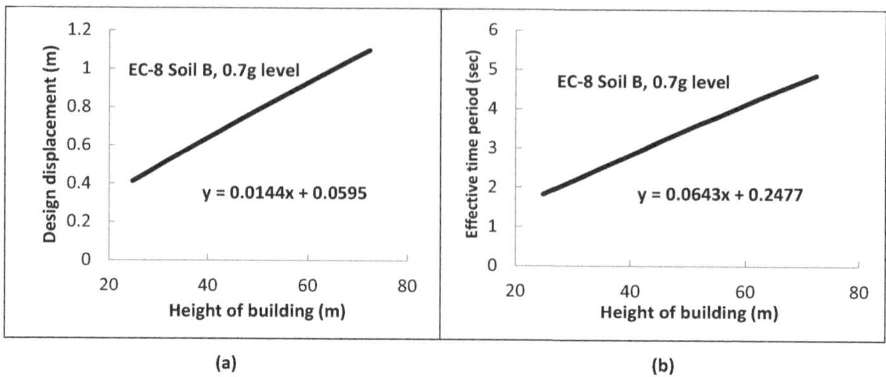

FIGURE 10.21　Variation with building height: (a) design displacement and (b) effective time period.

slightly increases with height of the building because the total mass increases with increase of building height (number of storeys).

It is interesting to see that the design displacement and effective time period increases linearly with height of the building (Figure 10.21), and it possible to get equations of such straight lines.

10.9　EFFECT OF VARIATION OF PERCENTAGE BASE SHEAR CARRIED BY FRAME

The common practice is to allow 25–40% of base shear to be carried by the frame, and the remaining part of base shear is carried by the wall. For studying the effect of such variation of base shear carried by frame, we take up an example here.

Example 10.9.1 *Take a 15-storey frame-wall building. Wall length is 10 m, beam is 450 mm × 900 mm in size and 6 m in length, column size is constant, diameter of rebar is 25 mm, yield strength of rebar is 500 MPa, storey height is 3.3 m, ground storey height is 4.4 m, floor mass as 800 kN-s²/m, roof mass as 700 kN-s²/m, post yield stiffness ratio is 01, design drift 2%. EC-8 spectrum for soil type B at seismicity level 0.6g is taken. Vary the percentage frame shear from 25% to 45% @ 2% increment. Calculate all system properties and plot them.*

Figure 10.22 shows the variation of ductility and damping with percent frame shear. It is found that the wall ductility increases with percent frame shear. This is due to the reason that, as frame shear increases the wall shear decreases. This raises the height of inflection, which is a parameter in wall ductility. Figure 10.22 shows that the frame ductility and system ductility are insensitive to variation of frame shear.

Figure 10.23 shows the variation of effective stiffness, effective mass and base shear percent frame shear. It is found that the effective stiffness slightly decreases

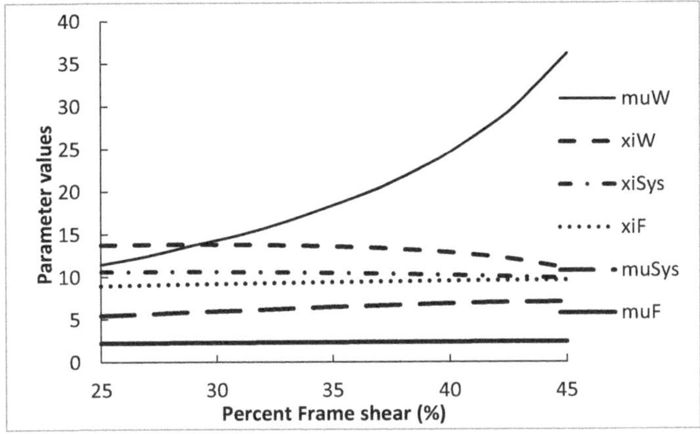

FIGURE 10.22   Variation of ductility and damping with frame shear.

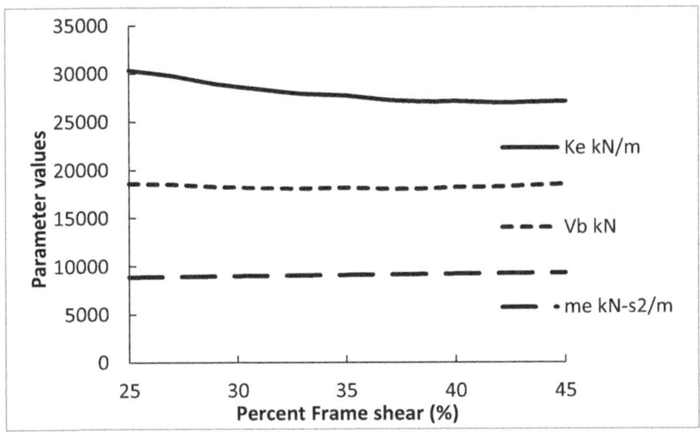

FIGURE 10.23   Variation of effective stiffness and mass, and base shear with frame shear.

with percent frame shear. The effective mass and base shear are insensitive to the variation of percent frame shear. Figure 10.24(a) shows that effective time period slightly increases with percent frame shear, while Figure 10.24(b) shows that the design displacement increases with percent frame shear.

10.10 CRITICAL CASES

In some cases, the given design drift may be less than the plastic rotation corresponding to the target PL. In such cases, the Eq. (10.2.4) does not work. However, the case is not an impractical one altogether. We discuss this aspect below.

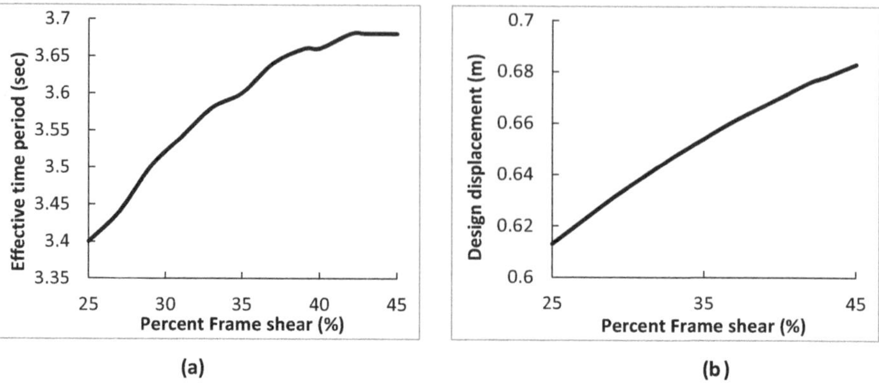

FIGURE 10.24   Variation with percent frame shear of: (a) T_e and (b) Δ_d.

Let us consider Eqs. (10.2.2) and (10.2.3).

$$\theta_{yW} = \frac{\phi_{yW} h_{inf}}{2}$$

$$\phi_{yW} = \frac{2\varepsilon_y}{L_W}$$

Combining these two equations, we get Eq. (10.9.1) for yield rotation of frame.

$$\theta_{yW} = \frac{\varepsilon_y h_{inf}}{L_W} \tag{10.10.1}$$

This equation is represented in Figure 10.25 with $\dfrac{h_{inf}}{L_W}$ on horizontal axis and θ_{yW} on vertical axis.

On the same figure, we show the limits of average plastic rotations for IO, LS and CP as horizontal lines. If we take f_y as 415 MPa and h_{inf}/L_W ratio as 8 (point A), then AC is the amount of yield rotation. If we design for LS performance level, the plastic rotation allowable is AB. Total rotation allowable is AB + AC. In the figure AB + AC = 0.00975 +0.0166 = 0.0264 radian (= AD). So, for this combination of h_{inf}/L_W ratio and steel, a system can be designed.

Now, take up the case when $\theta_d - \theta_{pW}$ is negative. Here, θ_d is less than θ_{pW}. It indicates we are below the horizontal line corresponding to the PL considered. P is

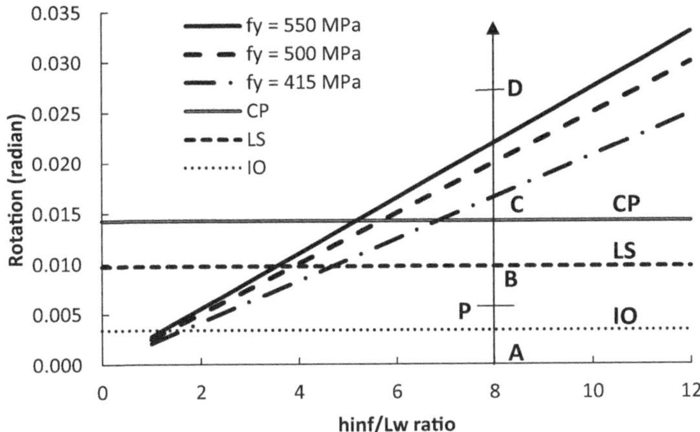

FIGURE 10.25 Interpretation of yield rotation and plastic rotation.

such a point for LS PL and Fe415 steel. In such a case the shear wall will remain elastic. If h_{inf}/L_W is lower (say 4, or less), the scenario will be different.

10.11 CLOSURE

The UPBD method for frame–shear wall buildings has been presented in this chapter. The theoretical background of the method was discussed. The interpretation of the basic equation was done. The step-wise procedure for design was described. Design examples have been given to discuss various aspects. An empirical expression has been given for inflection height of the wall. The variation of different parameters involved, with length of wall and, height of building, was discussed.

The advantage of UPBD method is that it can be used to design the building for both target drift and target PL. It also gives the wall size and beam size at the beginning, so that iteration for size is not required. The method has been found to give consistent results.

Readers are to note that only computation of floor forces arising out of seismic action have been computed here. The implementation in the computer model requires more work. The shear wall can be modelled as wide column or layered shell element. The analysis and design are to be done with the expected strength. The nonlinear analyses are to be carried out with reduced stiffness as per actual strength. This will need patience and time. Capacity design is to be carried out so that columns do not yield. Adequate ground motions (generally spectrum compatible ground motions) are to be used for dynamic analysis. The achieved drift and performance levels are to be compared with those of the target values. The achieved performance level should never exceed the target value. Achieved drift may lie within ± 20% of target value.

10.12 EXERCISES

Q 10.12.1 Give a brief outline of UPBD method for dual system.

Q 10.12.2 Derive an expression of wall length in dual system building.

Q 10.12.3 How do we find the thickness of shear wall?

Q 10.12.4 Write down the steps in design of dual system buildings using UPBD method.

Q 10.12.5 Discuss the (a) effect of wall length on design parameters, (b) effect of building height on design parameters and (c) effect of percentage base shear carried by frame on design parameters.

Q 10.12.6 A 16-storey dual system building plan is 20 m ×15 m in size. The design drift is 2.7% and target performance level is LS. The floor thickness is 120 mm, superimposed DL in floor is 1 kN/m^2 in floor. Exterior infill walls are 250 mm thick and interior wall thickness are 125 mm thick. Column size is 600 mm × 600 m all through. Find the lateral seismic design forces. Take as f_y as 550 MPa for rebar, and concrete strength as 30 MPa. Use EC-8 spectrum for soil type B at 0.6g level. Interstorey height is 3.3 m and ground storey height is 4.3 m. Consider 20% opening in infill walls. Assume that frame carries 33% of base shear. Diameter of longitudinal bar in wall is 20 mm.

Q 10.12.7 A 12-storey dual system building plan is 20 m ×15 m in size. The design drift is 1.5% and target performance level is IO. The floor thickness is 120 mm, superimposed DL in floor is 1 kN/m^2 in floor. Exterior infill walls are 250 mm thick and interior wall thickness are 125 mm thick. Column size is 600 mm × 600 m all through. Find lateral seismic design forces. Take as f_y as 550 MPa for rebar, and concrete strength as 30 MPa. Use EC-8 spectrum for soil type B at 0.45g level. Interstorey height is 3.3 m and ground storey height is 4.2 m. Consider 20% opening in infill walls. Assume that frame carries 23% of the base shear. The diameter of the longitudinal bar in wall is 20 mm.

Q 10.12.8 A 13-storey dual system building plan is 25 m × 16 m in size. The design drift is 3% and target performance level is CP. The floor thickness is 120 mm, superimposed DL in floor is 1 kN/m^2 in floor. Exterior infill walls are 250 mm thick and interior wall thickness are 125 mm thick. Column size is 600 mm × 600 m all through. Find the lateral seismic design forces. Take f_y as 550 MPa for rebar, and concrete strength as 30 MPa. Use EC-8 spectrum for soil type B at 0.55g level. Interstorey height is 3.3 m and ground storey height is 4.2 m. Consider 20% opening in the infill walls. Assume that frame carries 25% of the base shear. The diameter of the longitudinal bar in wall is 20 mm.

Q 10.12.9 A 19-storey dual system building plan is 35 m ×30 m in size. The design drift is 2.7% and target performance level is LS. The seismic load in floor is 13 kN/m^2 in floor and 10 kN/m^2 at roof. Take f_y as 550 MPa for rebar, and concrete strength as 30 MPa. Use EC-8 spectrum for soil type B at 0.7g level. Interstorey height is 3.5 m and ground storey height is 4.3 m. Assume that the frame carries 25% of the base shear. Find the lateral seismic design forces along both the directions of the building. Beam length and depth are to be decided. The rebar diameter of vertical steel in wall is 20 mm.

FURTHER READINGS

Choudhury, S. (2007) Performance-Based Seismic Design of Hospitals, Ph.D. Thesis, Indian Institute of Rooorkee, India.

Choudhury, S. and Singh, S.M. (2013) A Unified Approach to Performance-Based Design of RC Frame Buildings, *Journal of Institution of Engineers (I)-Series-A*, May, 94(2), pp. 73–82, DOI 10.1007/s40030-013-0037-8

Das, T. K. and Choudhury, S. (2021) Comparison of the Cost of Reinforced Concrete Frame Buildings Designed Using IS Code and Unified Performance-Based Design Method, *Book Proceedings of Springer on International Conference on Advances in Structural Mechanics and Applications* (ASMA-2021), 6–8 October 2021, conducted by the Department of Civil Engineering, NIT Silchar, Paper ID 136.

Debnath, P.P. and Choudhury, S. (2015) Effect of Unreinforced Masonry Infill on Floor Amplification and Other Parameters in Frame-Wall Buildings, *Journal of Mechanics Based Design of Structures and Machines*, 43, 450–465, DOI: 10.1080/15397734.2015.1025961

Debnath, P.P. and Choudhury, S. (2017) Nonlinear Analysis of Shear Wall in Unified Performance-Based Seismic Design of Buildings, *Asian Journal of Civil Engineering*, 18(4), 633–642.

Mayengbam, S.S. and Choudhury, S. (2014) Determination of Column Size for Displacement-Based Design of Reinforced Concrete Frame Buildings, *Journal of Earthquake Engineering & Structural Dynamics*, July, 43(8), pp. 1149–1172. Article first published online: 21 Nov 2013 | DOI: 10.1002/eqe.2391

Mibang, D. (2022) Performance of Reinforced Concrete Frame–Shear Wall Building Designed Using Unified Performance-Based Design and Assessment of Damage Index, Ph.D. Thesis, Department of Civil Engineering, National Institute of Technology Silchar, India.

Mibang, D. and Choudhury, S. (2019a) Performance-Based Design of Dual System, International Conference on Recent Development in Sustainable Infrastructure (Materials and Management) (ICRDSI-2019), Kalinga Institute of Technology, Bhubaneswar, 11–13 July 2019, Paper ID 132.

Mibang, D. and Choudhury, S. (2019b) Performance of Dual System Designed Using UPBD Method, 2nd International Conference on Recent Advancements in Interdisciplinary Research (ICRAIR-2019), Asian Institute of Technology Conference Centre, Thailand, 1–2 June 2019, Paper ID-69, pp. 327–333.

Mibang, D. and Choudhury, S. (2020) Nonlinear Performance of Frame–Shear Wall Building, *Challenges of Resilient and Sustainable Infrastructure Development in Emerging Economies (CRSIDE 2020)*, Organized by ASCE India Section in Association with IIEST, Shibpur and IEI, 2–4 March p. 352.

Priestley, M.J.N. (2003) Myths and Fallacies in Earthquake Engineering, Revisited, *European School for Advanced Studies in Reduction of Seismic Risk*, 9th Mallet-Milne Lecture.

Soumen, N.S. and Choudhury, S. (2016) Comparative Study on Layered-Shear Wall and Wide-Column Shear Wall Modeling, Demystifying Global Trends in Technology & Management, Organised by Cyber Times and Indira Gandhi Delhi Technical University for Women, 29–30 April 2016, pp. 129–134.

Sullivan, T.J., Priestley, M.J.N. and Calvi, G.M. (2005) Development of an Innovative Seismic Design Procedure for Frame–Wall Structures, *Journal of Earthquake Engineering,* 9(Special Issue 2), 279–307.

Sullivan, T.J., Priestley, M.J.N. and Calvi, G.M. (2006a) Direct Displacement-Based Design of Frame–Wall Structures, *Journal of Earthquake Engineering,* 10(Special Issue 1), 91–124.

Sullivan, T.J., Priestley, M.J.N. and Calvi, G.M. (2006b) Seismic Design of Frame-Wall Structures, Research Report No. ROSE-2006/02.

11 UPBD for Steel Frame Buildings

11.1 INTRODUCTION

We are now familiar with the UPBD method of design. We have applied this method to the design of RC frame buildings in Chapter 9 and to RC frame–wall buildings in Chapter 10. In this chapter, we apply this method to steel frame buildings. The additional intricacy in applying the UPBD method to steel buildings arises out of the fact that the plastic rotation for a performance level of the system is not available as a numerical value. It depends on the yield rotation, which in turn depends on many parameters, including geometry of the section. Unless a section is chosen, the yield rotation cannot be obtained. And the chosen section may not satisfy the requirement, thus needing many iterations.

11.2 DEVELOPMENT OF UPBD METHOD FOR STEEL BUILDINGS

A multiple degree of freedom (MDOF) system can be converted to an equivalent single degree of freedom (ESDOF) system by utilizing available dynamical equations. Figure 11.1 explains this. The properties of an ESDOF system are given below.

Effective mass (m_e)
Equivalent height (H_e)
Equivalent damping (ξ_e)
Design displacement (Δ_d).

Let the target design drift be θ_d and target design plastic rotation corresponding to the desired performance level be θ_p. In the UPBD method, we wish to design for these two target design criteria.

Figure 11.2(a) shows an MDOF system that can be converted to an ESDOF system as shown in Figure 11.2.1(b). The total rotation that the ESDOF system undergoes is equal to the target angular design drift (θ_d). The total angular rotation is the sum of elastic rotation (θ_e) and plastic rotation (θ_p). So, we can write Equation (11.2.1) describing this relationship.

DOI: 10.1201/9781003441090-11

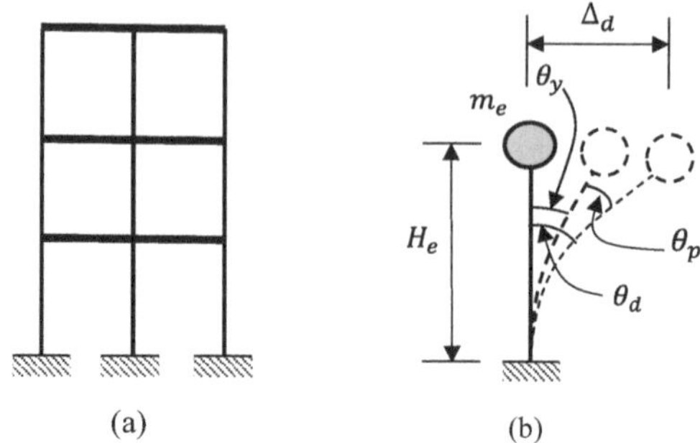

(a)　(b)

FIGURE 11.1　(a) MDOF system (b) ESDOF system.

$$\theta_d = \theta_e + \theta_p \tag{11.2.1}$$

The maximum elastic rotation of ESDOF system (θ_e) is equal to the yield rotation of the 3D steel frame (θ_{yF}). Priestley et al. (2007) gave the expression for yield rotation of structural steel frame as Eq. (11.2.2).

$$\theta_{yF} = \frac{0.65\varepsilon_y l_b}{h_b} \tag{11.2.2}$$

where ε_y = yield strain of rebar $(\varepsilon_y = f_y / E_s)$

f_y = yield strength of steel and E_s is modulus of elasticity of steel

l_b = length of the beam in the direction of seismic load under consideration
h_b = depth of the beam in the direction of seismic load under consideration.

In a steel frame building, we apply capacity design in the form of weak-beam strong-column concept. This implies that the beams will attain a plastic state while columns will remain elastic. So, the entire plastic rotation of the system arises out of the beam alone. In Eq. (11.2.1), we can write $\theta_p = \theta_{pb}$, where θ_{pb} is the plastic rotation in the beams. Now we can rewrite Eq. (11.2.1) as Eq. (11.2.3).

$$\theta_d = \theta_{yF} + \theta_{pb} \tag{11.2.3}$$

Combining Eq. (11.2.3) and Eq. (11.2.2), we get Eq. (11.2.4).

$$h_b = \frac{0.65\varepsilon_y l_b}{\theta_d - \theta_{pb}} \tag{11.2.4}$$

Eq. (11.2.4) gives a beam depth that satisfies both drift and PL in terms of plastic rotation. This beam depth is obtained at the start of the design. Hence, iteration is not necessary for determining beam depth. The designer is required to match from the available standard sections of structural steel the one that matches nearly to this depth.

11.3 PLASTIC ROTATION OF STEEL BEAMS

Unlike that in RC beams, the plastic rotations for steel beams are not available directly in tabular numerical values for any PL. It is available in terms of chord rotation of beam (θ_y) (see Figure 11.2). Such values are available in FEMA-356 (2000) as well as in ASCE-SEI-41-13 (2013), as shown in Table 11.1. The ASCE-SEI-41-17 (2017) seems to have some typological error in this provision.

Table 11.1 gives the plastic rotations corresponding to IO, LS and CP in terms of chord yield rotation (θ_y). The yield rotation of chord (θ_y) in Table 11.1 is given by Eq. (11.3.1) and is explained in Figure 11.2. The chord yield rotation is a function of beam length, beam moment of inertia about the transverse axis and material properties. Now it is clear to the readers that until a section is chosen, the value of θ_y and hence plastic rotations at different PLs cannot be found out.

$$\theta_y = \frac{Z_p f_{ye} l_b}{6EI_b} \tag{11.3.1}$$

where Z_p = plastic modulus of section of the beam

f_{ye} = expected yield strength of steel

l_b = length of beam under consideration

E = modulus of elasticity of steel

I_b = moment of inertia of beam section about the axis of bending.

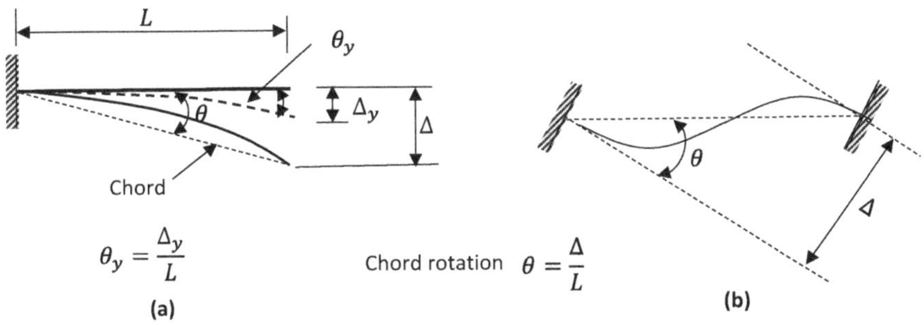

FIGURE 11.2   Chord rotation explained: (a) cantilever and (b) frame element.

TABLE 11.1
Plastic rotations for different performance levels as per FEMA-356 and ASCE-SEI-41-13

| | Plastic rotation for different performance levels | | | | | |
| | IO | | LS | | CP | |
Condition	FEMA-356	ASCE-SEI-41-13	FEMA-356	ASCE-SEI-41-13	FEMA-356	ASCE-SEI-41-13
$\dfrac{b_f}{2t_f} \le \dfrac{52}{\sqrt{f_{ye}}}$ and $\dfrac{h}{t_w} \le \dfrac{418}{\sqrt{f_{ye}}}$	θ_y	θ_y	$6\theta_y$	$9\theta_y$	$8\theta_y$	$11\theta_y$
$\dfrac{b_f}{2t_f} \ge \dfrac{65}{\sqrt{f_{ye}}}$ or $\dfrac{h}{t_w} \ge \dfrac{640}{\sqrt{f_{ye}}}$	$0.25\theta_y$	$0.25\theta_y$	$2\theta_y$	$3\theta_y$	$3\theta_y$	$4\theta_y$

11.4 METHODOLOGY

The lumped floor masses (m_i) are supposed to be known in the beginning. Shape profiles are obtained from available equations (Priestley and Calvi, 1997). The critical storey is that storey where the mass is largest or, the inter-storey height of storey is largest. Displacement profiles are now obtained from which the ESDOF system properties are computed. System ductility is obtained from yield displacement and target displacement. From system ductility system, damping is computed, which enables one to compute the base shear utilizing the displacement spectra. Beam section is selected by computing beam depth from Eq. (11.2.4). In doing so, the chord rotation angle is to be utilized. The chord rotation angle again depends on the section properties. So, selecting a beam section for a particular target design criteria requires some search in the steel table. The process has been explained with a numerical example in Section 11.5.

The design steps are given below.

Step 1: Decide target design criteria, namely, drift and PL; decide the hazard level.

Step 2: Compute the beam depth from Eq. (11.2.4) and select a beam section (see numerical example in Section 11.6).

Step 3: Compute the shape profile from Eqs. (11.5.1) and (11.5.2).

$$\text{For number of stories} \leq 4 \quad \phi_i = h_i/H \tag{11.5.1}$$

$$\text{For number of series} > 4 \quad \phi_i = \frac{4}{3}\frac{h_i}{H}\left(1 - 0.25\frac{h_i}{H}\right) \tag{11.5.2}$$

Here, ϕ_i is the shape coefficient for the i-th floor and h_i is the height of the i-th floor from the base of the building.

Step 4: Identify the critical storey. Find critical storey displacement from Eq. (11.5.3). In this step, the design drift comes into picture. In Eq. (11.5.3), h_{sc} is critical storey height, Δ_c is critical storey displacement. In Eq. (11.5.4), Δ_i is displacement of the i-th floor.

$$\Delta_c = \theta_d h_{sc} \tag{11.5.3}$$

The floor displacement is found out from Eq. (11.5.4).

$$\Delta_i = \phi_i \frac{\Delta_c}{\phi_c} \tag{11.5.4}$$

Step 5: Compute ESDOF system properties from Eq. (11.5.5). In Eq. (11.5.5), H_e is the effective height of the ESDOF system.

$$\Delta_d = \frac{\sum_{i=1}^{n} m_i \Delta_i^2}{\sum_{i=1}^{n} m_i \Delta_i} \tag{11.5.5a}$$

$$m_e = \frac{\sum_{i=1}^{n} m_i \Delta_i}{\Delta_d} \tag{11.5.5b}$$

$$H_e = \frac{\sum_{i=1}^{n} m_i \Delta_i h_i}{\sum_{i=1}^{n} m_i \Delta_i} \tag{11.5.5c}$$

Step 6: Compute system ductility and damping from Eq. (11.5.6) where, Δ_y is the yield displacement of ESDOF system, μ is ductility and ξ_e is equivalent damping.

$$\mu = \frac{\Delta_d}{\Delta_y} \tag{11.5.6a}$$

$$\Delta_y = \theta_{yF} H_e \tag{11.5.6b}$$

$$\xi_e = 2 + 120\left(\frac{1 - \dfrac{1}{\sqrt{\mu}}}{\pi}\right)\% \tag{11.5.6c}$$

Step 7: Find the effective time period (T_e) from displacement spectra corresponding to the design spectrum. Find base shear from Eq. (11.5.7), where k_e is effective stiffness of ESDOF system and V_b is design base shear.

$$k_e = 4\pi^2 \frac{m_e}{T_e^2} \tag{11.5.7a}$$

$$V_b = k_e \Delta_d \tag{11.5.7b}$$

Step 8: Distribute the base shear by Eq. (11.5.8).

$$F_i = V_b \frac{m_i \Delta_i}{\sum_{i=1}^{n} m_i \Delta_i} \quad \text{(when the building height is up to 10 stories)} \tag{11.5.8a}$$

$$F_i = 0.9V_b \frac{m_i \Delta_i}{\sum_{i=1}^{n} m_i \Delta_i} \quad \text{(when the building height is more than 10 stories)}$$

$$\tag{11.5.8b}$$

Put the remaining 10% base shear at roof level (to take care of higher mode effects).

Step 9: Load combination

The computation of floor level lateral force in step 8 is carried out in both directions of the building. Denote F_x for F_i values the in x-direction and, F_y for F_i values in y-direction. The load combinations to be used are as below:

DL + LL
DL + LL ± F_x
DL + LL ± F_y

Step 10: Design the building using expected strength of steel ($1.25 f_y$).

Step 11: Evaluate the building performance through nonlinear analyses.
Step 12: If design targets are not satisfied, go to step 2 with a new beam section.

11.5 NUMERICAL EXAMPLES ON THE SELECTION OF BEAM SECTION

We are taking here the structural steel section conforming to Indian Standard. The principle is general and is not restricted to code of any country.

Example 11.6.1 *Arrive at a beam section to accommodate a design drift of 2% and PL LS. The beam length is 3 m.*

Solution: Prepare an Excel sheet with all available steel sections for beam (I-sections) in increasing order of depth. In Table 11.2, only one part of such Excel sheet is shown, where solution was obtained at the condition $(h \sim h_b)$. Table 11.1 is used to get value of θ_{pb} (FEMA-356). For LS PL, $\theta_{pb} = 2\theta_y$. When h and h_b are found to be approximately equal, one can stop and select that beam section.

From Table 11.2, beam section ISMB400 is adopted (out of some more viable sections).

11.6 CLOSURE

The UPBD method for steel frame buildings is more round about than that of RC frame buildings. Search is required in the available beam sections. It is also possible that in some cases solution may not come. So, a personal judgement is also warranted.

A shape profile for steel frame building was proposed by Kotapaty as given by Eq. (11.7.1).

$$\phi_i = 1.2\frac{h_i}{H} - 0.2\left(\frac{h_i}{H}\right)^2 \tag{11.7.1}$$

where h_i is height of i-th storey from base of building; H is overall height of the building.

FURTHER READINGS

ASCE-SEI-7 Minimum Design Loads and Associated criteria for Buildings and Other Structures.

ASCE-SEI-41-17 (2017) Seismic Evaluation and Retrofit of Existing Buildings.

Azeel, P.K. (2022) Unified Performance-Based Design of Steel Frame Buildings with Masonry Infill Walls, M.Tech. Thesis, Department of Civil Engineering, NIT Silchar, India.

FEMA-273 (1996) NEHRP Guidelines for the Seismic Rehabilitation of Buildings, Seismic Safety Council, Washington DC.

FEMA-356 (2000) Prestandard and Commentary for the Seismic Rehabilitation of Buildings, US Federal Emergency Management Agency.

FEMA-368 (2001) NEHRP Recommended Provisions for Seismic Regulations for New Buildings and other Structures, US Federal Emergency Management Agency.

TABLE 11.2
Search for appropriate depth of beam ($l_b = 3$ m, PL = LS, $\theta_d = 2\%$)

Section	h (mm)	Z_P (mm^3)	I_b (mm^4)	b_f (mm)	t_f (mm)	$\dfrac{b_f}{2t_f}$	$\dfrac{b_f}{2t_f} \leq \dfrac{52}{\sqrt{f_{ye}}}$	$\dfrac{b_f}{2t_f} \leq \dfrac{65}{\sqrt{f_{ye}}}$	θ_y (rad) Eq. (11.3.1)	$\theta_{pb} = 2\theta_y$ (rad)	h_b (mm) Eq. (11.2.4)	$h \sim h_b$?
ISWB-350	350	995484	155217000	200	11	8.77	No	Yes	0.006915	0.013829	545	No
ISHB-350	350	1213551	191597000	250	12	10.77	No	Yes	0.006829	0.013657	530	No
ISHB-350	350	1268676	198028000	250	12	10.77	No	Yes	0.006907	0.013814	544	No
ISMB-400	400	1176163	204584000	140	16	4.37	No	Yes	0.006198	0.012396	442	Yes
ISWB-400	400	1320176	234267000	200	13	7.69	No	Yes	0.006076	0.012151	429	Yes
ISHB-400	400	1566357	280835000	250	13	9.84	No	Yes	0.006013	0.012026	422	Yes
ISHB-400	400	1626357	288235000	250	13	9.84	No	Yes	0.006083	0.012167	429	Yes
ISMB-450	450	1553347	303908000	150	17	4.31	No	Yes	0.005511	0.011021	375	No

FEMA-450 (2004) NEHRP Recommended Provisions for Seismic Regulations for New Buildings and Other Structures, US Federal Emergency Management Agency.

Kotapatty, V.K., Unified Performance-Based Seismic Design of Steel Frame Buildings, M.Tech thesis, Department of Civil Engineering, National Institute of Technology, Silchar.

Priestley, M.J.N. and Calvi, G.M. (1997) Concepts and Procedures for Direct Displacement-based Design and Assessment", *Proc. of International Workshop on Seismic Design Methodologies for Next Generation of Codes*, Bled, Slovenia.

Priestley, M.J.N., Calvi, G.M. and Kowalsky (2007) *Displacement-Based Design*, IUSS Press, Pavia, Italy.

Seismic Evaluation and Retrofit of Existing Buildings (2013) ASCE.

12 Effect of Infill

12.1 INTRODUCTION

In this chapter, the effect of infill on buildings and modelling of the infill are discussed. Infill is a nonstructural component, as it is not designed to carry any force. In this chapter, infill refers to masonry infills. Infill walls are provided in RC buildings and steel buildings to serve the following purposes:

    (i)   It protects the dwellers from sun, wind, rain, frost, etc.
    (ii) It protects the dwellers from theft and burglary.
    (iii) It protects the dwellers from heat wave and cold wave.
    (iv) It gives privacy to the dwellers.
    (v) It allows partitioning of space.

Masonry infill consists of bricks and mortar joining the bricks in a specified pattern (bonding). Both bricks and mortar are weak in tension but good against compressive forces. Under tensile force, the brick masonry cracks, generally along the mortar joints. During shaking from earthquake, the frame sways and causes bounded infill panel to experience compression along one diagonal and tension along the other. The tension and compression along the two diagonals toggle during reversible shaking. Separation or tearing may take place along the tension diagonal. But the compression diagonal will take a large compressive force before failure. This sets a strut action along the compression diagonal. The strut action increases the stiffness of the system. In fact, a large additional stiffness is set into the building from infills. It may be mentioned that masonry is a brittle material with very limited ductility.

The normal design of the buildings considers the weight of infill but ignores its stiffness. This leads to an unrealistic assessment of the structural behaviour and hence, leads to improper design.

12.2 STRUT ACTION AND STRUT MODEL OF MASONRY

A typical masonry infill panel is shown in Figure 12.1. Figure 12.2 shows the same panel when subjected to frame sway. Various strut models are available for unreinforced masonry (Figure 12.3). The strut model is an equivalent strut that can offer

DOI: 10.1201/9781003441090-12

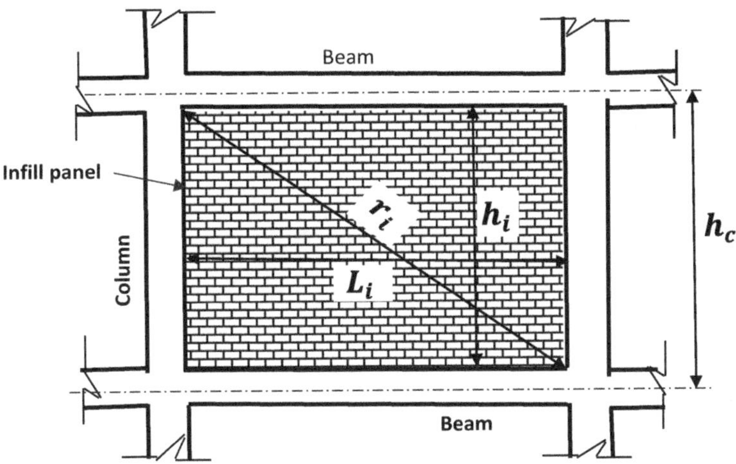

FIGURE 12.1 A typical infill panel.

same stiffness as that of real masonry panel. The thickness of the strut is same as that of the masonry. The width of the strut is given by empirical formula (Eq. 12.2.1). The modulus of elasticity of the strut material is same as that of the masonry. In the computer, strut is modelled as a brace element (axially loaded element) with thickness as that of masonry and width as given by Eq. (12.2.1). The expected strength of masonry is taken as 1.3 times the nominal strength.

FEMA-356 gives Eq. (12.2.1) for equivalent width of the strut.

$$a = 0.175\left(\lambda h_c\right)^{-0.4} r_i \tag{12.2.1}$$

where

 a = width of equivalent strut
 h_c = height of column between centerlines of beams
 r_i = diagonal length of infill panel

$$\lambda = \left[\frac{E_m t_i \sin 2\theta}{4 E_f I_c h_i}\right]^{0.25}$$

E_m = modulus of elasticity of masonry material (taken as 550 times compressive strength of masonry)

E_f = modulus of elasticity of frame material

t_i = thickness of infill panel
h_i = height of infill panel
I_c = moment of inertia of column
θ = angle of inclination of infill strut with horizontal

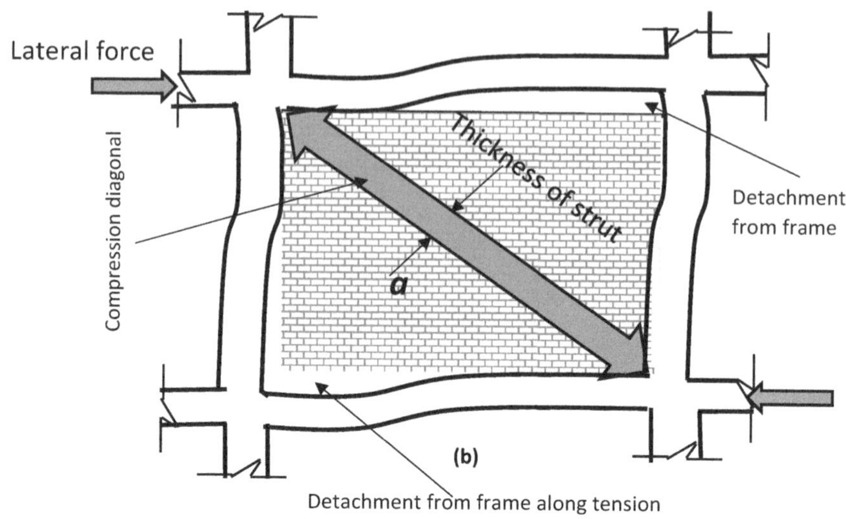

FIGURE 12.2   Masonry infill panel under frame sway.

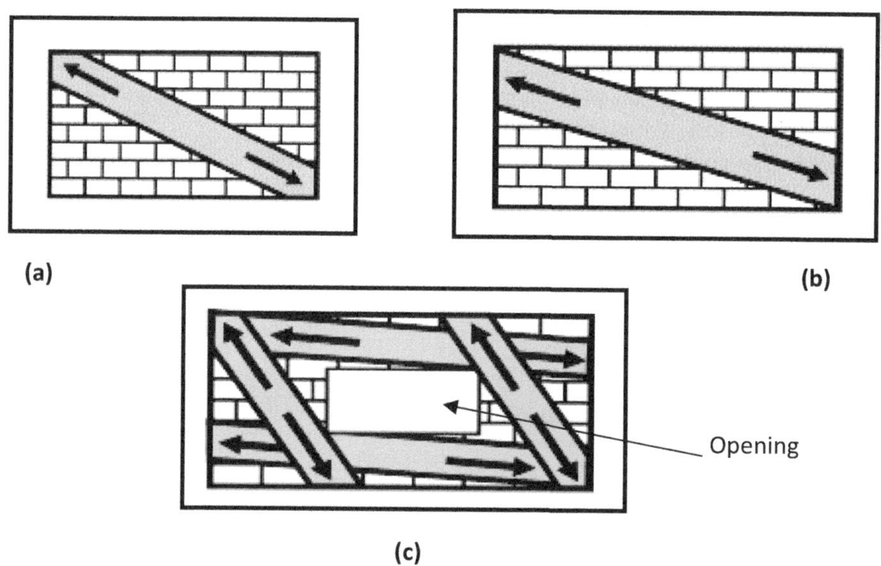

FIGURE 12.3   Infill strut modelling: (a) concentric placement, (b) eccentric placement and (c) with opening.

12.2.1   Infill Strut Modelling

The infill is modelled as an axially loaded pin-connected member. The connection may be placed in a concentric way as shown in Figure 12.3(a), in which case the forces from the infill panel acts at the corner joint. The strut may be placed eccentrically as shown in Figure 12.3(b), in which case the strut force is transferred to the columns only. In many practical cases, there shall be openings in the infill wall. The strut model in such a case is shown in Figure 12.3(c).

12.3 MASONRY STRENGTH

The strength of masonry depends on the strength of bricks and strength of the mortar that binds the bricks. The compressive strength of masonry is given by Eq. (12.3.1).

$$\sigma_m = 0.433 \left(\sigma_{brick} \right)^{0.64} \left(\sigma_{mortar} \right)^{0.36} \tag{12.3.1}$$

where
σ_m = compressive strength of masonry strength in MPa
σ_{brick} = compressive strength of brick in MPa
σ_{mortar} = compressive strength of mortar in MPa.

Various strengths of masonry are tabulated in Table 12.1.

12.3.1 LATERAL SHEAR STRENGTH OF MASONRY

The lateral strength of masonry arises out of shear strength along bed joints. The lateral strength is given by Eq. (12.3.1.1).

$$Q_{CE} = v_e \left(Lt \right) \tag{12.3.1.1}$$

In Eq. (12.3.1.1), Q_{CE} is the expected lateral strength of infill panel, v_e is the expected shear strength of masonry, L is the horizontal length of infill panel clear between columns, t is the thickness of infill wall. Q_{CE} can be used to find the axial capacity of the infill strut.

TABLE 12.1
Strength of masonry (FEMA-356)

Masonry property	Masonry condition		
	Good	Fair	Poor
Compressive strength	900 psi	600 psi	300 psi
(σ_m)	(6.2 MPa)	(4.1 MPa)	(2.07 MPa)
Flexural tensile strength	20 psi	10 psi	0
	(0.138 MPa)	(0.07 MPa)	
Shear strength with bond	27 psi	20 psi	13 psi
	(0.186 MPa)	(0.138 MPa)	(0.09 MPa)
Modulus of elasticity	550 σ_m	550 σ_m	550 σ_m

12.3.2 Force–Deformation Behaviour of Infill Strut

The force–deformation response of infill strut as per FEMA-356 is shown in Figure 12.4. The parameters used in the response (namely, *d, e, c*) and performance levels (IO, LS, CP) are shown Table 12.2. The values are applicable for *nonlinear static procedure for URM infill.*

12.3.2.1 Axial Force on Infill Strut

The shear failure of masonry takes place along the bed joint planes. The bed joint shear strength is given by Eq. (12.3.1.1). This can be utilized to calculate the axial force in the infill strut. Consider Figure 12.5. The axial force is given by Eq. (12.3.2.2).

$$F = \frac{Q_{CE}}{\cos \theta} \tag{12.3.2.2}$$

where F is the axial force in strut, θ is angle inclination of strut with horizontal.

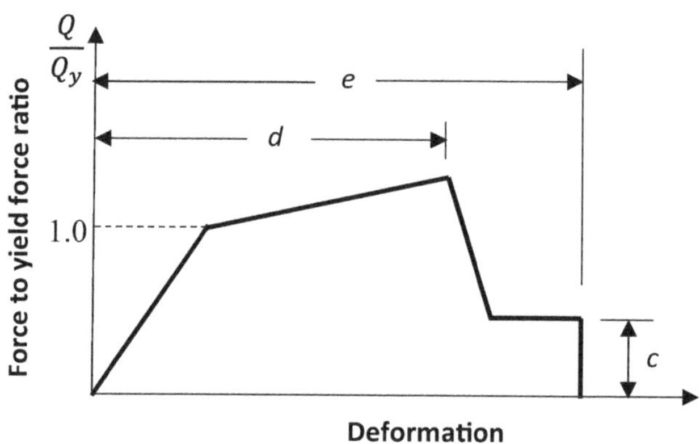

FIGURE 12.4 Force–deformation behaviour of infill strut element of URM for nonlinear static procedure.

TABLE 12.2

Force–deformation parameters and PLs of infill strut element for nonlinear static procedure (FEMA-356)

	Force-deformation parameters			Acceptance criteria				
					Performance levels			
					Primary		Secondary	
Item	c	d	e	IO	LS	CP	LS	CP
Bed-joint sliding	0.6%	0.4%	0.8%	0.1%	0.3%	0.4%	0.6%	0.8%

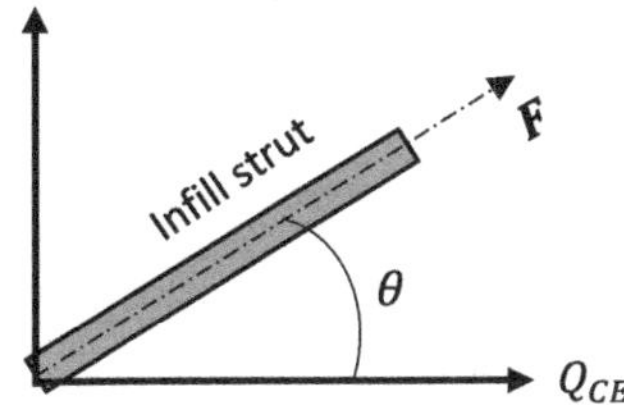

FIGURE 12.5   Force along the infill strut.

As infill is made up of brittle material, the modelling should be a force-controlled one. For ductile members, deformation-controlled behaviour is modelled.

12.3.2.2   Effect of Infill on Building

Infill imparts additional stiffness to the building. It is common with the designers to consider the weight of the infill in the model but ignore the stiffness part. As a result, the computed dynamic time period of building will be erroneous. We have the following three situations that may be encountered:

(i)   Bare frame building where the weight of infill is considered but stiffness from infill is disregarded: time period of building T_{bare} (obtained from dynamic analysis).
(ii)   Infill stiffness considered: time period T_{actual} (obtained from dynamic analysis)
(iii) Codal approximate time period T_a (obtained from code).

In all the cases, we have $T_{actual} < T_{bare}$.

If, $T_{actual} > T_a$, base shear correction as per code is necessary.
If, $T_{actual} < T_a$, base shear correction is not necessary.

Providing infill strut in building model also needs incorporating plastic hinges for the infill. It is a bit clumsy to model such plastic hinges. Because of its brittle nature, infill strut will fail much earlier than the frame members.

12.4   EXAMPLES

Example 12.4.1 *In a building, the columns are spaced at 4 m centre to centre. The storey height is 3.3 m. The beam size is 300×450 mm and column size is 400 × 400 mm. Find the thickness of infill strut in the panel and find the axial force capacity of the strut. Masonry is of fair quality. The concrete grade is M30. Thickness of infill wall is 200 mm.*

Solution: Consider Figure 12.6 The length of infill = 4000 −2 × (400/2) = 3600 mm.
   Height of infill panel = 3300 −2 × (450/2) = 3075 mm.
   Diagonal length of infill panel = $\sqrt{3600^2 + 3075^2}$ =4734.5 mm.

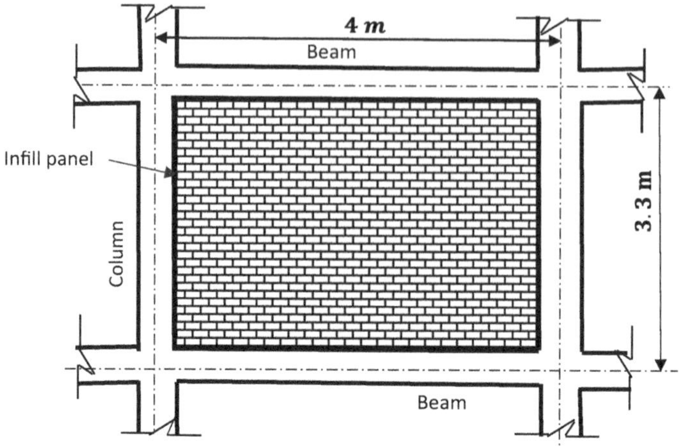

FIGURE 12.6   Infill panel in Example Ex.12.4.1.

$$a = 0.175\left(\lambda h_c\right)^{-0.4} r_i$$ Eq. (12.2.1)

h_c = height of column between centerlines of beams = 3075 mm.
r_i = diagonal length of infill panel = 4734.5 mm.

$$\lambda = \left[\frac{E_m t_i \sin 2\theta}{4 E_{fe} I_c h_i}\right]^{0.25}$$

For fair quality of masonry, σ_m = 4.1 MPa [Table 12.3]

$E_m = 550\,\sigma_m = 550 \times 4.1 = 2255$ MPa
$\theta = \tan^{-1}\left(3075 / 3600\right) = 40.5$ degree
E_{fe} = modulus of elasticity of concrete = $5000\sqrt{30} = 27386$ MPa
I_c = moment of inertia of column = $400 \times 400^3 / 12 = 2.133 \times 10^9$ mm^4
h_i = height of infill = 3075 mm
t_i = thickness of infill wall = 200 mm

$$\lambda = \left[\frac{E_m t_i \sin 2\theta}{4 E_{fe} I_c h_i}\right]^{0.25} = \left[\frac{2255 \times 200 \times \sin(2 \times 40.5 \times \pi/180)}{4 \times 27386 \times 2.133 \times 10^9 \times 3075}\right]^{0.25} = 2.73$$

$$a = 0.175\left(\lambda h_c\right)^{-0.4} r_i = 0.175 \times \left(8.87 \times 10^{-4} \times 3075\right)^{-0.4} = 554.5 \text{ mm.}$$

So, width of equivalent infill panel is 554.5 mm.

12.5 CLOSURE

This chapter chiefly refers to FEMA-356. Some additional information on masonry is available in ASCE-SEI-41-17, but this document appears to lack simplicity. Incorporation of masonry infill in the building model increases the stiffness and hence reduces the time period. But this is the 'true' time period of the building. More discussion on the real model can be found in Chapter 4.

12.6 EXERCISES

Q12.6.1 Describe the effect of infill in buildings.

Q12.6.2 Describe the strut model of infill in buildings.

Q12.6.3 Describe how the lateral strength of masonry is computed.

Q12.6.4 In a building the columns are spaced at 5.5 m centre to centre. The storey height is 3.5 m. The beam size is 400×550 mm and column size is 500×500 mm. Find the thickness of infill strut in the panel and find the axial force capacity of the strut. Masonry is of fair quality. The concrete grad is M30. Thickness of infill wall is 250 mm.

FURTHER READINGS

ASCE-SEI-7 Minimum Design Loads and Associated criteria for Buildings and Other Structures.

ASCE-SEI-41-17 (2017) Seismic Evaluation and Retrofit of Existing Buildings.

FEMA-273 (1996) NEHRP Guidelines for the Seismic Rehabilitation of Buildings. US Federal Emergency Management Agency, Building Seismic Safety Council, Washington DC.

FEMA-356 (2000) Prestandard and Commentary for the Seismic Rehabilitation of Buildings, US Federal Emergency Management Agency.

FEMA-368 (2001) NEHRP Recommended Provisions for Seismic Regulations for New Buildings and other Structures, US Federal Emergency Management Agency.

FEMA-389 (2004) Premier for Design Professionals, US Federal Emergency Management Agency.

FEMA-450 (2004) NEHRP Recommended Provisions for Seismic Regulations for New Buildings and other Structures, US Federal Emergency Management Agency.

13 DDBD of RC Frame Buildings with Viscous Dampers

13.1 INTRODUCTION

Damping in structure reduces its response and hence it is to our advantage. Structural response can be reduced by increasing the amount of damping. We can increase the damping in a system by adding damping devices, called dampers. There are various types of dampers, but passive damper and active damper are the basic categories. Active damper needs some real-time feedback to activate it. Hence, these require some electrical/electronic triggering. Passive dampers are activated by movement of the structure. There are also semi-active dampers. Passive dampers may be working with the orifice action of some fluid (when it is called fluid viscous damper (FVD)). Passive dampers may also work on the principle of mechanism, friction, rolling-bending or yielding of metal. There are also magneto-rheological dampers and shape memory alloy dampers. A damper acts as a fuse and itself may get damaged during earthquake, thereby saving the main structure.

An FVD works on the principle of orifice action on a fluid. The fluid is filled in a cylinder with piston and two chambers. The fluid used should be: (i) highly viscous, (ii) stable over long life, (iii) chemically non-reactive and (iv) non-inflammable. Generally, silicone fluid is used in FVD.

Figure 13.1 shows a rough longitudinal section of an FVD. Silicone fluid remains partially in chambers 1 and 2. As the piston moves, the fluid changes chamber through the orifice. While passing through orifice, high velocity is generated in the compressed fluid, which does work, thereby generating heat. The kinetic work done and dissipated heat are the sources of damping. An accumulator chamber temporarily stores fluid.

The force acting in the damper due to fluid movement is given by Eq. (13.1.1).

$$F_D = C |v|^{\alpha} . \mathrm{sign}(v) \tag{13.1.1}$$

where F_D is the damper force, C is a constant, v is fluid velocity, $\mathrm{sign}(v)$ is a sign of velocity and α is the index signifying the degree of nonlinearity in the fluid action. When $\alpha = 1$, it is a linear damper. As α reduces, nonlinearity increases. A real damper is shown in Figure 13.2.

DOI: 10.1201/9781003441090-13

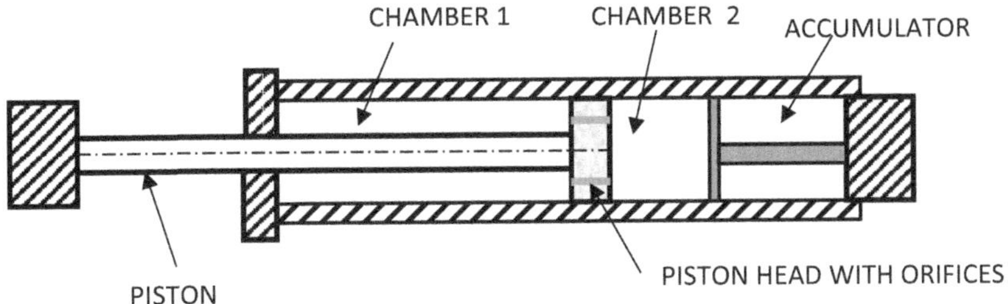

FIGURE 13.1 Longitudinal section of a fluid viscous damper.

FIGURE 13.2 FVD in real structure (*Source*: internet).

Viscous damping response can be expressed by Eq. (13.1.2) (Chopra, 2008):

$$\left(\frac{u}{u_{max}}\right)^2 + \left(\frac{F_D}{c\omega u_{max}}\right)^2 = 1 \tag{13.1.2}$$

where u is the displacement response of damper, c is damping coefficient, ω is the frequency of damper movement and u_{max} is the maximum damper displacement response. Obviously, it is an equation of ellipse and the response curve is shown in Figure 13.3(a). The elastic structural response is shown in Figure 13.3(b), and structure–damper combined response is shown in Figure 13.3(c).

The placements of damper in structure can be of various forms as shown in Figure 13.4.

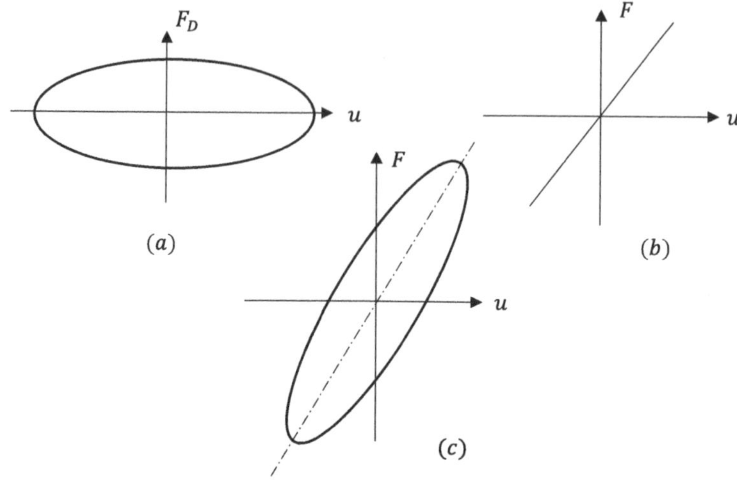

FIGURE 13.3　Damper response and structure response: (a) damper response, (b) elastic structural response and (c) combined structure–damper response.

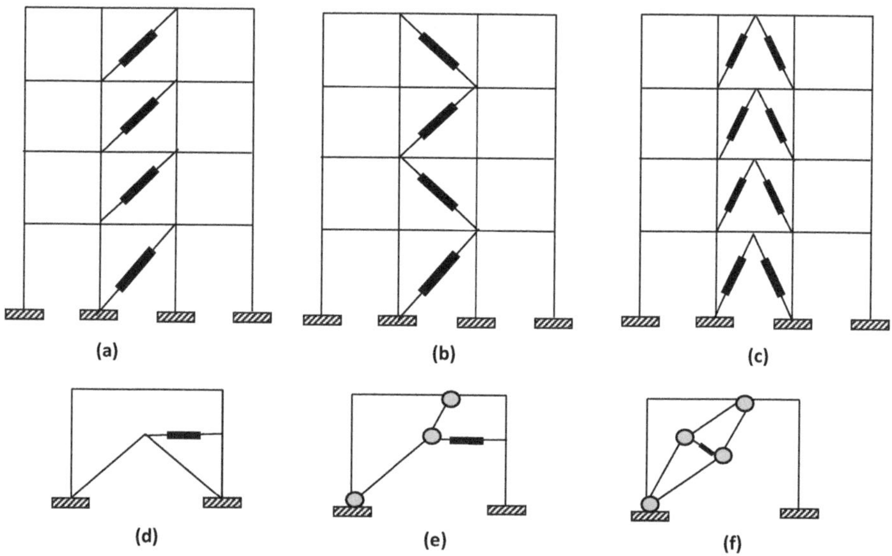

FIGURE 13.4　Various arrangements of placements of dampers in building.

13.2　EXISTING METHODS OF DESIGN USING FVD

The design of structure with passive dampers like friction dampers and metallic yielding dampers have been available in the literature for three decades. But performance-oriented design approach is comparatively new. FEMA-356 (2000) described evaluation of buildings with added dampers. Subsequent FEMA documents continued to include a similar treatment. Lin et al. (2008) proposed a DDBD method for design of structures with viscoelastics and FVD. The damping of the damper was added to the structure's damping and design was done following iterative procedure by converting the MDOF system to ESDOF system. Sullivan and Lago (2012) introduced a DDBD method for design of buildings with FVD considering damping force as a constant proportion to storey shears and applied it to linear FVD. Moradpour and Dehestani (2019) proposed a DDBD method accommodating nonlinear dampers and optimization of the dampers. In this chapter, both linear and nonlinear dampers have been treated.

13.3 DDBD METHOD FOR RC FRAME BUILDINGS WITH DAMPERS

Sullivan and Lagos (2012) developed a DDBD method for RC frame buildings with supplemental dampers of FVD type. The steps involved in the design are described below.

Step 1: *Decide design drift and hazard level.*
Step 2: *Find out displacement profile* from Eq. (13.3.1).

$$\Delta_i = \omega_\theta \theta_d h_i \frac{4H - h_i}{4H - h_1} \tag{13.3.1a}$$

$$\omega_\theta = 1.15 - 0.0034H \leq 1.0 \tag{13.3.1b}$$

where
    Δ_i = profile displacement at i-th floor level
    ω_θ = factor taking care of dynamic amplification
    H = total height of building
    h_i = height of i-th floor from base of building
    h_1 = height of ground floor
    θ_d = design drift

Step 3: *The ESDOF system properties are found out* from Eqs. (13.3.2 to 13.3.4) with already familiar notations.

$$\Delta_d = \frac{\sum_{i=1}^{n} m_i \Delta_i^2}{\sum_{i=1}^{n} m_i \Delta_i} \tag{13.4.2}$$

$$m_e = \frac{\sum_{i=1}^{n} m_i \Delta_i}{\Delta_d} \tag{13.4.3}$$

$$H_e = \frac{\sum_{i=1}^{n} m_i \Delta_i h_i}{\sum_{i=1}^{n} m_i \Delta_i} \tag{13.4.4}$$

Step 3: *Decide the fraction of storey shear that will be taken by the dampers*. This is expressed by Eq. (13.3.5).

$$F_{di} = \beta_i V_i \tag{13.3.5}$$

where

F_{di} = damper force at i-th floor

β_i = fraction, showing storey shear taken by damper at i-th floor; this value may be also kept constant over the floors.

V_i = storey shear at i-th floor

Step 4: *Compute system effective damping* Using Eqs. (13.3.6).

$$\xi_{sys} = \xi_F + \xi_D \tag{13.3.6a}$$

where $\xi_D = \beta/2$ = damping in dampers. Here, β is average value of β_i, ξ_F is frame damping.

$$\xi_F = 5 + 71\frac{\mu-1}{\mu\pi}\% \tag{13.3.6b}$$

Here, μ is frame displacement ductility given by Eq. (13.3.6c).

$$\mu = \frac{\Delta_d}{\Delta_y} \tag{13.3.6c}$$

$$\Delta_y = \frac{0.5\varepsilon_{ye} l_b}{h_b} H_e \tag{13.3.6d}$$

ε_{ye} = yield strain of rebar at expected strength level [$f_{ye} = 1.25 f_y$ as per FEMA-356; $f_{ye} = 1.1 f_y$ as per Sullivan and Lago (2012).

Step 5: *Generate displacement spectra corresponding to design spectrum and scale it down for damping* ξ_{sys}.

$$S_{d,\xi sys\%} = S_{d,5\%}\sqrt{\frac{10}{5+\xi sys\%}} \tag{13.3.6e}$$

Step 6: *Compute base shear*

$$k_e = 4\pi^2 \frac{m_e}{T_e^2} \tag{13.3.7a}$$

$$V_b = k_e \Delta_d \tag{13.3.7b}$$

Step 7: *Distribute the base shear (n is number of storey)*

$$\text{For } n \le 10 \quad F_i = V_b \frac{m_i \Delta_i}{\sum_{i=1}^{n} m_i \Delta_i} \tag{13.3.8a}$$

For $n > 10$

$$\text{Below roof level } F_i = kV_b \frac{m_i \Delta_i}{\sum_{i=1}^{n} m_i \Delta_i} \tag{13.3.8b}$$

$$\text{Roof level } F_{roof} = \left(1-k\right)V_b + kV_b \frac{m_{roof} \Delta_{roof}}{\sum_{i=1}^{n} m_i \Delta_i} \tag{13.3.8c}$$

k is generally 0.9.

Step 8: *Compute damping in damper (for damper manufacture purpose)*

The damping in i-th damper is given by Eq. (13.3.9a). F_{di} is given in Eq. (13.3.5).

$$C_i = \frac{F_{di} T_e}{2\pi u_i} \tag{13.3.9a}$$

$$u_i = (\Delta_i - \Delta_{i-1})\cos\delta \tag{13.3.9b}$$

where u_i is axial deformation of i-th damper, which is inclined at angle δ with the horizontal axis. Eq. (13.3.9b) is explained in Figure 13.5.

In Figure 13.5, OA is a damper-element inclined at angle δ with horizontal. The storey displacement is BC = Δ_i at i-th floor. The axial deformation of the damper is AC = BCcos δ. So, $u_i = (\Delta_i - \Delta_{i-1})\cos\delta$.

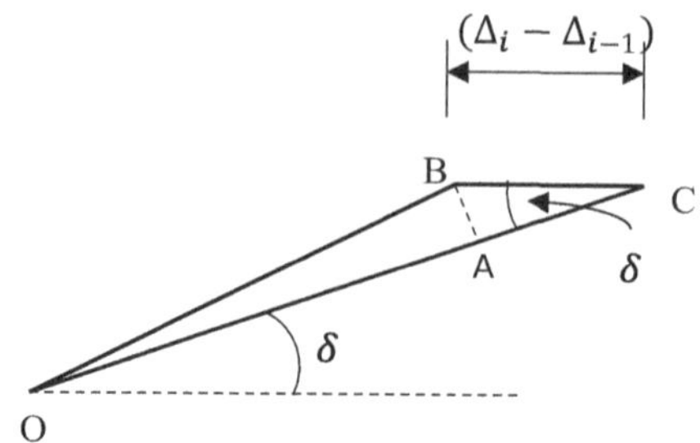

FIGURE 13.5 Damper axial deformation.

13.4 EXAMPLES

Example 13.4.1 *A 12-storey RC frame building is fitted with viscous dampers over all floors. The ground storey height is 4.4 m and other storey height is 3.3 m. The floor mass is 700 kN-s²/m and roof mass is 600 kN-s²/m. The building is to be designed for a drift of 2.6%. Use yield strength of rebar as 500 MPa. Take the length of beam as 5 m and depth of the beam as 0.8 m. Use EC-8 spectrum for Soil type B at 0.6g level. Angle made by dampers with horizontal is 35°. Find storey lateral forces, storey shear and damper constants. Assume that dampers carry 20% of storey shears.*

Solution: *Compute ESDOF system properties*

The total height of the building is $H = 40.7$ m. Factor for taking care of dynamic amplification factor is,

$$\omega_\theta = 1.15 - 0.0034H \le 1.0 = 1.15 - 0.0034 \times 40.7 = 1.012; \text{ so, } \omega_\theta = 1.0.$$

$\Delta_i = \omega_\theta \theta_d h_i \dfrac{4H - h_i}{4H - h_1}$, using this equation, the values are found out (Column 4 of

Table 13.1). Now ESDOF system properties calculated using Table 13.1.

$\Delta_d = 2425/4100 = 0.591$ m; $m_e = 4100/0.591 = 6932$ kN-s²/m; $H_e = 112037/4100 = 27.326$ m.

Compute Base shear

$$\Delta_y = \frac{0.5\varepsilon_{ye} l_b}{h_b} H_e = \frac{0.5 \times \left(1.1 \times 500/2 \times 10^5\right) \times 5}{0.8} \times 27.326 = 0.229 \text{ m.}$$

$$\mu = \frac{\Delta_d}{\Delta_y} = 0.591/0.229 = 2.58.$$

TABLE 13.1
ESDOF system properties for Ex. 13.4.1

Floor	Ht. of floor (h_i) (m)	Mass (m_i) (kN-s²/m)	Profile Displacement (Δ_i) (m)	$m_i\Delta_i$	$m_i\Delta_i^2$	$m_i\Delta_i h_i$
12	40.7	600	0.816	489	399	19919
11	37.4	700	0.770	539	415	20154
10	34.1	700	0.720	504	363	17195
9	30.8	700	0.667	467	312	14388
8	27.5	700	0.611	428	261	11757
7	24.2	700	0.551	385	212	9326
6	20.9	700	0.487	341	166	7122
5	17.6	700	0.419	294	123	5168
4	14.3	700	0.349	244	85	3489
3	11	700	0.274	192	53	2110
2	7.7	700	0.196	137	27	1057
1	4.4	700	0.114	80	9	352
			Sum	4100	2425	112037

$$\xi_F = 5 + 71\frac{\mu-1}{\mu\pi} = 5 + 71\frac{2.58-1}{2.58\pi} = 18.85\%$$

$$\xi_{sys} = \xi_F + \xi_D = \xi_F + \beta/2 = 18.9 + 20/2 = 28.85\%.$$

Corresponding to $\Delta_d = 0.591$ m and $\xi_{sys} = 28.85\%$ in EC-8 displacement spectra for Soil type B at 0.6g seismicity level, $T_e = 4.84$ sec (Figure 13.6).

$$k_e = 4\pi^2 \frac{m_e}{T_e^2} = 11683\,\text{kN/m}.$$

$$V_b = k_e\Delta_d = 6910 \text{ kN}.$$

The base shear is now distributed over floors and tabulated in Table 13.2.

Compute damping in dampers

$$C_i = \frac{F_{di}T_e}{2\pi u_i}$$

$$F_{di} = \beta V_i$$

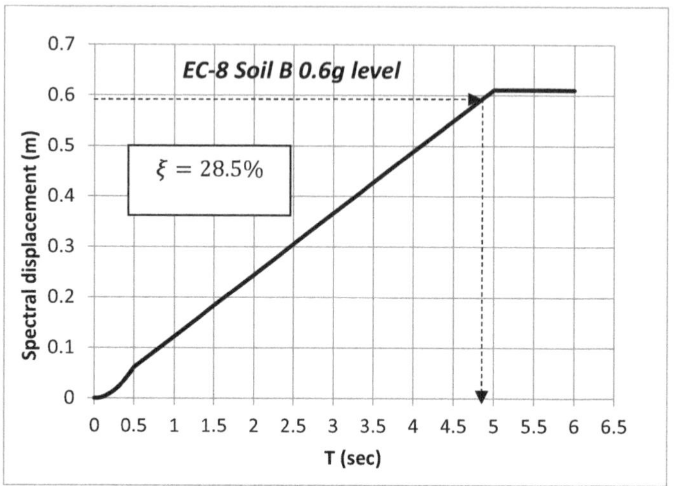

FIGURE 13.6   Scaled displacement spectra for Ex. 13.4.1.

TABLE 13.2
Floor force and damping constant for Ex. 13.4.1 ($\alpha = 1$, $\beta = 0.2$, $\xi_D = 10\%$)

Floor	h_i	m_i	$m_i h_i$	V_b (kN)	F_i (kN)	V_i (kN)	F_{di} (kN)	C_i (kN-sec/m)
12	40.7	600	24420		1510	1510	302	6191
11	37.4	700	26180		878	2389	478	9084
10	34.1	700	23870		801	3190	638	11312
9	30.8	700	21560		723	3913	783	13001
8	27.5	700	19250	6910	646	4559	912	14247
7	24.2	700	16940		568	5127	1025	15125
6	20.9	700	14630		491	5618	1124	15693
5	17.6	700	12320		413	6031	1206	15998
4	14.3	700	10010		336	6367	1273	16079
3	11	700	7700		258	6626	1325	15965
2	7.7	700	5390		181	6806	1361	15682
1	4.4	700	3080		103	6910	1382	11360
		sum	185350	check	6908		sum	159737

$$u_i = (\Delta_i - \Delta_{i-1})\cos\delta$$

Calculation of C_i at *roof level*: $u_i = (\Delta_i - \Delta_{i-1})\cos\delta = (0.816 - 0.770) \times \cos(35°) = $

$0.03776\,\text{m}$. $F_{di} = \beta_i V_i = 0.2 \times 1510 = 302\,\text{kN}$. $C_i = \dfrac{F_{di} T_e}{2\pi u_i} = \dfrac{302 \times 4.84}{2\pi \times 0.03776} = 6161\,\text{kN}$. A

little difference with table value (6191) is due to rounding off effects. Values of C are given in Table 13.2.

13.5 NONLINEAR DAMPER

Damper nonlinearity is reflected by the parameter λ, which is a function of α and Γ, as given in Eq. (13.5.1) (Moradpour and Dehestani, 2019) where Γ is the Gamma function.

$$\xi_D = \frac{\lambda}{2} \frac{\sum_{i=1}^{n} \beta_i V_i \left(\Delta_i - \Delta_{i-1} \right)}{V_b \Delta_d} \tag{13.5.1a}$$

$$\lambda = 2^{2+\alpha} \frac{\Gamma^2 \left(1 + \alpha/2 \right)}{\pi \Gamma \left(2 + \alpha \right)} \tag{13.5.1b}$$

Typical values of α are 1.0 (linear damper), 0.6, 0.3 and 0.15; the value of $\alpha = 0.15$ gives highest nonlinearity out of the four values mentioned.

Let us plot Eq. (13.5.1b) to see the variation of λ with α (Figure 13.7). For this purpose, values of Γ are required. The values of Γ for various values of α are tabulated in Table 13.3. We take the help of mathematical relation $\Gamma(n+1) = n\Gamma(n)$ and $\Gamma(n) = (n-1)!$ Fractional values of Γ are available from Gamma tables (see books on mathematics).

Figure 13.7 shows that, as α increases, λ decreases, which implies nonlinearity in damper decreases so also the damping (Eq. 13.5.1a). At $\alpha = 1$, $\lambda = 1$ (linear damper).

13.5.1 INTERPRETATION OF EQUATION $\beta = 2\xi_D / \lambda$

We may keep the value of β (fraction of storey shear taken by damper) to be constant, or we can vary its value storey wise. The relationship between β and ξ_D given by Eq. (13.5.1.1) after Moradpour and Dehestani (2019).

TABLE 13.3
Relevant Gamma values

α	$2 + \alpha$	$\Gamma(2 + \alpha)$	$1 + \alpha/2$	$\Gamma^2(1 + \alpha/2)$	λ
1.0	3	2.0	1.5	0.785	1
0.9	2.9	1.827355	1.45	0.784	1.020
0.8	2.8	1.676491	1.4	0.787	1.041
0.7	2.7	1.544686	1.35	0.794	1.063
0.6	2.6	1.429626	1.3	0.805	1.087
0.5	2.5	1.329341	1.25	0.822	1.113
0.4	2.4	1.24217	1.2	0.843	1.140
0.3	2.3	1.166712	1.15	0.871	1.170
0.2	2.2	1.101803	1.1	0.905	1.201
0.15	2.15	1.072997	1.075	0.925	1.218

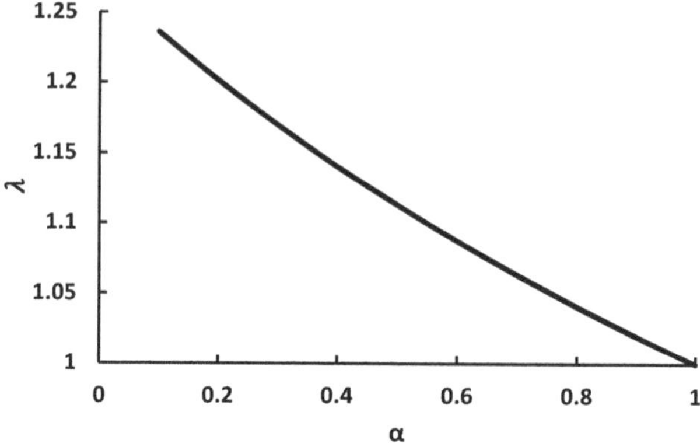

FIGURE 13.7 Variation of λ with α (Eq. 13.5.1b).

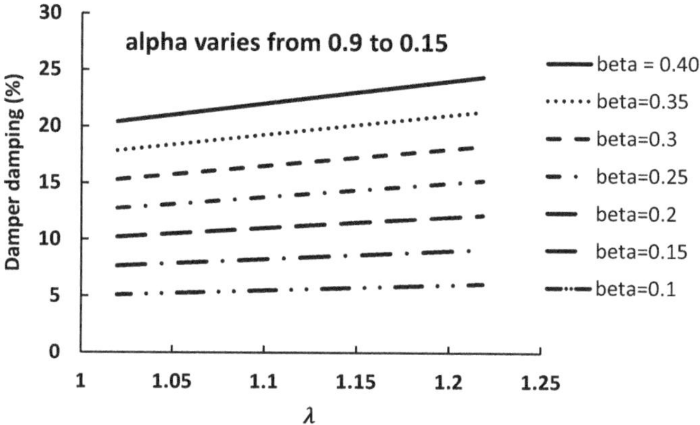

FIGURE 13.8 Variation of ξ_D with β and λ (α).

$$\beta = \frac{2}{\lambda}\xi_D \tag{13.5.1.1}$$

This can be written as Eq. (13.5.1.1a).

$$\xi_D = \frac{\beta\lambda}{2} \tag{13.5.1.1a}$$

Eq. (13.5.1.1a) is plotted in Figure 13.8 for various values of β. In this plot, the value of α is varied from 0.9 to 0.15. These all are nonlinear dampers. As α decreases, nonlinearity of damper increases, but the total damping constants decreases. The damping in the dampers is taken as a fixed value (around 15%) in design. The value of ξ_D more than 20% may not be practical. It is evident that ξ_D increases with β.

It simply implies that as the damper has to carry more storey shear, more damping is demanded in the damper.

For linear damper ($\lambda = 1$) and when β reamins constant over floors, we get Eq. (13.5.1.1b).

$$\xi_D = \frac{\beta}{2}$$

(13.5.1.1b)

If β is kept constant over the storeys, ξ_D also becomes constant. But it is possible to vary β over the storeys. One possibility is to vary β in proportion to storey drift. When β varies over storeys, the damping in damper is given by Eq. 13.5.1.2,

$$\xi_D = \frac{\lambda}{2} \sum_{i=1}^{n} \beta_i$$

(13.5.1.2)

13.6   DRIFT PROPORTIONAL DAMPING FORCE

As discussed in Section 13.5.1, the damper force can be proportional to storey shear. Eq. (13.3.5) expresses damper force in that way. Damper is inclined to horizontal by angle δ. So, actual damper force along its axis is given by Eq. (13.6.1).

$$F_{di} = \frac{\beta_i V_i}{cos\delta}$$

(13.6.1)

Eq. (13.3.5) and (13.6.1) are similar, only values of β_i will be different.

Also, Eq. (13.3.9a) is valid for linear damper ($\alpha = 1$). For nonlinear dampers, the damper force is given by Eq. (13.6.2) and consequently, damping constant is given by Eq. (13.6.3).

$$F_{d,i} = C_i \left(\frac{2\pi u_i}{T_e} \right)^{\alpha}$$

(13.6.2)

$$C_i = F_{di} \left(\frac{T_e}{2\pi u_i} \right)^{\alpha}$$

(13.6.3)

$$u_i = (\Delta_i - \Delta_{i-1})cos\delta$$

[(13.3.9b)]

Values of β_i can be constant or be varying with storeys. But other possibility is to vary it proportional to storey drift. The storey drift can be computed based on displacement profile (Eq. (13.3.1a)) or based on drifts after carrying out dynamic analysis. As damper properties are not known in the beginning, dynamic drift of building without damper may be used.

Now consider Eq. (13.5.1.2). The designer wishes to have a fixed percentage of damping arising out of dampers. So left hand side ξ_D is fixed by the designer; let this be $\xi_{D,decided}$. Let θ_i denote interstorey drift at i-th storey.

Now, the drift proportionality values of β will be given by,

$$\beta_i = k\theta_i \tag{13.6.4}$$

Also as per Moradpour and Dehestani (2019) we have,

$$\xi_D = \frac{\lambda}{2} \frac{\sum_{i=1}^{n} \beta_i V_i \left(\Delta_i - \Delta_{i-1}\right)}{V_b \Delta_d} \tag{13.6.5}$$

So, we can write,

$$\xi_{D,\,decided} = \frac{\lambda}{2} \frac{\sum_{i=1}^{n} k\theta_i V_i \left(\Delta_i - \Delta_{i-1}\right)}{V_b \Delta_d}$$

$$= \frac{k\lambda}{2} \frac{\sum_{i=1}^{n} \theta_i V_i \left(\Delta_i - \Delta_{i-1}\right)}{V_b \Delta_d}$$

$$\text{Or}, k = \frac{2 V_b \Delta_d \xi_{D,decided}}{\lambda \sum_{i=1}^{n} \theta_i V_i \left(\Delta_i - \Delta_{i-1}\right)} \tag{13.6.6}$$

Damper forces are given by Eq. (13.6.7).

$$F_{di} = \frac{k\theta_i V_i}{\cos\delta} \tag{13.6.7}$$

Finally,

$$C_i = F_{di} \left(\frac{T_e}{2\pi u_i}\right)^{\alpha} \tag{[(13.6.3)]}$$

$$u_i = (\Delta_i - \Delta_{i-1})\cos\delta \tag{[(13.3.9b)]}$$

Example 13.6.1 *Solve Ex. 13.4.1 by considering drift-proportional damping. Use drift based on displacement profile. Comment on the results. Use $\beta = 0.2$, that is $\xi_D = 10\%$.*

Solution: We use the displacement profile of Table 13.1, and compute storey drifts as shown Table 13.4. Let us decide (in line with Ex. 13.4.1) that damper supply 10% damping ($\xi_{D,decided} = 10\%$). Eq. (13.6.6) gives proportionality constant. But this is dependent on λ, which in turn depends upon α. So, Table 13.4 shows computation of damper force (as per Eq.(13.6.6)) for various values of α. From Ex. 13.4.1, $\Delta_d = 0.591$, $V_b = 6910$ kN. K used in Table 13.4 is obtained from Eq. (13.6.6).

From Table 13.4 and Figure 13.9, it is found that the damping constant decreases towards upper storeys. Also, the damping constant requirement decreases with higher nonlinearity of dampers (lower values of α). For linear damper ($\alpha = 1$), the damping constant demand is highest. Hence linear damper will not be economical.

Example 13.6.2 *With data of Ex13.6.1 and by considering drift-proportional damping, and taking $\beta = 0.3$ and 0.4 (that is, $\xi_D = 15\%$ and 20%), get the damping constants. Use drift based on displacement profile. Comment on the results.*

Solution: Tables 13.5 (for, $\xi_D = 15\%$) and 13.6 (for, $\xi_D = 20\%$) are prepared. The results of Example 13.6.1 and 13.6.2 are put in condensed form in Table 13.7.

The data in Table 13.7 has been plotted in Figure 13.10. From the figure, it is quite evident that the sum of damper constants is very high for linear dampers, thereby indicating that the linear dampers are costly. Nonlinear dampers by the same argument,

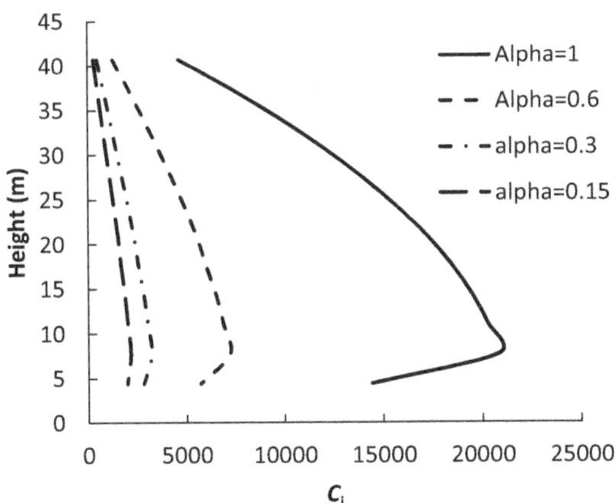

FIGURE 13.9 Variation of damping constant in Ex. 13.6.1.

TABLE 13.4

Damping constants for Ex. 13.6.1 ($\xi_D = 10\%$, proportional to drift as per profile)

level	Profile Δ_i	drift θ_i	V_i (kN)	$\theta_i V_i(\Delta_i - \Delta_{i-1})$	$\alpha = 1$ $k = 7.9954$		$\alpha = 0.6$ $k = 7.3535$		$\alpha = 0.3$ $k = 6.8357$		$\alpha = 15$ $k = 6.5627$	
					F_{di}	C_i	F_{di}	C_i	F_{di}	C_i	F_{di}	C_i
12	0.816	0.015	1510	1.07	226	4620	208	1271	193	478	185	292
11	0.770	0.017	2389	1.99	389	7309	357	2079	332	801	319	495
10	0.720	0.018	3190	2.99	550	9760	506	2841	470	1114	452	695
9	0.667	0.019	3913	4.09	713	11972	656	3563	610	1421	585	893
8	0.611	0.020	4559	5.47	890	13948	819	4267	761	1737	730	1104
7	0.551	0.021	5127	7.00	1068	15686	982	4924	913	2044	876	1311
6	0.487	0.023	5618	8.66	1243	17189	1143	5528	1063	2337	1020	1513
5	0.419	0.023	6031	9.85	1374	18452	1263	6004	1174	2560	1127	1665
4	0.349	0.025	6367	11.94	1554	19480	1429	6516	1328	2836	1275	1863
3	0.274	0.026	6626	13.44	1682	20273	1547	6888	1438	3034	1380	2005
2	0.196	0.027	6806	15.25	1816	20823	1670	7218	1552	3227	1490	2149
1	0.114	0.026	6910	20.41	1747	14415	1607	5700	1494	2814	1434	1968
			Sum	102.15		**173927**		**56798**		**24404**		**15954**

TABLE 13.5

Damping constants for Ex.13.6.2 ($\xi_D = 15\%$, proportional to drift as per profile)

	Profile Δ_i	drift θ_i	V_i (kN)	$\theta_i V_i (\Delta_i - \Delta_{i-1})$	$\alpha = 1$ $k = 7.9954$		$\alpha = 0.6$ $k = 7.3535$		$\alpha = 0.3$ $k = 6.8357$		$\alpha = 15$ $k = 6.5627$	
					F_{di}	C_i	F_{di}	C_i	F_{di}	C_i	F_{di}	C_i
12	0.816	0.015	1510	1.07	339	6930	312	1906	290	717	278	438
11	0.770	0.017	2389	1.99	583	10964	536	3118	498	1202	478	743
10	0.720	0.018	3190	2.99	825	14640	759	4262	705	1672	677	1043
9	0.667	0.019	3913	4.09	1069	17958	984	5344	914	2131	878	1340
8	0.611	0.020	4559	5.47	1335	20923	1228	6400	1141	2606	1096	1656
7	0.551	0.021	5127	7.00	1601	23529	1473	7386	1369	3066	1314	1967
6	0.487	0.023	5618	8.66	1864	25783	1715	8292	1594	3505	1530	2269
5	0.419	0.023	6031	9.85	2060	27678	1895	9006	1761	3840	1691	2497
4	0.349	0.025	6367	11.94	2330	29220	2143	9773	1992	4255	1913	2795
3	0.274	0.026	6626	13.44	2522	30409	2320	10332	2156	4551	2070	3008
2	0.196	0.027	6806	15.25	2724	31235	2505	10827	2329	4841	2236	3223
1	0.114	0.026	6910	20.41	2621	21622	2411	8550	2241	4220	2151	2953
			Sum	102.15		**260891**		**85196**		**36606**		**23931**

TABLE 13.6
Damping constants for Ex.13.6.2 (ξ_D = 20%, proportional to drift as per profile)

	Profile Δ_i	drift θ_i	V_i (kN)	$\theta_i V_i(\Delta_i - \Delta_{i-1})$	$\alpha = 1$, $k = 7.9954$		$\alpha = 0.6$, $k = 7.3535$		$\alpha = 0.3$, $k = 6.8357$		$\alpha = 15$, $k = 6.5627$	
					F_{di}	C_i	F_{di}	C_i	F_{di}	C_i	F_{di}	C_i
12	0.816	0.015	1510	1.07	452	9240	416	2542	386	955	371	583
11	0.770	0.017	2389	1.99	777	14619	715	4157	665	1603	638	991
10	0.720	0.018	3190	2.99	1100	19520	1012	5682	941	2229	903	1390
9	0.667	0.019	3913	4.09	1426	23944	1311	7125	1219	2842	1170	1787
8	0.611	0.020	4559	5.47	1780	27897	1637	8534	1522	3475	1461	2208
7	0.551	0.021	5127	7.00	2135	31373	1964	9848	1825	4088	1753	2623
6	0.487	0.023	5618	8.66	2486	34377	2286	11056	2125	4674	2040	3026
5	0.419	0.023	6031	9.85	2747	36904	2527	12007	2349	5120	2255	3329
4	0.349	0.025	6367	11.94	3107	38960	2858	13031	2657	5673	2550	3727
3	0.274	0.026	6626	13.44	3363	40545	3093	13776	2875	6068	2760	4010
2	0.196	0.027	6806	15.25	3632	41647	3340	14436	3105	6455	2981	4298
1	0.114	0.026	6910	20.41	3495	28829	3214	11401	2988	5627	2869	3937
			Sum	102.15		**347854**		**113595**		**48808**		**31908**

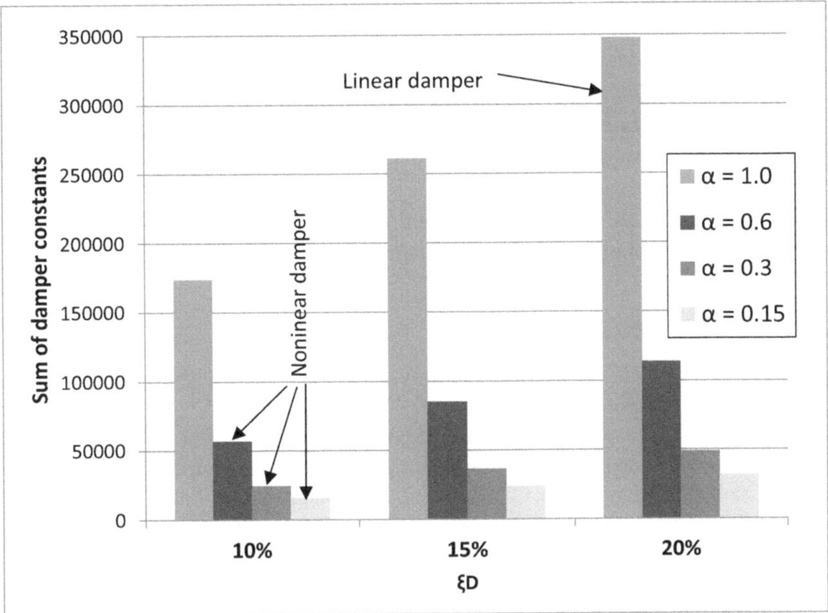

FIGURE 13.10   Variation of sum of damping constant in example Ex.13.6.2.

TABLE 13.7
Damping constants for Ex.13.6.1. and, Ex.13.6.2

Sum of damper constants ΣC_i											
Linear damper			Nonlinear dampers								
$\alpha = 1$			$\alpha = 0.6$			$\alpha = 0.3$			$\alpha = 0.15$		
ξ_D values			ξ_D values			ξ_D values			ξ_D values		
10%	15%	20%	10%	15%	20%	10%	15%	20%	10%	15%	20%
173927	260891	347854	56798	85196	113595	24404	36606	48808	15954	23931	31908

are cheaper. The cost decreases as the nonlinearity increases. However, to attain high nonlinearity may lead to practical problems in the manufacturing of dampers.

13.7   CLOSURE

The use of FVD in PBD has been highlighted in this chapter. The cost of damper is proportional to the damping coefficient. So, to reduce cost, this coefficient to be minimized. Examples are given to explain the application of the concept. Further research can be motivated through this chapter.

13.8 EXERCISES

Q13.8.1 What is a viscous damper? What are the advantages and disadvantages of viscous dampers?

Q13.8.2 Which parameter is to be optimized in fluid viscous dampers?

Q13.8.3 Give an outline of DDBD for RC frame buildings with fluid viscous damper.

Q13.8.4 Differentiate between linear and nonlinear dampers.

Q13.8.5 Explain what is meant by drift-proportional damping.

FURTHER READINGS

Chopra, A.K. (2007) *Dynamics of Structures – Theory and Applications to Earthquake Engineering*, Prentice Hall of India.

Kim, J., and Choi, H. (2006) Displacement-based Design of Supplemental Dampers for Seismic Retrofit of a Framed Structure, *Journal of Structural Engineering*, 132(6), 873–883.

Lin, Y. Y., Chang, K. C., and Chen, C. Y. (2008) Direct Displacement-based Design for Seismic Retrofit of Existing Buildings Using Nonlinear Viscous Dampers, *Bulletin Earthquake Engineering*, 6, 535–552, March,.

Malu, A. and Choudhury, S. (2020) Direct Displacement-Based Design of a Building Incorporating Viscous Dampers, *Proceedings of International Structural Engineering and Construction*, 7(2), doi 10.14455/ISEC.2020.7(2).STR-12. (SCOPUS).

Moradpour, S., and Dehestani, M. (2019) Optimal DDBD Procedure for Designing Steel Structures with Nonlinear Fluid Viscous Dampers, *Structures*, 22, 154–174.

Pettinga., J. D., and Priestley, M. J. N. (2010) Dynamic Behavior of Reinforced Concrete Frames Designed with Direct Displacement-based Design, *Journal of Earthquake Engineering*, 9(2), 309–330,.

Sullivan, T. J., and Lago, A. (2012) Towards a Simplified Direct DBD Procedure for the Seismic Design of Moment Resisting Frames with Viscous Dampers, *Engineering Structures*, 35, 140–148.

Symans, M.D. and Constantinou, M.C. (1998) Passive Fluid Viscous Damping Systems for Seismic Energy Dissipation, *ISET Journal* .

14 Displacement-Based Design for Bridge Piers

14.1 INTRODUCTION

Displacement-based design (DBD) method can be applied to the design of bridge piers. Unlike a building, a bridge has generally one mass at top of the idealized structure. Hence, the computation for ESDOF system properties is not required in the bridge design. A bridge has limited elements like piers, abutments and superstructure (deck). The piers are generally of dissimilar heights following the topography of the channel, which the bridge crosses. The piers can of different shapes like circular, oval, chamfered, square and rectangular (Figure 14.1). Mostly the sections are hollow. There can be more than one column in a pier system (multi-column pier). Circular columns are easy to construct (though form work is a bit costly). Circular columns have equal stiffness and strength in all directions. Also note that for bridge piers the design displacement parameter is *strain*.

14.2 DBD BASICS FOR CIRCULAR BRIDGES PIERS

Consider the circular pier section with circular ties as shown in Figure 14.2(a). The ties make a confining effect on the concrete. Under compressive load in column the confined inner core develops tendency to bulge out; this is prevented by ties with the development of tensile forces in the ties. This is the mechanism of confinement in concrete. Confinement in concrete increases its compressive strength and ductility. The confinement effect depends on the ratio of volume of transverse ties to the volume of core concrete, called transverse steel volumetric ratio (ρ_v). To find an expression of ρ_v, take the hatched portion of Figure 14.2(b). The volume of concrete in this portion is given by Eq. (14.2.1a).

$$V_c = \frac{\pi}{4} D_c^2 s \qquad (14.2.1a)$$

where s is spacing of the circular tie, V_c is volume of core concrete within height s, D_c is core diameter (diametrical distance between centres of tie).

The volume of tie in this hatched portion is given by Eq. (14.2.1b).

DOI: 10.1201/9781003441090-14

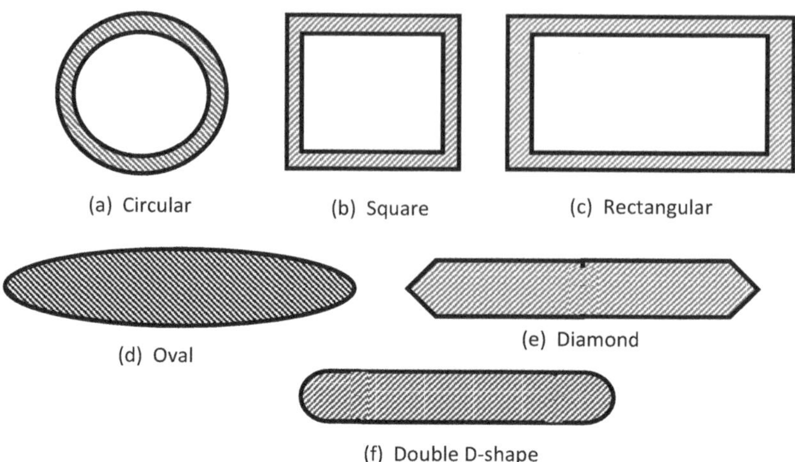

FIGURE 14.1 Various sections of bridge piers.

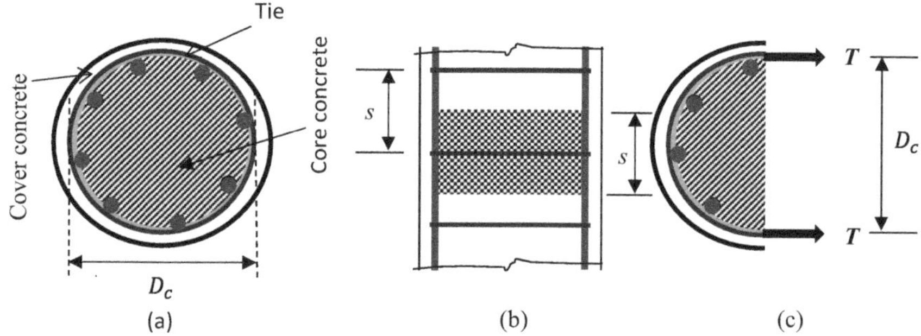

FIGURE 14.2 Circular pier: (a) cross-section (b) longitudinal section.

$$V_h = \frac{\pi}{4} d_h^2 \left(\pi D_c \right) \tag{14.2.1b}$$

where d_h is the diameter of tie. ρ_v is the ratio of V_h and V_c.

$$\rho_v = \frac{V_h}{V_c} = \frac{\dfrac{\pi}{4} d_h^2 \left(\pi D_c \right)}{\dfrac{\pi}{4} D_c^2 s} = \frac{\pi d_h^2}{s D_c} = \frac{4 \left(\dfrac{\pi}{4} d_h^2 \right)}{s D_c}$$

$$\rho_v = \frac{4 A_h}{s D_c} \tag{14.2.2}$$

where A_h is area of cross-section of tie bar.

The maximum confining pressure is exerted in concrete when the tie yields. Let T be the yield strength of tie. With yield strength of tie bar as f_{yh}, T is given by Eq. (14.2.3).

$$T = f_{yh}A_h \qquad\qquad (14.2.3a)$$

Referring to Figure (14.2c), the total tension in tie at yield is $2T$. Symbolizing radial stress in confined concrete by f_r, the radial compression force in height s of column is sD_cf_r. Equalizing compression and tension we get,

$$2f_{yh}A_h = sD_cf_r$$

$$\text{Or,} \quad f_r = \frac{2f_{yh}A_h}{sD_c} \qquad\qquad (14.2.3b)$$

Using Eq. (14.2.2) and (14.2.3b) we get,

$$f_r = 0.5\rho_v f_{yh} \qquad\qquad (14.2.4)$$

With as characteristic compression strength of concrete in MPa, the confined compression strength f_{con} is given by Eq. (14.2.5) (Priestley et al., 2007).

$$f_{con} = f_{ck}\left(2.254\sqrt{1+\frac{7.94f_r}{f_{ck}}} - 2\frac{f_r}{f_{ck}} - 1.254\right) \qquad\qquad (14.2.5)$$

The expected strength of concrete f_{ce} and of steel f_{ye} given by Priestley et al. (2007) is given by Eq. (14.2.6a) and (14.2.6b) respectively.

$$f_{ce} = 1.3f_{ck} \qquad\qquad (14.2.6a)$$

$$f_{ye} = 1.1f_y \qquad\qquad (14.2.6b)$$

Rearranging Eq. (14.2.5) and putting Eq. (14.2.4) we get,

$$\frac{f_{con}}{f_{ck}} = \left(2.254\sqrt{1+\frac{7.94\left(0.5\rho_v f_{yh}\right)}{f_{ck}}} - 2\frac{0.5\rho_v f_{yh}}{f_{ck}} - 1.254\right) \qquad\qquad (14.2.7a)$$

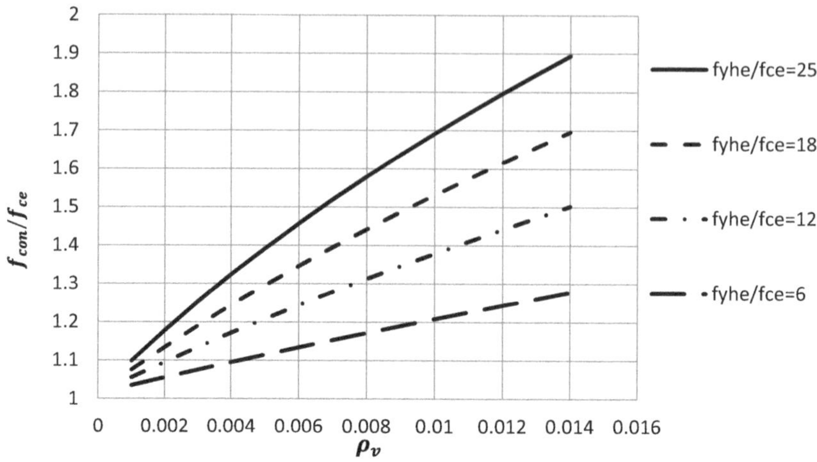

FIGURE 14.3   Confined concrete strength (Eq. 14.2.7b).

Noting that in displacement-based design we use expected strength and denoting expected strength of hoop ties as f_{yhe}, Eq. (14.2.7a) is written as

$$\frac{f_{con}}{f_{ce}} = \left(2.254\sqrt{1 + 3.97\rho_v \frac{f_{yhe}}{f_{ce}}} - \rho_v \frac{f_{yhe}}{f_{ce}} - 1.254 \right) \qquad (14.2.7b)$$

Equation (14.2.7b) is plotted in Figure 14.3 for various values of ρ_v and $\dfrac{f_{yhe}}{f_{ce}}$ ratios.

In the construction of this figure, concrete of grades M20, M25, M30, M35, M40 and steel of grades Fe415, Fe500 and Fe550 have been considered.

14.3   DBD FOR CIRCULAR BRIDGE PIERS

Priestley et al. (2007) has developed a damage control displacement-based design for bridge piers. We present here the steps in the method in our own language. In the beginning, the designer will select pier diameter and core diameter. The axial load in column has to be supplied.

> **Step 1:** The transverse steel volumetric ratio ρ_v is found out from Eq. (14.2.2). While using this equation, the designer has already chosen the yield strength of tie bar, tie diameter and tie spacing.
>
> **Step 2:** The confined compressive strength of concrete f_{con} is calculated by using Eq. (14.2.7b) or by using Figure 14.3.

Step 3: Calculate the damage control strain (ε_d) in compression concrete by using Eq. (14.3.1) (Priestley et al., 2007). In this equation, (ε_{su}) is ultimate strain in rebar.

$$\varepsilon_d = 0.004 + 1.4 \frac{\rho_v f_{yh} \varepsilon_{su}}{f_{con}} \tag{14.3.1}$$

Step 4: The column axial stress ratio (axial force to expected strength) is computed by using Eq. (14.3.2). Here A_g is gross area of section of pier, r is axial force ratio

$$r = \frac{P}{f_{ce} \times A_g} \tag{14.3.2}$$

Step 5: Calculate the neutral axis depth by using Eq. (14.3.3) (Priestley et al., 2007). Here, x is neutral axis depth and D is diameter of pier.

$$\frac{x}{D} = 0.2 + 0.65r \tag{14.3.3}$$

Step 6: The limit state curvatures can be expressed as the ratio of extreme fibre strain and neutral axis depth. Denoting limit state curvature for concrete by ϕ_{lc} and limit state curvature for steel as ϕ_{ls}, we can write,

$$\phi_{lc} = \frac{\varepsilon_c}{x} \tag{14.3.4a}$$

$$\phi_{ls} = \frac{\varepsilon_s}{d-x} \tag{14.3.4b}$$

Step 7: Find the plastic hinge length from Eq. (14.3.5) (Priestley et al., 2007). Here L_p is plastic hinge length, L_{sp} is strain penetration length, L_c is the distance from the critical section to the point of inflection, f_{ye} is expected yield strength of longitudinal bars, d_i is diameter of longitudinal bars, f_y and f_u are yield strength and ultimate tensile strength of longitudinal bars respectively.

$$L_p = kL_c + L_{sp} \geq 2L_{sp} \tag{14.3.5a}$$

$$L_{sp} = 0.022 f_{ye} d_l \quad \left[f_{ye} \text{ in MPa}, d_l \text{ in mm} \right] \tag{14.3.5b}$$

$$k = 0.2 \left(\frac{f_u}{f_y} - 1 \right) \leq 0.08 \tag{14.3.c5cc}$$

Step 8: Find the design displacement from Eq. (14.3.6) (Priestley et al., 2007). Here, is design displacement, is yield displacement, is inelastic displacement,

$$\Delta_D = \Delta_y + \Delta_p = \frac{2.25\varepsilon_y}{3D}\phi_y\left(H + L_{sp}\right)^2 + \left(\phi_{ls} - \frac{2.25\varepsilon_y}{D}\right)L_pH \qquad (14.3.6)$$

Step 9: Find the displacement by Eq. (14.3.7) as usual.

$$\mu = \frac{\Delta_D}{\Delta_y} \qquad (14.3.7)$$

Step 10: Find the pier damping from Eq. (14.3.8) (Priestley et al., 2007).

$$\xi = 0.05 + 0.444\left(\frac{\mu - 1}{\mu\pi}\right) \qquad (14.3.8)$$

Step 11: If there are multiple piers, find the equivalent damping from Eq. (14.3.9). Here, V_i is lateral shear in i-th pier.

$$\xi_{eq} = \frac{\Sigma\,\xi_i V_i}{\Sigma\,V_i} \qquad (14.3.9)$$

Step 12: Reduction factor for spectral displacement is given by Eq. (14.3.10) (Priestley et al., 2007).

$$\text{reduction factor} = \left(\frac{0.07}{0.02 + \xi_{eq}}\right)^{0.5} \qquad (14.3.10)$$

Example 14.3.1 *For a circular bridge pier with fixed base and top with bearing, the diameter is 1.8 m and clear height is 10 m. The axial load on the pier in pier including self-weight is 8000 kN. Concrete grade is M30 and steel grade is Fe415. The longitudinal bars are 30 mm in diameter with ultimate strain as 10%. The ties are 20 mm in diameter spaced at 150 mm c/c. The ultimate strain of steel is 15% and ultimate strength is 520 MPa. Clear cover is 40 mm. Find the effective damping in the pier.*

Solution: Preliminary data: $D = 1800$ m, $H = 10000$ mm, $L_c = 10000$ mm, $P = 8000$ kN, $f_{ce} = 1.3 \times 30 = 39$ MPa, $f_y = 415$ MPa, $f_{se} = 1.1 \times 415 = 465$ MPa, $D_c = 1800 - 2 \times 40 = 1720$ mm, $d_l = 30$ mm, $d_h = 20$ mm, $A_h = \frac{\pi}{4}d_h^2 = 314$ mm², $s = 150$ mm,

$$\varepsilon_{su} = 0.1, \ A_g = \frac{\pi}{4}D^2 = \frac{\pi}{4}1800^2 = 2544688 \text{ mm}^2, d = D - 40 - 20 - 30/2 = 1725 \text{ mm.}$$
$$f_u = 550 \text{ MPa}$$

Step 1: The transverse steel volumetric ratio ρ_v is found –

$$\rho_v = \frac{4A_h}{sD_c} = \frac{4 \times 314}{150 \times 1780} = 0.00047.$$

Step 2: The confined compressive strength of concrete f_{con} is calculated:

$$f_r = 0.5\rho_v f_{yh} = 0.5 \times 0.0047 \times 415 = 0.975.$$

$$f_{con} = f_{ck}\left(2.254\sqrt{1 + \frac{7.94f_r}{f_{ck}}} - 2\frac{f_r}{f_{ck}} - 1.254\right)$$

$$= 30\left(2.254\sqrt{1 + \frac{7.94 \times 0.975}{30}} - 2\frac{0.975}{30} - 1.254\right)$$

$$= 36.3 \text{ MPa}$$

Step 3: Calculate the damage control strain (ε_d) in compression -

$$\varepsilon_d = 0.004 + 1.4\frac{\rho_v f_{yh} \varepsilon_{su}}{f_{con}}$$

$$= 0.004 + 1.4\frac{0.00465 \times 415 \times 0.10}{36.2}$$

$$= 0.01146.$$

Step 4: The column axial stress ratio (r)

$$r = \frac{P}{f_{ce} \times A_g} = \frac{8000 \times 1000}{39 \times 2544688} = 0.081$$

Step 5: Calculate the neutral axis depth, $\frac{x}{D} = 0.2 + 0.65r = 0.2 + 0.65 \times 0.081 = 0.2527$. So, $= x = 0.2527 \times 1800 = 455$ mm.

Step 6: The limit state curvatures

$$\phi_{lc} = \frac{\varepsilon_d}{x} = 0.01146/455 \text{ per mm} = 0.0252 \text{ per m.}$$

$$\varepsilon_s = 0.6\varepsilon_{su} = 0.6 \times 0.1 = 0.06.$$

$$\phi_{ls} = \frac{\varepsilon_s}{d-x} = \frac{0.06}{1725-455} \text{ per mm} = 0.047 \text{ per m.}$$

Step 7: Find the plastic hinge length

$$k = 0.2\left(\frac{f_u}{f_y} - 1\right) = 0.2\left(\frac{550}{415} - 1\right) = 0.065 \leq 0.08$$

$$L_{sp} = 0.022 f_{ye} d_l = 0.022 \times 465 \times 30 = 306.9 \text{ mm}$$

$$L_p = kL_c + L_{sp} = 0.065 \times 10000 + 306.9 = 966.9 \text{ mm} \geq 2L_{sp}$$

Step 8: Find the design displacement

$$\phi_y = \frac{2.25\varepsilon_y}{D} = \frac{2.25 \times 415/200000}{1800} \text{ per mm} = 0.0026 \text{ per m.}$$

$$\Delta_y = \frac{\phi_y \left(H + L_{sp}\right)^2}{3} = \frac{0.009(10 + 0.3069)^2}{3} \text{ m} = 0.092 \text{ m.}$$

$$\Delta_p = \left(\phi_{ls} - \phi_y\right)L_p H = (0.047 - 0.0026) \times 0.9669 \times 10 = 0.429 \text{ m.}$$

$$\Delta_D = \Delta_y + \Delta_p = 0.092 + 0.429 = 0.521 \text{ m.}$$

Step 9: Find the ductility

$$\mu = \frac{\Delta_D}{\Delta_y} = 0.521/0.092 = 5.7.$$

Step 10: Find the pier damping

$$\xi = 0.05 + 0.444\left(\frac{\mu - 1}{\mu\pi}\right) = 0.05 + 0.444\left(\frac{5.7 - 1}{5.7\pi}\right) = 0.167 \text{ or } \mathbf{16.7\%.}$$

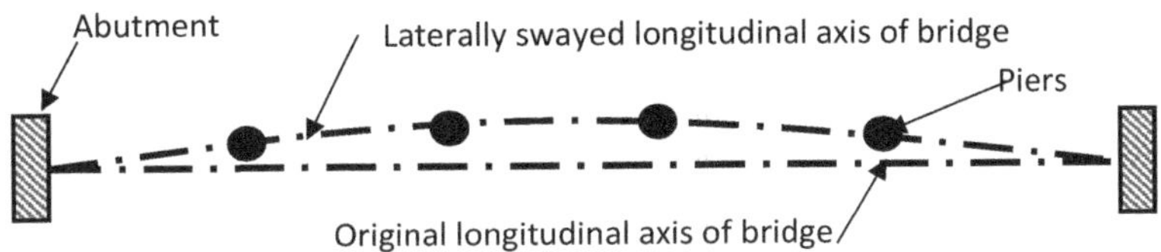

FIGURE 14.4 Top view of longitudinal axis of bridge.

14.4 DDBD FOR BRIDGES PIERS

Bridge has only one vertical level of mass, namely, at bearing/deck level. The question arises – can we apply DDBD method in bridge? Well, in vertical direction DDBD method is not applicable, but it can be applied in horizontal direction, considering all piers together, and their lateral sway. The top view of a bridge is shown in Figure 14.4. Interested readers may see Priestley et al. (2007) for further information.

14.5 CLOSURE

A brief outline of displacement-based design of bridge piers has been discussed in this chapter. Interested readers may see Priestley et al. (2007).

14.6 EXERCISES

Ex14.6.1 Give an outline of displacement-based design of bridge piers.

Ex14.6.2 Write down the steps for DBD of circular bridge piers.

Ex14.6.3 For a circular bridge pier with fixed base and bearing at the top, the diameter is 2.0 m and clear height is 15 m. The axial load on the pier including self-weight is 10000 kN. Concrete grade is M30 and steel grade is Fe415. The longitudinal bars are 30 mm in diameter with ultimate strain as 10%. The ties are 16 mm in diameter spaced at 150 m c/c. The ultimate strain of steel is 15% and ultimate strength is 520 MPa. Clear cover is 50 mm. Find effective damping in the pier.

FURTHER READING

Priestley, M.J.N., Calvi,G.M. and Kowalsky, M.J. (2007) *Displacement-Based Seismic Design of Structures*, IUSS Press..

15 UPBD for Bridge Piers

15.1 INTRODUCTION

We have seen the DDB for bridge piers in Chapter 14, where only drift could be satisfied as a target performance objective. In this chapter, we shall see how PL along with the drift can also be accommodated in the UPBD method for bridge piers. The UPBD method for circular bridge piers was proposed by Banerjee (2020) and later on further strengthened by Banerjee and Choudhury (2020a, 2020b). This chapter will be mainly reflecting on these three works. We confine ourselves to RC circular bridge piers.

Figure 15.1 shows a typical bridge section. The effective mass over an individual bridge pier is the sum of: (i) mass from superstructure of length stretched to half of the spans on either side of pier and, (ii) one third of the mass of pier (Priestley et al., 2007). This mass works over the equivalent single degree of freedom (ESDOF) system which the pier assembly is converted to. Figure 15.2 shows a pier assembly and the ESDOF system.

15.2 THEORETICAL BACKGROUND

In the UPBD method, we wish to design the structure for a given drift and PL under a given hazard level. Let the target design drift for the bridge pier be θ_d and the PL is indicated by plastic rotation of pier to angle θ_p. Let the yield rotation of pier by θ_y. The angular drift consists of the yield rotation and plastic rotation. This can be expressed by Eq. (15.2.1) and explained through Figure 15.3.

$$\theta_d = \theta_y + \theta_p \tag{15.2.1}$$

Now we look at finding the expression for θ_y.

DOI: 10.1201/9781003441090-15

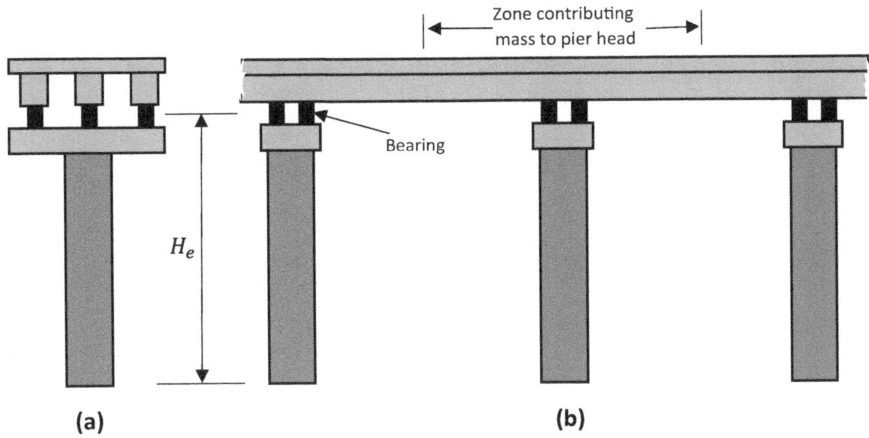

FIGURE 15.1 Typical bridge section: (a) cross-sectional view and (b) longitudinal view.

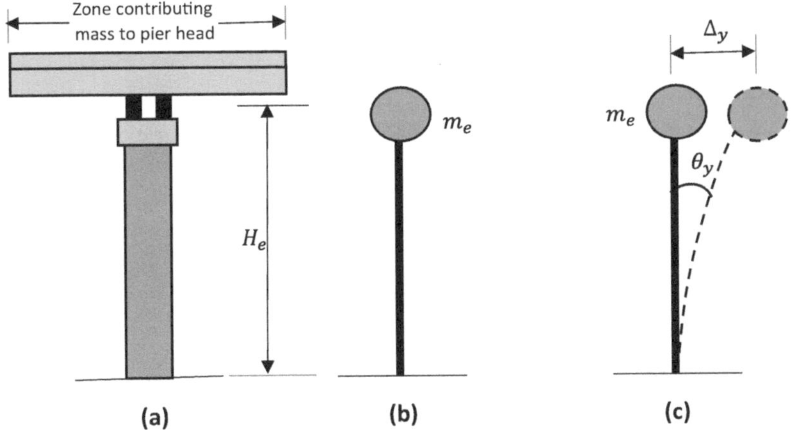

FIGURE 15.2 (a) Bridge section, (b) ESDOF system and (c) yield displacement.

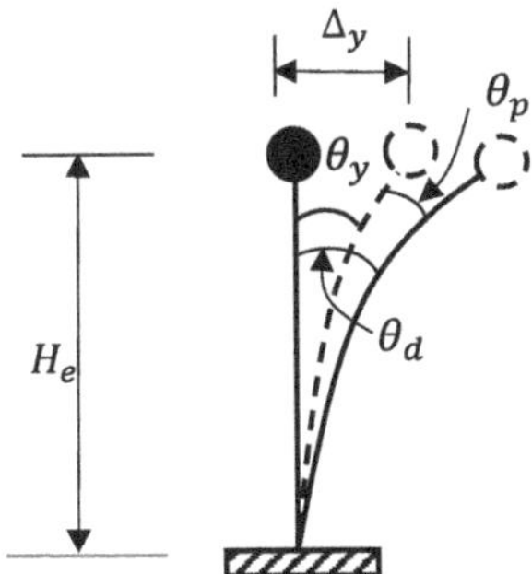

FIGURE 15.3 (a) Bridge section, (b) ESDOF system and (c) yield displacement.

15.2.1 EXPRESSION FOR THE YIELD ROTATION OF PIER

Referring to Figure 15.3, we can write the relationship between θ_y and Δ_y as Eq. (15.2.1.1).

$$\theta_y = \Delta_y / H_e \tag{15.2.1.1}$$

Expression for yield displacement for pier is available from Priestley et al. (2007) as Eq. (15.2.1.2).

$$\Delta_y = C\phi_y \left(H_e + L_{sp}\right)^2 \tag{15.2.1.2}$$

where

Δ_y = yield displacement at top of pier

ϕ_y = yield curvature given by Eq. (15.2.1.4) (Priestley et al., 2007)

C = a coefficient for end condition of pier, typically 1/3 when top end is pinned and bottom end of pier is fixed

H_e = effective height of pier

L_{sp} = strain penetration length, given by Eq. (15.2.1.3) (Priestley et al., 2007)

$$L_{sp} = 0.022 f_{ye}\, d_{bl} \tag{15.2.1.3}$$

f_{ye} = expected strength of rebar = $1.1 \times f_y$

d_{bl} = diameter of longitudinal rebar in m.

$$\phi_y = \frac{2.25\varepsilon_y}{D} \tag{15.2.1.4}$$

ε_y = yield strain of rebar = f_y / E_s, E_s = modulus of elasticity of rebar = 2×10^5 MPa

D = diameter of pier in m

Substituting the above expressions in Eq. (15.2.1.2) we get Eq. (15.2.1.5).

$$\Delta_y = \frac{1}{3} \times \frac{2.25\varepsilon_y}{D}\left(H_e + 0.022 f_{ye} d_{bl}\right)^2 \tag{15.2.1.5}$$

Combing Eq. (15.2.1.1) and (15.2.1.5), and noting that $\theta_y = \theta_d - \theta_p$, we get Eq. (15.2.1.6).

$$D = \frac{0.75\varepsilon_y\,(H_e + 0.022 f_{ye} d_{bl})^2}{H_e\left(\theta_d - \theta_p\right)} \tag{15.2.1.6}$$

Eq. (15.2.1.6) gives the diameter of the pier which will satisfy both drift and PL for the pier. The plastic rotations for piers (taken at bottom of pier) are available from FEMA-356 or SEI documents. The values of θ_p for various PLs, as per ASCE-SEI-41-13 are given below.

    (a) Average Plastic rotation for IO: 0.00375 radian
    (b) Average Plastic rotation for LS: 0.02125 radian
    (c) Average Plastic rotation for CP: 0.02725 radian

15.2.2 MODIFIED EXPRESSION FOR PIER DIAMETER

It was observed by Banerjee and Choudhury (2020a, 2020b) that the bridge system is very stiff compared to buildings and achieved drift by using Eq. (15.2.1.6) was much less than the target drift. This is due to the reason that abutment makes the bridge laterally very stiff. So, drift correction factor (λ) was introduced in the equation leading to the Eq. (15.2.2.1).

$$D = \frac{0.75\,\varepsilon_y\,(H_e + 0.022 f_{ye} d_{bl})^2}{H_e\left(\theta'_d - \theta_p\right)} \tag{15.2.2.1}$$

where θ'_d is modified drift, given by Eq. (15.2.2.2).

$$\theta'_d = \lambda\theta_d \tag{15.2.2.2}$$

As suggested by Banerjee and Choudhury (2020a, 2020b)

    $\lambda = 2.0$ for IO PL
    $\lambda = 3.0$ for LS PL.

15.2.3 SIZE OF NON-CIRCULAR PIERS

So far, we have discussed about circular pier. If square or rectangular pier is considered Eq. (15.2.1.2) can be used with vales of C and ϕ_y for rectangular section.

15.3   DESIGN STEPS FOR UPBD METHOD FOR BRIDGE PIERS

The design steps under UPBD method for bridge pier design are given below.

Step 1: The supplied (that is, obtained from client) data are: the design drift (θ_d), and PL (θ_p) and hazard level.

Modified drift (θ_d') is obtained from Eq. (15.2.2.2).

Step 2: Diameter of the per is obtained from Eq. (15.2.2.1).

Step 3: The bridge pier is idealized as an ESDOF system. The ESDOF system properties are calculated. Equivalent mass (m_e) is the summation of the mass of half the length of the superstructure on both the sides of the concerned pier and one third the mass of the pier. The equivalent height (H_e) equals to the distance between the base of the pier to the centre of the bearing.

Step 3: Yield rotation (θ_y) is obtained from Eq. (15.2.1.1). Ductility is obtained as shown below.

$$\mu = \frac{\Delta_d'}{\Delta_y} = \frac{\theta_d' H_e}{\theta_y H_e} = \frac{\theta_d'}{\theta_y}$$

Step 4: The effective damping is given by Eq. (15.3.1) (Priestley et al., 2007).

$$\xi_e = 0.05 + 0.444 \left(\frac{\mu - 1}{\mu \pi} \right) \tag{15.3.1}$$

Step 5: Displacement spectra corresponding to the design acceleration spectrum is generated for various damping values (Figure 15.4).

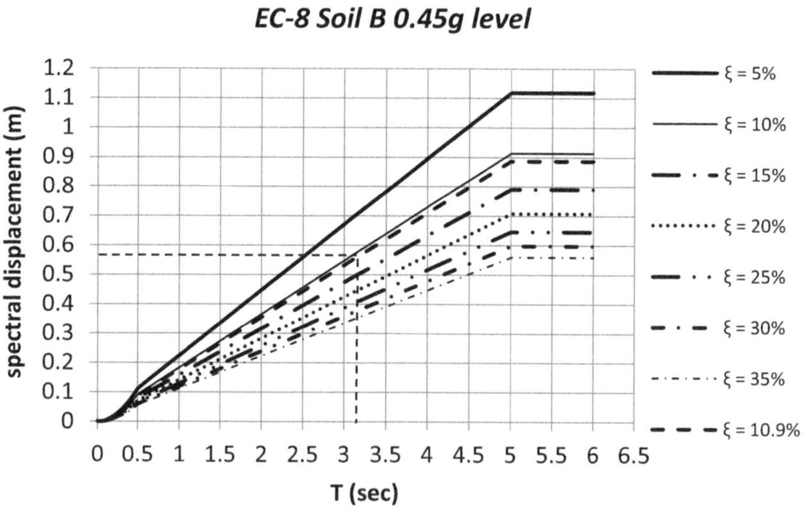

EC-8 Soil B 0.45g level

FIGURE 15.4   Displacement Spectra for EC-8 at 0.45g level for soil type B.

Step 6: Effective time period (T_e) of the system is read from the curve corresponding to the effective damping (ξ_e) and modified design displacement $(\Delta'_d = \theta'_d H_e = \lambda \theta_d H_e)$.

Step 7: Effective stiffness (K_e) for the system is calculated from Eq. (15.3.2). The base shear (V_b) is given by Eq. (15.3.3).

$$K_e = 4\pi^2 \frac{m_e}{T_e^2} \tag{15.3.2}$$

$$V_b = K_e \times \Delta'_d \tag{15.3.3}$$

Unlike that in building bridge has only one mass level at top. The base shear (V_b) is applied at the top of the effective height of the pier.

Step 8: Design is carried out with expected strength of the materials. For concrete, it is taken as 1.5 times the 28-day characteristic strength, and that of steel it is as 1.25 times the yield strength of reinforcement bars (FEMA-356). The load combinations are as follows:

$$D + L$$

$$D + L \pm F_x$$

$$D + L \pm F_y$$

Here, D stands for dead load, L stands for live load, and F_x, F_y stand for the seismic loads in two mutually perpendicular directions of the bridge.

15.4 EXAMPLES

Example 15.4.1 *An RC bridge pier is 10.5 m high up to the bottom of bearing. Bearing height is 400 mm and deck system thickness is 200 mm. Weight of segment of superstructure including half spans on each side of pier is 1000 kN. The pier is to be designed for 1.8% drift and LS PL in seismic zone V of India. Find the diameter of the pier. Take Fe415 grade rebar of 20 mm diameter for longitudinal bars.*

Solution: Effective height of pier $H_e = 10.5 + 0.4/2 = 10.7$ m.

Effective mass of the pier head $m_e = 1000/9.81 = 101.94$ kg.

Design drift $= 1.8\% = 0.018$

Modified drift $= \theta'_d = \lambda \theta_d = 3 \times 0.018 = 0.054$ (for LS PL $\lambda = 3$).

Yield strain of rebar $\varepsilon_y = 415/200000 = 0.002075$.

Diameter of bar $= 20$ mm $= 0.02$ m.

Expected yield strength of rebar $f_{ye} = 1.1 \times 415 = 456.5$ MPa.

Plastic rotation for LS PL $\theta_p = 0.02125$

Diameter of pier =

$$D = \frac{0.75\varepsilon_y(H_e + 0.022f_{ye}d_{bl})^2}{H_e(\theta'_d - \theta_p)} = \frac{0.75 \times 0.002075(10.7 + 0.022 \times 456.5 \times 0.02)^2}{10.7(0.054 - 0.02125)}$$

$$= 0.528 \text{ m}$$

A diameter of 550 mm is recommended.

$$\Delta_y = \frac{1}{3} \times \frac{2.25\varepsilon_y}{D}\left(H_e + 0.022f_{ye}d_{bl}\right)^2$$

$$= \frac{1}{3} \times \frac{2.25 \times 0.002075}{0.55}(10.7 + 0.022 \times 456.5 \times 0.02)^2$$

$$= 0.336 \text{ m.}$$

$$\mu = \frac{\Delta'_d}{\Delta_y} = \frac{\theta'_d H_e}{\Delta_y} = \frac{0.054 \times 10.7}{0.336} = 1.72.$$

$$\xi_e = 0.05 + 0.444\left(\frac{\mu - 1}{\mu\pi}\right) = 0.05 + 0.444\left(\frac{1.72 - 1}{1.72\pi}\right)$$

$$= 0.109 \ (10.9\%).$$

$$\Delta'_d = \theta'_d H_e = 0.054 \times 10.7 = 0.578\text{m.}$$

From Figure 15.4, corresponding to spectral displacement 0.578 m and effective damping 10.9%, we get $T_e = 3.2$ s.

$$K_e = 4\pi^2 \frac{m_e}{T_e^2} = 4\pi^2 \frac{101.94}{3.2^2} = 393\text{kN/m.}$$

$$V_b = K_e \times \Delta'_d = 393 \times 0.578 = 227.2 \text{ kN.}$$

This base shear will be put at effective height level for design.

15.5 CLOSURE

UPBD treatment has been done in this chapter for a circular pier. Similar procedure can be applied to square or rectangular piers with a little modifications.

15.6 EXERCISES

Q15.6.1 Discuss the theoretical background of UPBD method for bridge pier design.

Q15.6.2 Write down the steps involved in the UPBD method for bridge pier design.

Q15.6.3 An RC bridge pier is 15 m high up to the bottom of the bearing. Bearing height is 800 mm and deck system thickness is 300 mm. Weight of segment of superstructure including half spans on each side of pier is 3000 kN. The pier is to be designed for 2% drift and LS PL in seismic zone V of India. Find the diameter of the pier. Take Fe500 grade rebar of 25 mm diameter for longitudinal bars.

FURTHER READINGS

ASCE/SEI 41-13 (2014) Seismic Evaluation and Retrofit of Existing Buildings. 2014

Banerjee, S. (2020) Unified Performance-Based Design of Bridge Piers, M.Tech. Thesis (under the supervision of S. Choudhury), Department of Civil Engineering, National Institute of Technology Silchar, India, May.

Banerjee, S. and Choudhury S. (2020a) An Introduction to Unified Performance-Based Design of Bridge Piers, 17th WCEE, Sendai, Japan, 13–18 September.

Banerjee S. and Choudhury S. (2020b) Design of Bridge Pier using Unified Performance-Based Design Method, Second ASCE India Conference on Challenges of Resilient and Sustainable Infrastructure Development in Emerging Economies (CRSIDE2020), 2–4 March.

Priestley, M.J.N., Calvi G. M. and Kowalsky M. J. (2007) *Displacement-Based Seismic Design of Structures*, IUSS Press.

16 Energy Dissipating Devices

16.1 INTRODUCTION

Damping reduces the response of the structures and as such, it works as a protection against damage to the structures. The damping in the structure can be increased by adding additional damping devices. The devices so added are known as energy dissipation devices (EDD). The damping obtained through EDDs is known as added damping. The total damping (ξ_{total}) arises out of: (i) material damping (ξ_m), (ii) hysteretic damping (ξ_h) and (iii) added damping (ξ_a), as shown in Eq. (16.1.1a). In normal buildings, the elements are damaged during earthquakes of a very high magnitude. In most of the cases, the damage may be beyond repair. The EDDs act like fuses and may get damaged during a strong earthquake, thereby saving the main structural components. The damaged EDD can be replaced after the earthquake. There are various types of EDDs, such as friction dampers, metallic yielding dampers, fluid viscous dampers, viscoelastic dampers, shape memory alloys and magneto-rheological dampers. Each variety of EDD has many sub-variants. Some of the dampers are discussed in this chapter.

$$\xi_{total} = \xi_m + \xi_h + \xi_a \tag{16.1.1a}$$

Energy-wise, the balance equation can be written as (16.1.1b).

$$E_t = E_s + E_k + E_D + E_h + E_a \tag{16.1.1b}$$

where

E_t = total energy in the vibrating system = $\int \left(m\ddot{u}(t) \right) d\ddot{u}_g(t)$

m is mass of the system, u is displacement, $\ddot{u}_g$ is acceleration of ground

E_s = strain energy = $\dfrac{1}{2} Fu = \dfrac{1}{2} F \left(\dfrac{F}{k} \right) = \dfrac{1}{2} F^2 / k$; k is stiffness

E_k = kinetic energy = $\dfrac{1}{2} m\dot{u}^2$

DOI: 10.1201/9781003441090-16

E_D = viscous damping energy = $\int c\dot{u}^2 dt$

E_h = hysteretic energy arising out of hysteretic action after damage

E_a = energy dissipated by added dampers.

The equation of motion of a normal structure is given by Eq. (16.1.2). The equation of motion of a structure with additional damper is given by Eq. (16.1.3a).

$$[m]\{\ddot{u}(t)\}+[c]\{\dot{u}(t)\}+[k]\{u(t)\}=\{F(t)\} \qquad (16.1.2)$$

$$[m]\{\ddot{u}(t)\}+[c]\{\dot{u}(t)\}+[k]\{u(t)\}+[k_d]\{u_d(t)\}=\{F(t)\} \qquad (16.1.3a)$$

where $[m]$ is the mass matrix for structure, $\{\ddot{u}(t)\}$ is the acceleration vector for structure, $[c]$ is the inherent damping matrix for the structure, $\{\dot{u}(t)\}$ is the velocity vector for the structure, $\{u(t)\}$ is the displacement vector for the structure, $\{F(t)\}$ is the excitation vector, $[k_d]$ is the stiffness matrix for dampers and $\{u_d(t)\}$ is the displacement vector for dampers. For earthquake excitation, $\{F(t)\}$ may be replaced by $-[m]\{\ddot{u}_g(t)\}\{\iota\}$, where $\{\ddot{u}_g(t)\}$ is the ground acceleration record and $\{\iota\}$ is the unit vector. So, Eq. (16.1.3a) is rewritten as Eq. (16.1.3b).

$$[m]\{\ddot{u}(t)\}+[c]\{\dot{u}(t)\}+[k]\{u(t)\}+[k_d]\{u_d(t)\}$$
$$=-[m]\{\ddot{u}_g(t)\}\{\iota\} \qquad (16.1.3b)$$

It may be noted that Eq. (16.1.3) is applicable when the structure is considered as an MDOF system and there are multiple dampers in the system. If the system is idealized to a single SDOF system for the structure and an SDOF system for the damper, then the dynamic equation reduces to (16.1.4).

$$m\ddot{u}(t)+c\dot{u}(t)+ku(t)+k_d u_d(t)=F(t) \qquad (16.1.4a)$$

$$m\ddot{u}(t)+c\dot{u}(t)+ku(t)+k_d u_d(t)=-m\ddot{u}_g(t) \qquad (16.1.4b)$$

Dampers can be classified into three broad categories:

(i)   Passive dampers
(ii) Semi-active dampers
(iii) Active dampers.

The passive dampers do not require any external power to operate; the motion of the structure sets them in action. Active dampers require an external power to get

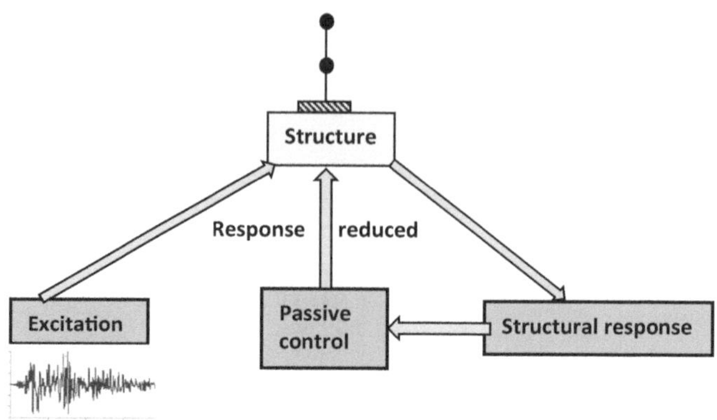

FIGURE 16.1   Working of a passive control system.

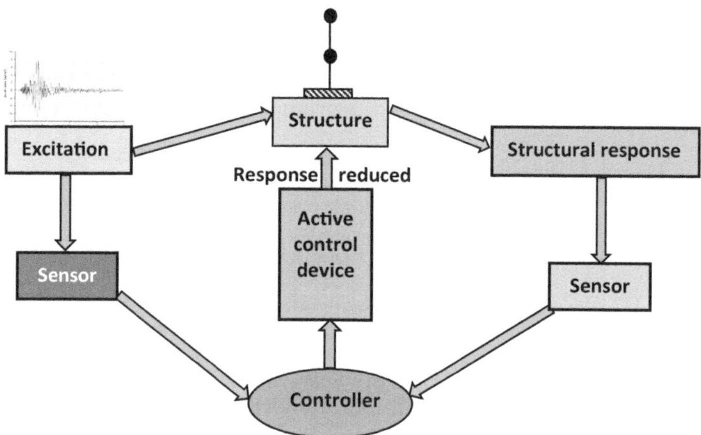

FIGURE 16.2   Working of an active control system.

activated during the earthquakes. Semi-active dampers are in between the passive and active types, requiring some power to operate.

In active dampers, the external power supply triggers the motion of the damping system such that it opposes the motion of the main structure, induced by the ground motion. Active dampers are generally sophisticated and efficient. The problems with the active dampers are as follows: (i) they are costly, (ii) they require specialized skill for installation and (iii) they may not work in case of power failure, unless there is a huge power backup system. Dampers are also called control systems, especially in case of active systems.

Active control devices operate on the principle of hydraulic actuators, which produce force opposite to that of the excitation. Active control devices are of many types, some of these being active mass damper (AMD), active tendons (AT) and active variable stiffness (AVS). AMDs are also effective against wind vibration.

The diagrammatic working of passive dampers and active dampers is shown in Figures 16.1 and 16.2, respectively.

The force–deformation behaviour of friction damper is rectangular. The movement of friction surfaces start at some fixed force. The displacement reaches a maximum value and the system becomes standstill for a moment. When the displacement takes place in the opposite direction, it follows a similar path with negative (reversed) force (Figure 16.3a).

The yielding metallic damper initially undergoes elastic deformation and finally yields. After yielding, the force may slightly increase due to strain hardening. The reverse movement occurs with some lag and it forms a typical hysteresis loop (Figure 16.3b).

Fluid viscous damper also shows movement practically at some finite force. As orifice action is just similar in forward and backward movements of the fluid, the force–deformation behaviour is elliptical (Figure 16.3c).

The viscoelastic damper has both elastic component and viscous fluid action component. So, the force–displacement relationship is inclined ellipse (Figure16.3d).

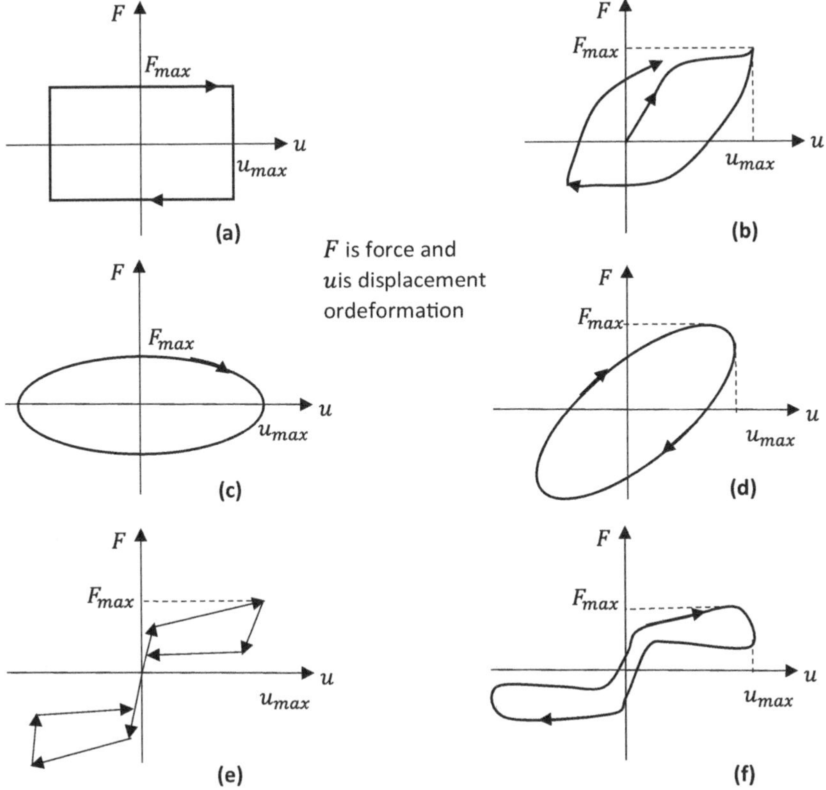

FIGURE 16.3 Force–displacement behaviour of dampers: (a) friction damper, (b) hysteretic damper, (c) fluid viscous damper, (d) viscoelastic damper, (e) friction damper with self-centring capability and (f) fluid viscous damper with self-restoring capability.

Friction damper with restoring capacity show force–displacement loop as two quadrilaterals in the positive and negative sectors (Figure 16.3e).

The fluid damper with a restoring capability has the loops that are continuous over the positive and negative sectors (Figure 16.3f).

Damper is generally placed in braces. The flexibility of the brace affects the performance of the damper. If the brace flexibility is not considered, it leads to an overestimation of damping.

The damping in a damper in a structure at k-th mode is given by Eq. (16.1.5) (Constantinou et al., 1998):

$$\xi_k = \frac{1}{2} \frac{\sum_{i=1}^{n} c_i \cos^2 \theta_i \phi_{ir}^2}{\omega_k \sum_{j=1}^{N} m_j \phi_j^2} \tag{16.1.5}$$

where ξ_k is the damping ratio for the k-th mode, n is number of EDDs, N is the number of lumped masses, θ_i is the angle of inclination of the i-th damper with horizontal inclination, ϕ_{ir} is the relative modal displacement of the i-th damper, c_i is the damping coefficient of the i-th damper, ω_k is the k-th modal frequency, m_j is the j-th lumped mass and ϕ_j is the mode shape coefficient of j-th mass.

16.2 FRICTION DAMPERS

Friction dampers works on the principle of friction between surfaces arising out of the movement of the surfaces in contact with each other. Friction pads are used in the plates to enhance friction. The amount of damping mainly depends on the quantum of movement of the surfaces, coefficient of friction between surfaces and normal reaction arising out of grip force between the surfaces. Figure 16.4 shows a friction damper where friction takes place between the flange of the beam and the friction pad. Figure 16.5 shows a double-action friction damper.

The frictional force in the damper can be expressed by Eq. (16.2.1) (Li and Reinhorn, 1995):

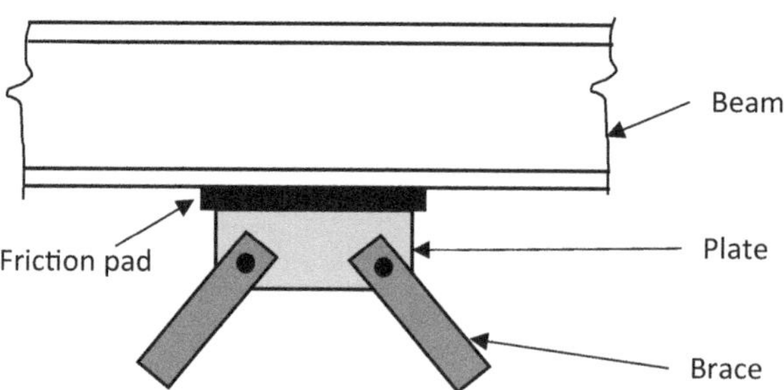

FIGURE 16.4 Friction damper.

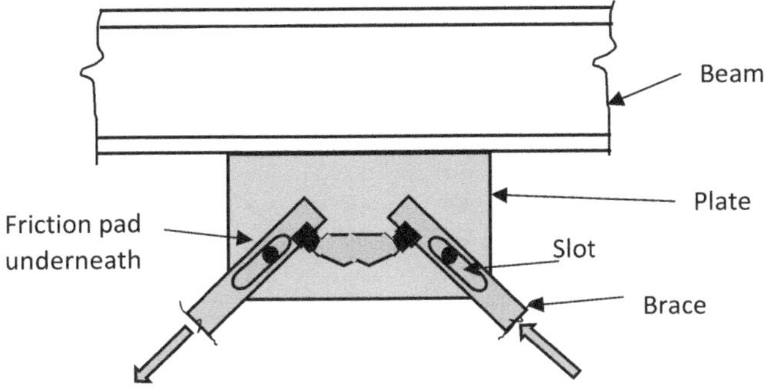

FIGURE 16.5   Double action friction damper.

$$F_f = ku \qquad \text{where} \quad \left|F_f\right| \le \mu_{max} R \qquad\qquad (16.2.1a)$$

$$F_f = \mu_{min} R \quad \text{where} \quad \mu_{min} R \le \left|F_f\right| < \mu_{max} R \qquad\qquad (16.2.1b)$$

Where F_f is the friction force in the damper, k is the stiffness of the damper, μ_{max} is the maximum coefficient of friction that occurs at the start of the movement, μ_{min} is the minimum coefficient of friction that occurs after sliding occurs and R is the normal reaction. It may be noted that friction force can be positive or negative depending on the direction of motion. It always opposes the applied force.

In order to have a normal reaction, there should be a grip between two sliding surfaces. This is achieved through bolting or spring action.

Figure 16.6 shows a friction damper in action with central links. Frictional energy dissipation takes place in the link due to friction among plates in the links. This damper can be easily fixed in the diagonal brace system.

Figure 16.7 shows Sumitomo friction damper reported by Aiken et al. (1992).

Figure 16.8 shows an arrangement of friction damper in a steel structure.

Slotted bolted connection (SBC) is a simple and low-cost friction damper (Figure 16.9). This type of damper was reported by Fitzgerald et al. (1989). The main feature of this damper is a slotted plate connected to another plate by bolting. The slot allows movement of the plate. The bolt gives necessary grip and normal force.

An arrangement of EDD in a building is shown in Figure 16.10.

16.3   YIELDING METALLIC DAMPERS

The yielding metallic dampers work on the principle of hysteresis of metals (generally mild steel). Hysteresis is the lag between the input force and output deformation. The energy dissipated in one hysteretic cycle is the area under the hysteresis loop. Larger hysteresis loop indicates more energy dissipation. A larger area of the

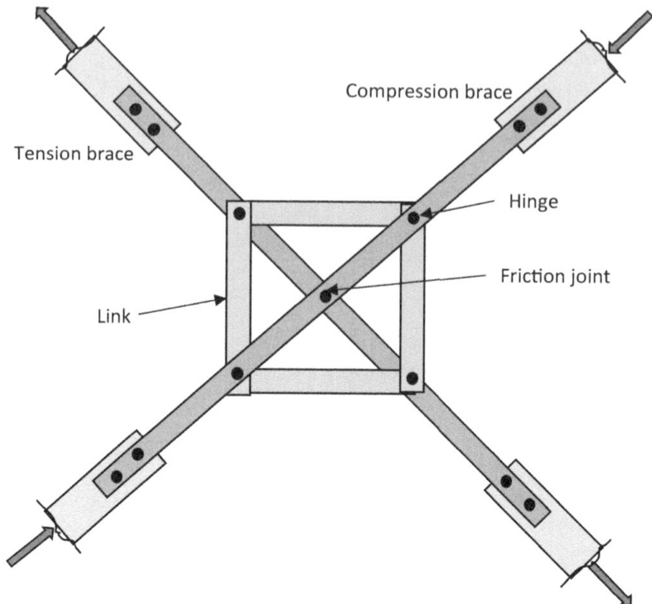

FIGURE 16.6 Friction damper with links.

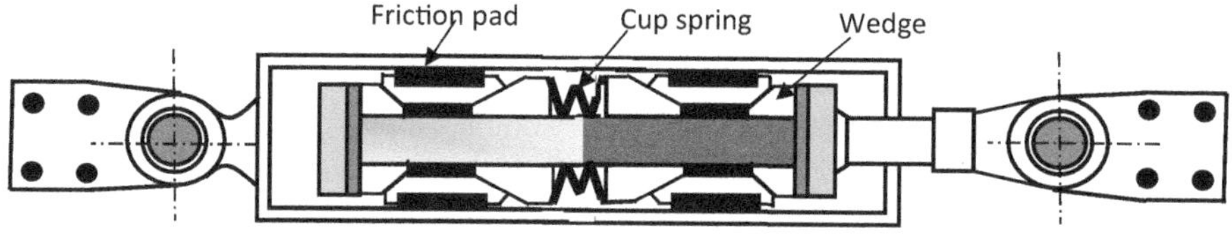

FIGURE 16.7 Sumitomo friction device (schematic).

loop will occur when the inelastic deformation is large, without failure. Some of the metallic plate dampers are explained briefly in the following.

16.3.1 Constricted Plate Damper

An yielding metallic damper with a dumbbell-shaped plates yields easily at the constricted area of the plate. Once inelasticity sets in, the hysteresis loops will be generated in to-and-fro motion caused by earthquakes. A schematic diagram of the damper is shown in Figure 16.11.

16.3.2 Rolling Bending Damper

Rolling bending dampers are made of plates bent in an elliptical shape. Two faces of the dampers are welded to moving parts of the structure. As the structural parts move relative to each other, the plate rolls and bends in rolling. The damper may also touch

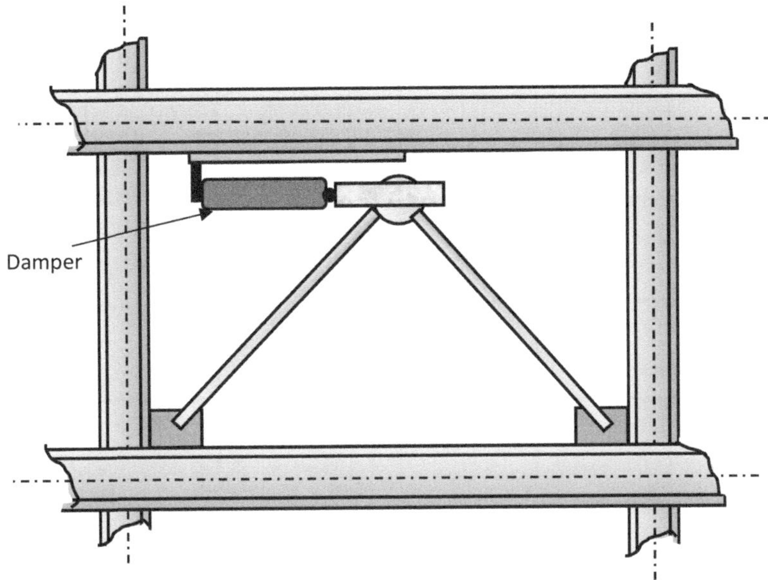

FIGURE 16.8   Typical placement of a Sumitomo friction device.

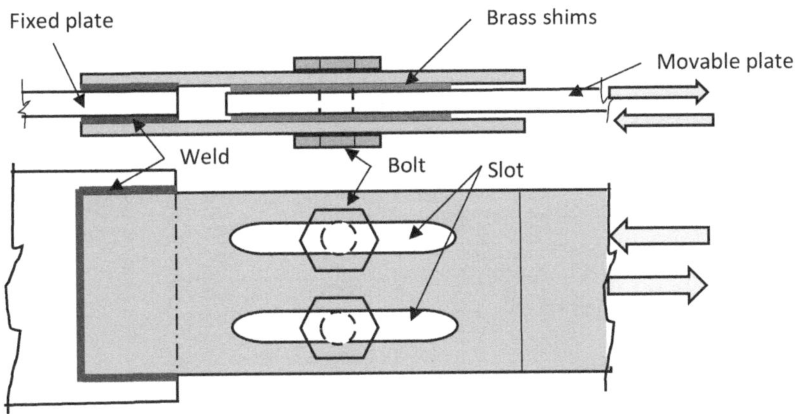

FIGURE 16.9   Slotted bolted connection.

the inelastic range. The energy is dissipated during rolling–bending and through hysteresis action. A schematic view of the damper is shown in Figure 16.12.

The demerit of yielding metallic dampers is during minor earthquakes, they may not yield and will act as an added stiffness.

16.4   FLUID VISCOUS DAMPERS

Fluid viscous damper (FVD) is being used since long in shock absorbing devices in the automobile industry. It has also been used in the airplane wheel shock absorbing

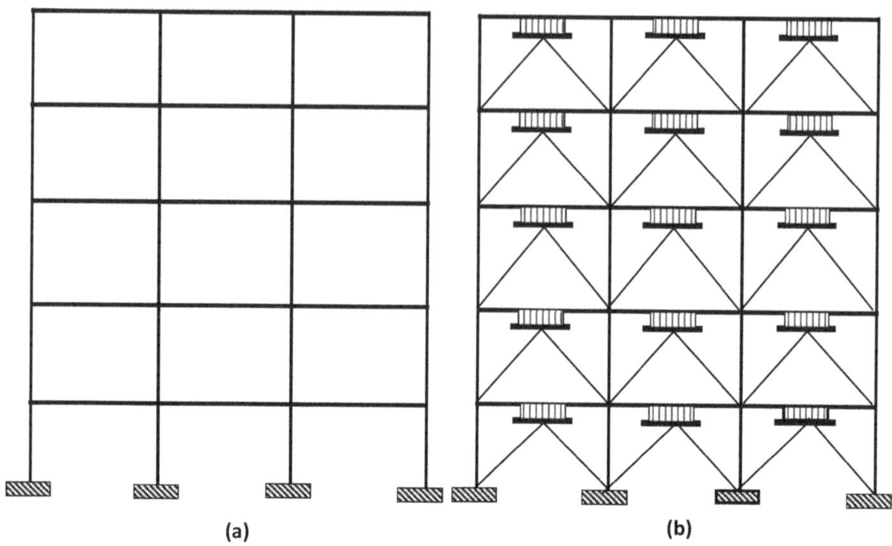

(a)   **(b)**

FIGURE 16.10   An arrangement of dampers in a steel frame building.

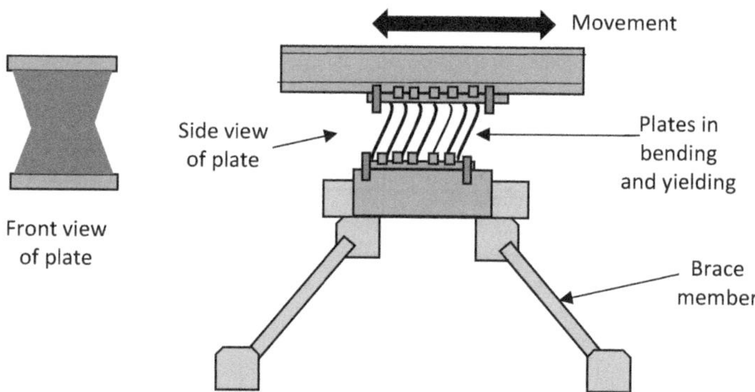

FIGURE 16.11   Yielding metallic damper with dumbbell-shaped plates.

mechanism. FVD acts on the principle of fluid flow through orifice. They are also known as hydraulic dampers.

The movement of fluid produces a drag or resistance to any object within it. This is because of viscous force that is generated during the movement. On the other hand, if the fluid passes through an orifice, it does some work and as a result energy gets dissipated. This is the principle of fluid viscous dampers.

The force acting in the damper due to fluid movement given by Eq. (16.4.1).

16.4.1   ORIFICE TYPE DAMPER

$$F_D = C|\dot{u}|^\alpha .\mathrm{sign}(\dot{u})$$

(16.4.1)

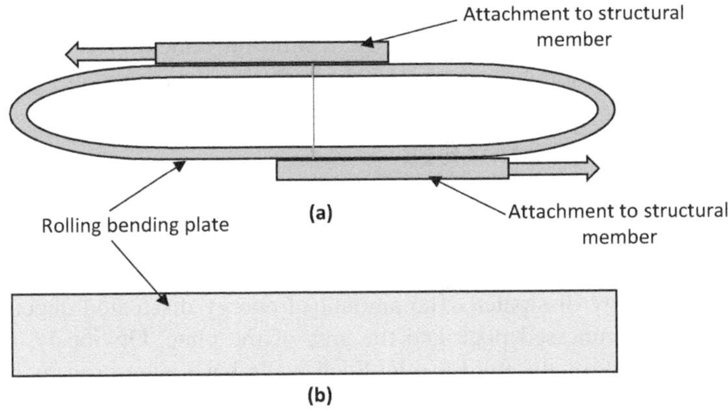

FIGURE 16.12   Rolling–bending damper: (a) elevation and (b) plan.

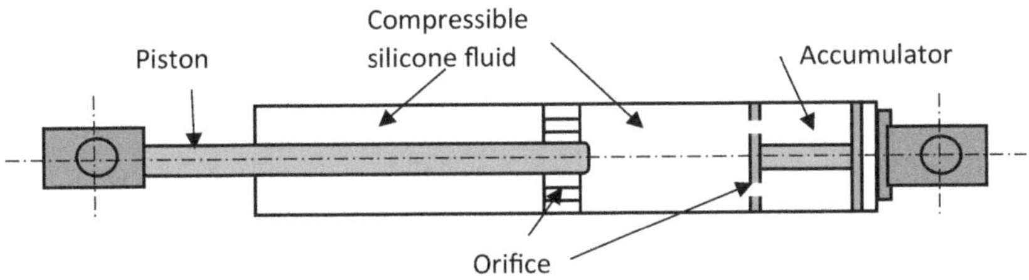

FIGURE 16.13   Fluid viscous damper.

where F_D is the damper force, $\dot{u}$ is fluid velocity, $\mathrm{sign}(\dot{u})$ is the sign of velocity and α is the index signifying the degree of nonlinearity in the fluid action. When $\alpha = 1$, it is a linear damper. As α reduces, nonlinearity increases. The force–deformation response of FVD depends on the frequency of excitation.

Where F_D is the force in the damper due to movement of the fluid, the displacement of fluid is the velocity of the fluid. The sign of the velocity decides the positive and negative nature of the force, that is, its direction of action. The FVD is used not only in structural response control, but also in gun recoil control, airplane shock absorber and various other hydraulic controls. A general picture of a FVD is shown in Figure 16.13. It consists of a cylinder and piston. The cylinder contains a highly viscous silicone fluid. The fluid moves by the movement of piston caused by the earthquake (when tied diagonally in braces). The fluid has an accumulator to accumulate the fluid when compressed. There are orifices through which fluid moves, and in the process, work is done. The work done is a measure of energy dissipated in the process.

The efficiency and usefulness of FVD depends on (i) fluid viscosity, (ii) stroke length, (iii) orifice action and (iv) velocity of fluid movement. The cost of FVD is proportional to the damping coefficient to be attained in the damper.

16.5 VISCOUS WALL DAMPER

A viscous wall damper consists of a steel plate and is submerged in viscous fluid contained in a wall. The submerged wall moves under the influence of the earthquake. The drag generates force and work is done by the earthquake motion. The work done is nothing but energy dissipated. The amount of energy dissipated depends on the movement of the immersed plate and the area of the plate. Obviously, it depends also on the viscosity of the fluid inside. Such walls have been used in Japan. The force generated in the viscous wall is given by Eq. (16.5.1). A viscous wall damper is shown in Figure 16.14.

$$F = \mu \dot{u} A \tag{16.5.1}$$

where F is the force of drag that acts in a direction opposite to the direction of the earthquake motion, μ is the coefficient of viscosity of the fluid, $\dot{u}$ is the velocity of the moving wall and A surface is the area of the moving wall.

For linear viscous response, Eq. (16.5.2) is valid.

$$\left(\frac{F}{c \omega u_{max}} \right)^2 + \left(\frac{u}{u_{max}} \right)^2 = 1 \tag{16.5.2}$$

where F is the wall force, c is the damping coefficient, ω is the angular frequency, u is the displacement parameter and u_{max} is the maximum displacement. Eq. (16.5.2) gives the response as an oval shape (Figure 16.15).

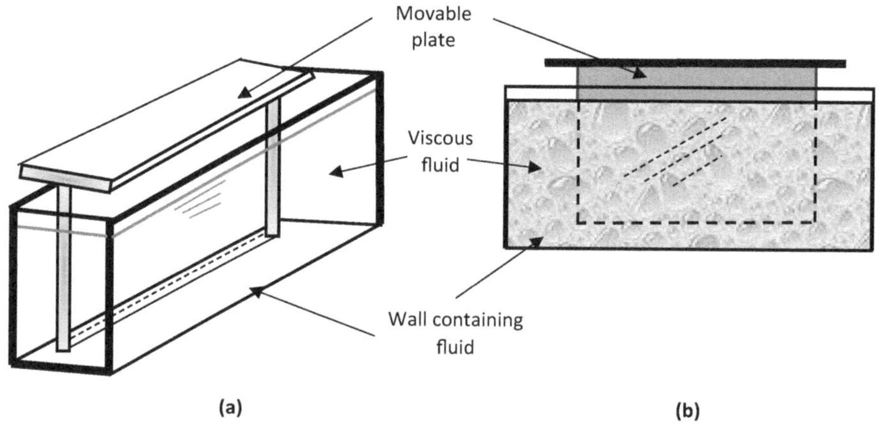

FIGURE 16.14 Viscous wall damper.

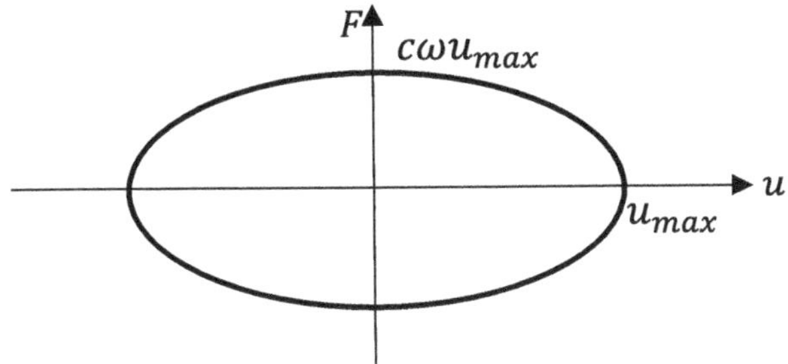

FIGURE 16.15 Response of a linear viscous wall damper.

16.6 VISCOELASTIC DAMPERS

Viscoelastic dampers (VED) consist of one or more layer of viscoelastic material such as an acrylic polymer sandwiched between metal elements. The movement in the metal element cause the VE material to undergo reversible shear deformation. Through this process, hysteresis energy is dissipated. VEDs are influenced by operating frequencies and temperature. VEDs have been used in many important buildings including the World Trade Centre (110 stories), Columbia SeeFirst Building (73 stories) and Union Square Building (60 stories) (Li and Reinhorn, 1995).

The relationship between the shear stress and shear strain in VEDs is given by Eq. (16.6.1).

$$\tau(t) = G_{storage}\,\gamma(t) \pm G_{loss}\sqrt{\gamma_0^2 - \gamma(t)^2} \qquad (16.6.1)$$

where

$\tau(t)$ = time dependent shear stress

$G_{storage}$ = shear storage modulus = $\tau_0\cos\delta/\gamma_0$

G_{loss} is shear loss modulus = $\tau_0\sin\delta/\gamma_0$

γ_0 = peak shear strain

$\gamma(t)$ = time varying shear strain

δ = phase angle between shear stress and shear strain.

Eq. (16.6.1) is graphically shown in Figure 16.16. The ratio of shear loss modulus and shear storage modulus is called the loss factor. A higher loss factor indicates more efficiency of the damper in dissipating energy.

$$\text{Loss factor} = G_{loss}/G_{storage}.$$

VE dampers consists of two outer plates and an inner plate with two layers of VE material on two sides of the inner plate (Figure 16.17). As VEDs have both stiffness and damping, they can be modellea d as spring-dashpot system.

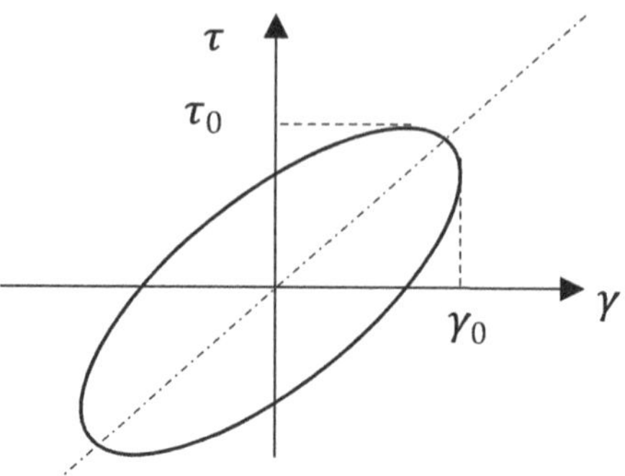

FIGURE 16.16   Shear stress–shear strain response of VE dampers.

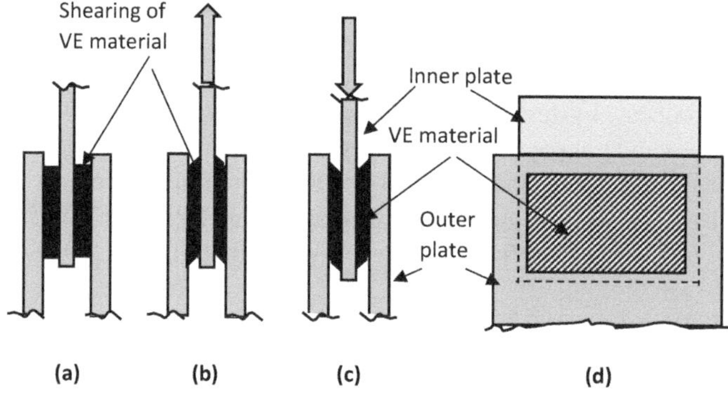

FIGURE 16.17   VE damper: (a) section, (b) force applied upwards, (c) force applied downwards and (d) side view.

The spring-dash pot model of a VED-structure is shown in Figure 16.18. The structure has its stiffness and material damping. The damper has its stiffness and damping. The VED acts along the brace, so the stiffness of brace comes in series with the VED.

With the notations k_s for stiffness of the structure, m for mass of the structure-damper-brace, k_b for stiffness of the brace, k_d for stiffness of the damper, k_{bd} the combined stiffness of the brace and damper and T_{sd} for time period of structure-damper system, And the brace and damper being in series, we can write

$$\frac{1}{k_{bd}} = \frac{1}{k_b} + \frac{1}{k_d}$$

$$\text{Or,} \quad k_{bd} = \frac{k_b k_d}{k_b + k_d} \tag{16.6.2}$$

The time period for structure-damper system is given by Eq. (16.6.3).

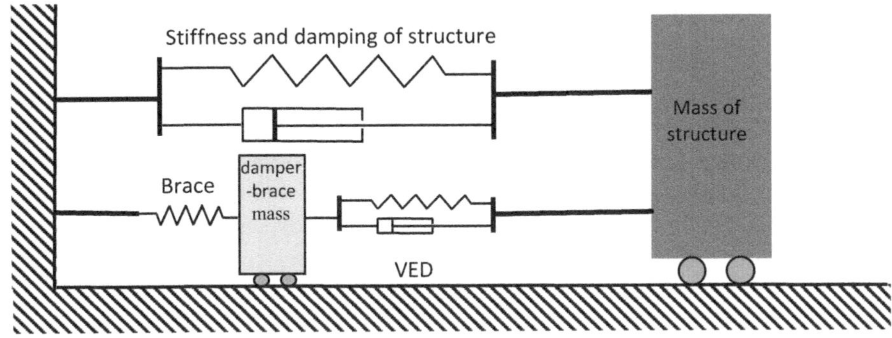

FIGURE 16.18   Modelling of VED with a brace.

$$T_{sd} = \sqrt{\frac{m}{k_s + k_{bd}}}$$
(16.6.3)

The damping ratio of the system is given by Eq. (16.6.4).

$$\xi = \frac{1}{4\pi}\frac{E_D}{E_{so}}$$
(16.6.4a)

$$E_{so} = \frac{1}{2}k_s u_{max}^2$$
(16.6.4b)

$$E_D = \frac{1}{2}\pi\gamma_0^2 G_{loss} V$$
(16.6.4c)

Here, V is volume of VE material in the damper. The volume of VE material is obtained from Eq. (16.6.5).

$$V = \frac{2\omega_1^2 [U]^T [m]\{U\}\xi_{target}}{G_{loss}\gamma_0^2}$$
(16.6.5)

where ω_1 is the first mode frequency, $\{U\}$ is the maximum amplitude vector in the structure, m is the mass matrix in the structure and ξ_{target} is the target damping to achieve from dampers. It may be noted that V is total volume of the VE material in all the dampers. This may be equally distributed in the dampers.

16.7 ECCENTRICALLY BRACED FRAMES

Eccentrically braced frames (EBF) are used to increase stiffness in steel frame buildings. The concentrically braced frame (CBF) also increases the stiffness but the EBF has better hysteretic energy dissipating capacity. The hysteretic behaviour is concentrated in EBF through a capacity design, so that other members remain elastic (see Figure 16.19).

    The joint between the brace and frame is through a shear link, which dissipates energy through nonlinear actions. The behaviour of EBF depends on the post-buckling cyclic behaviour of individual braces.

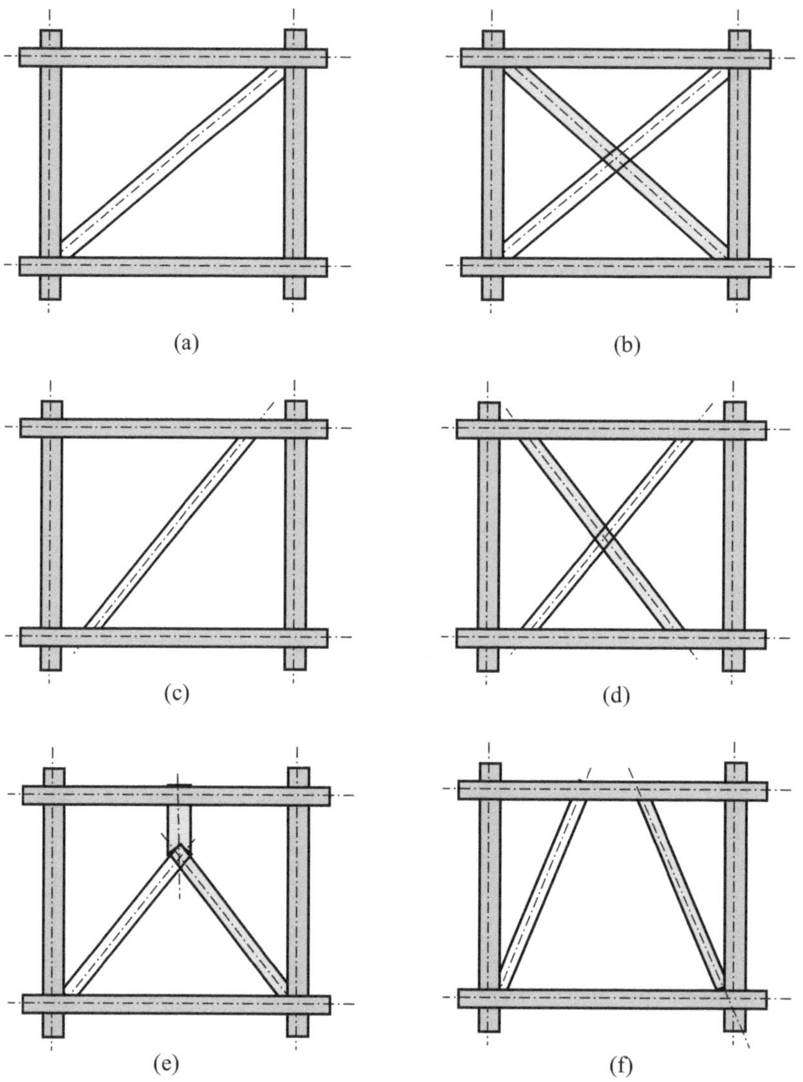

FIGURE 16.19 (a) CBF with one diagonal, (b) CBF with two diagonals, (c) EBF with one diagonal, (d) EBF with two diagonals, (e) inverted Y-brace and (f) EBF.

16.8   ELASTOMERIC SPRING DEVICE

Elastomer is a silicon-based material. An elastomeric spring damper utilizes pressurized compressible elastomer. The compressibility of the elastomer gives rise to spring action and the viscous property of the elastomer gives it the hysteretic property. The damper imposes damping and stiffness on the system. There is a load value up to which the elastomer remains stiff; beyond this force, the elastomer flows through the orifice in the piston head. On the removal of load, the pre-pressure takes the damper to the original state. Elastomeric spring devices are used in end-of rack buffer, car-to-car coupling system, shock absorbing device in missiles and torpedoes, steel mills, material handling units, etc. (Pekcan et al., 1995).

Eq. (16.8.1) was proposed by Pekcan et al. (1995).

$$F = k_2 u + \frac{\left(k_1 - k_2\right)u}{\sqrt{1 + \left(k_1 u/F_s\right)^2}} + c\,\mathrm{sign}\left(\dot{u}\right)\left|u\dot{u}/u_{max}\right|^\alpha \qquad (16.8.1)$$

where

F = damper force response
F_s = damper static force
k_1 = stiffness of the initial damper stiffness
k_2 = stiffness of the damper with elastomer
u = damper displacement response
u_{max} = damper maximum displacement
c = damping coefficient
α = exponent.

A schematic view of an elastomeric damper is shown in Figure 16.20. The force–deformation response of elastomeric spring damper is shown in Figure 16.21.

16.9   SHAPE MEMORY ALLOY

Shape memory alloys (SMA) are special alloys that can 'remember' their original shape. On removal of the load, the element goes back to its original shape. In other words, SMAs are self-centring. Thus, they can repeatedly yield without rupture. The material of the SMA undergoes a reversible phase transformation. The hysteresis loop of SMA is an inclined elongated shape.

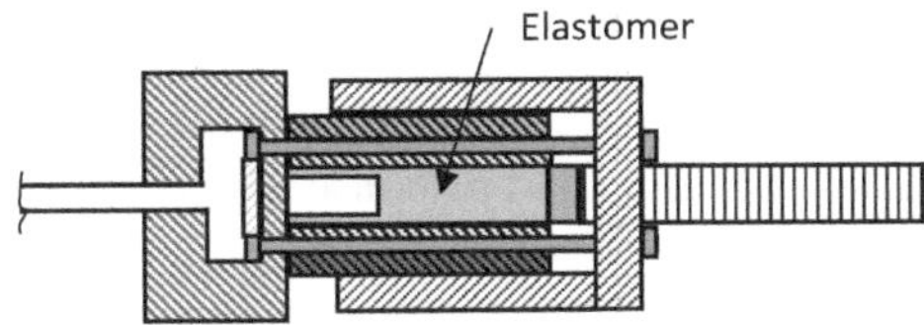

FIGURE 16.20   Schematic view of an elastomeric damper.

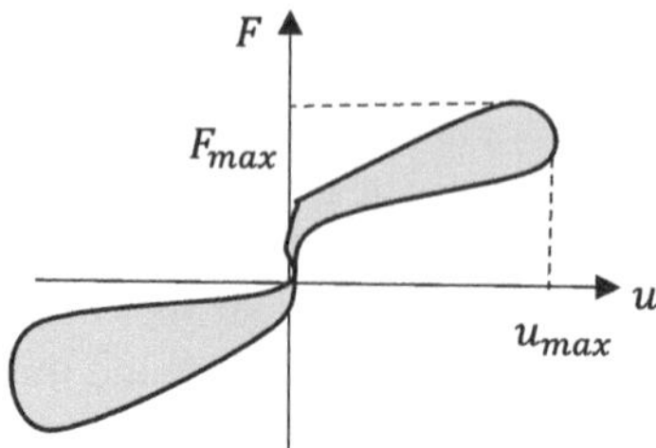

FIGURE 16.21   Force–deformation response of an elastomeric damper.

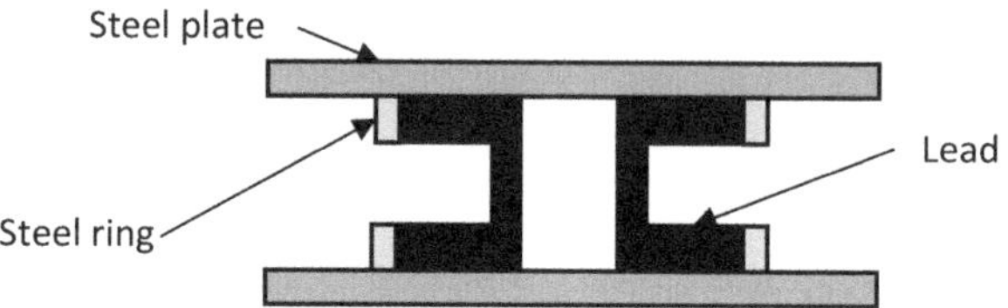

FIGURE 16.22   Lead extrusion device.

16.10   LEAD EXTRUSION DEVICES

Lead extrusion devices (LED) can dissipate energy through hysteresis. Their behaviour is like the friction devices. LEDs have been used in real buildings in New Zealand (Reinhorn and Li, 1995). LEDs are stable devices and can sustain a large number of cycles of operation. These are largely not affected by environmental factors such as temperature and hysteresis. A schematic diagram of lead extrusion device is shown in Figure 16.22.

16.11   CLOSURE

In this chapter, general descriptions of different types of energy dissipating devices are given. EDDs have found large-scale applications in structural response mitigation. They are passive damping devices. To apply the dampers in a real structure, readers may have to undertake a detailed study on the devices, their modelling and dynamic behaviours. There can be a mixture of two or more damper types in a structure.

16.12   EXERCISES

Q16.12.1 Give a brief account of EDDs utilizing fluids.

Q16.12.2 Describe different friction dampers and their working principle.

Q16.12.3 Describe different types of yielding metallic dampers.

Q16.12.4 Describe the modelling and response behaviour of an elastomeric spring damper.

Q16.12.5 Make a thorough literature review of passive dampers.

FURTHER READINGS

Aiken, I.D., Nims, D.K., Whittaker, A.S., Kelly, J.M. (1993) Earthquake Spectra, 9(3), 335–370. https://doi.org/10.1193/1.1585720

Chang K.C., Soong T.T., Oh S.T, and Lai, M.L. (1991) Seismic Response of a 2/5 Scale Steel Structure with Added Viscoelastic Dampers, Report No. NCEER-91-0012.

Constantinou, M.C., Soong, T.T. and Dargush, G.F. (1998) *Passive Energy Dissipation Systems for Structural Design and Retrofit*, MCEER Monograph,.

Fitzgerald, T.F., Anagnos, T., Goodson, M. and Zsutty, T. (1989) Slotted Bolted Connections in Aseismic Design for Concentrically Braced Connections, *Earthquake Spectra*, 5(2), 383–391.

Li, C. and Reinhorn, A.M. (1995) Experimental and Analytical Investigation of Seismic Retrofit of Structures with Supplemental Damping, Report No. NCEER-95-0009, 1995.

Pall, A.S. and Marsh, C. (1982) Seismic Response of Friction Damped Braced Frames, *Journal of Structural Division*, 108, 1313–1323.

Pall, A. and Pall, R.T. (2004) Performance-Based Design Using Pall Friction Dampers – An Economical Design Solution, 13th World Conference on Earthquake Engineering Vancouver, B.C., Canada, 1–6August, 2004, Paper No. 1955.

Pekcan, G., Mander, J.B. and Chen, S.S. (1995) Experimental Performance and Analytical Study of a Non-Ductile Reinforced Concrete Frame Structure Retrofitted with Elastomeric Spring Damper, Report No. NCEER-95-0010.

Pong, W.S., Tsai, C.S. and Lee, G.C. (1994) Seismic Study of Building Frames with Added Energy-Absorbing Devices, Report No. NCEER-94-0016, June.

Reinhorn, A.M. and Li, C. (1995) Experimental and Analytical Investigations of Seismic Retrofit of Structures with Supplemental Damping: Part III – Viscous Walls. Report No. NCEER-95-0013.

Robinson, W.H. (1998) Passive Control of Structures, the New Zealand Experience, ISET Journal.

17 Tuned Liquid Damper

17.1 INTRODUCTION

Any liquid when accelerated under an earthquake in a container experiences an inertia force in a direction opposite to the direction of the earthquake force, which causes the motion. So, if the liquid container is put in a structure, the opposite inertia force of the liquid will try to mitigate the earthquake response of the structure. The movement of the water in a container is called *sloshing*. Tuned liquid damper (TLD) is such a liquid container that has been engineered in some way that it is "tuned" to the structure to arrive at the maximum benefit of sloshing force. The motion of the liquid and structure are explained in Figure 17.1. The gravitational force generated by the liquid mass, the viscous interactive force out of liquid interaction with the rigid container and transitional force due to liquid movement within columns causes energy dissipation. The mechanism of TLDs in controlling the vibrations of the structures is based on the action of liquid sloshing and wave breaking. Both these actions produce damping in in the TLD through energy dissipation by the action of internal fluid viscous force and from breaking of the waves. The quantum of damping induced depends on the amplitude of the liquid motion in the tank and wave patterns of wave breaking. TLD is designed to be in tune with the natural frequency of the structure, but its action should be out of phase. In that case, TLD will help reduce the structural response like story drift, floor response and roof displacements.

The first installation of TLD on a real structure was on the 42 m high Nagasaki airport tower in 1987 (Tamura et al., 1995). This was followed by Yokohama Marine Tower and Tokyo international airport tower in Japan (Tamura et al., 1995).

With reference to Figure 17.1, as the structure moves to the right, the liquid sloshes to left and vice versa. Thus, the water sloshing action tries to diminish the motion of the building by acting in the opposite direction.

There are various types of TLDs as described below.

1. *Flat bottom TLD*: A flat bottom TLD is a rectangular tank with liquid (Figure 17.1(a)). The normal liquid is water. The viscosity of the liquid leads to damping in the system. Flat bottom TLD is called the shallow type when the depth to length of water is less than or equal to 0.15; otherwise, it is called deep water TLD. Shallow-type TLD exhibits nonlinearity in liquid movement

DOI: 10.1201/9781003441090-17

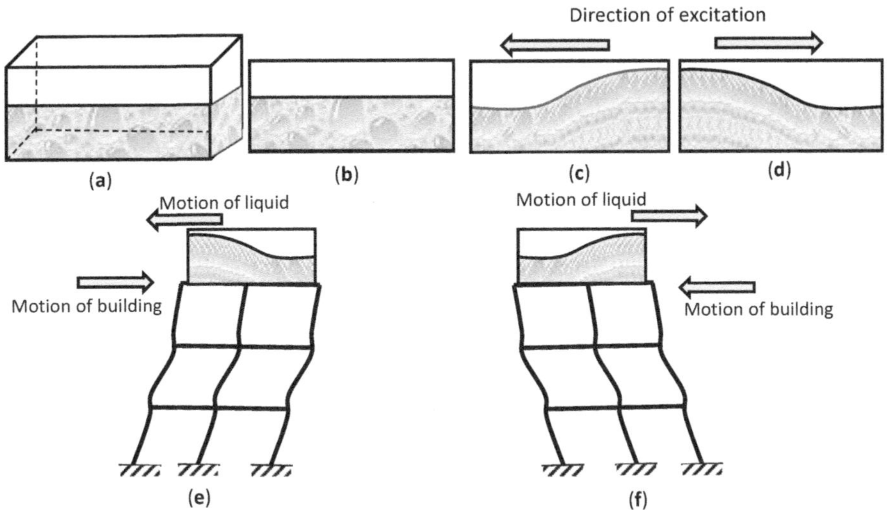

FIGURE 17.1 Motion of liquid in the tank: (a) tank with liquid, (b) front view of the tank, (c) liquid sloshing to the right, (d) liquid sloshing to the left, (e) building moving to the right with the tank and (f) building moving to the left with the tank.

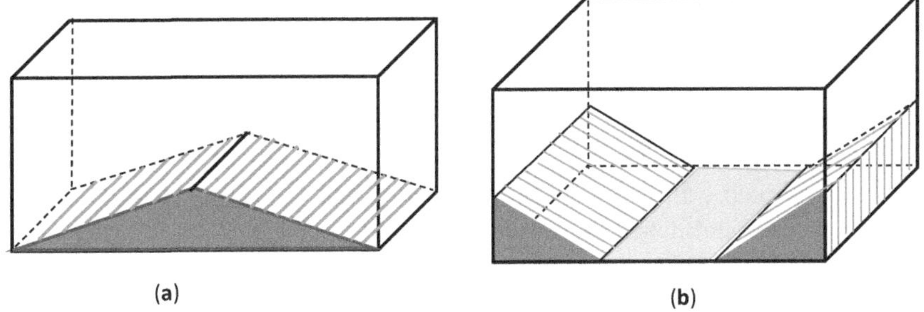

FIGURE 17.2 Slope bottom TLD: (a) V-shape and (b) trough shape.

and shows wave breaking. The mass ratio is the ratio of mass of liquid in the tank to the structural mass. The efficiency of the TLD depends on the mass ratio.

2. *Slope bottom TLD*: The concept of slope bottom TLD is analogous to seashore or tsunami waves near the shore. Near the shore, the sea bottom is sloped towards the sea. As such, the waves moving towards the shore confront an upward sloping bottom, thereby generating turbulence and the wave energy is dissipated. Slope bottom TLD may have V-shaped or trough-shaped bottom (Figure 17.2). As the liquid moves upgradient in the slope, sloshing action, turbulence and wave breaking cause energy dissipation. Slope bottom TLD induces more mass participation from the liquid. The frequency of the TLD

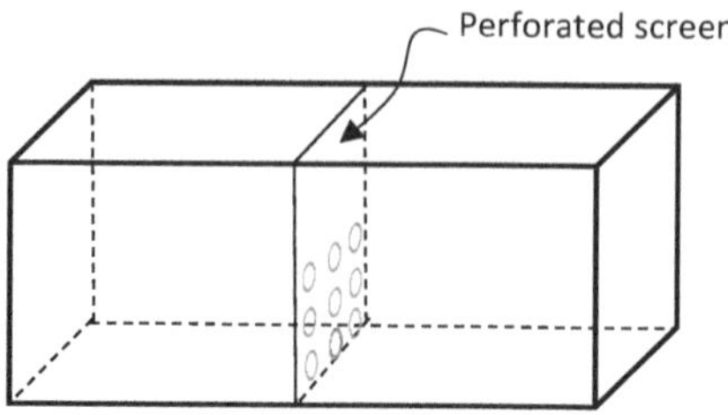

FIGURE 17.3   Perforated wall TLD.

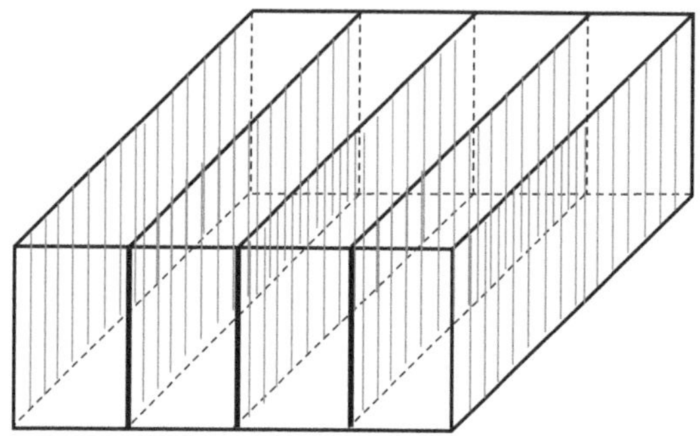

FIGURE 17.4   Multiple TLD.

should be slightly higher than the natural frequency of the structure for better performance.

3. *Perforated screen TLD*: The liquid in the tank may be allowed to pass through a perforated wall while sloshing (Figure 17.3). This will set an orifice action in motion and dissipate energy in the movement of water. It is possible to put multiple walls too. It is also possible to achieve the desired damping by adjusting the hole area to the total screen area (called solidity ratio).

4. *Multiple TLD*: A series of tanks with liquid can be used to increase the total energy dissipation through TLD. The number of TLDs can be decided based on the energy dissipation requirement. A schematic multiple tank is shown in Figure 17.4.

5. *TLD with baffle wall*: Baffle wall is a wall put at some angle in the TLD. This acts as a barrier to the flow of water, thereby dissipating energy. Multiple baffle walls can be put inside a tank. Baffle walls are also used to diminish the destructive action of sloshing in normal liquid retaining tanks.

6. *Tuned liquid column dampers (TLCD)*: Two arms of a continuous U-shaped tube can allow sloshing action during the motion of TLD (Figure 17.5). There is a constriction orifice in the horizontal arm that causes energy dissipation. The horizontal leg may be smaller in diameter to increase the efficiency (Figure 17.6). The efficiency of the TLCD depends on the diameter of the tube, orifice size, horizontal arm length and vertical height of the liquid in the vertical arms. The rise of liquid in arms against gravity also causes work to be done and energy dissipation.

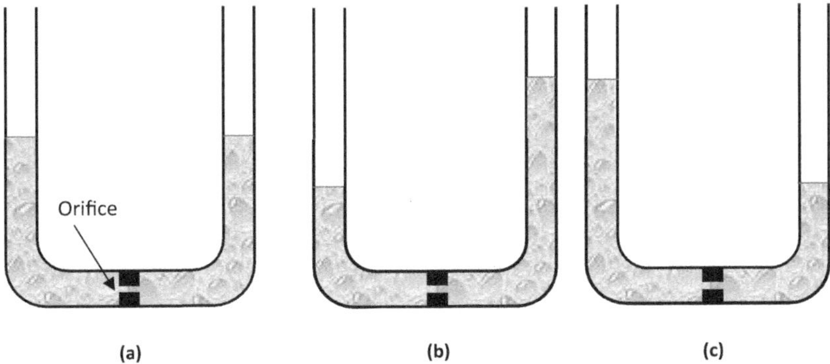

FIGURE 17.5 TLCD: (a) Static condition; (b) sloshing to the right and (c) sloshing to the left.

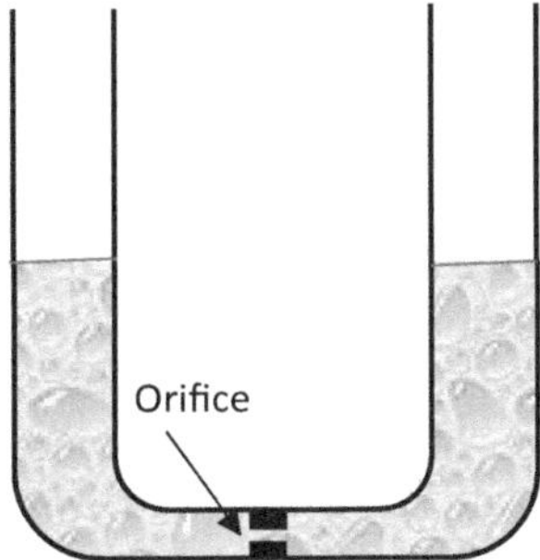

FIGURE 17.6 TLCD with a constricted horizontal arm.

TLD with water as liquid is the most common one in use. Water is also cheaply available. However, water has low viscosity and hence can dissipate less energy. A liquid with a higher viscosity may be used for better performance. Also, a liquid of a higher density works better. The density of water may be increased by mixing salt or sugar. The frictional force at tank bottom can be increased by putting non-soluble granular materials like clean sand. A number of floating balls may be used to increase damping and reduce wave beating. The TLD of other shapes may also be experimented with.

7. *Tuned liquid column ball dampers (TLCBD)*: Here a metallic ball is placed in the horizontal arm of the TLD (Figure 17.7). The provision of the ball leads to an orifice action and that increases the efficiency of the system in dissipating energy.

17.2 DYNAMICS OF TLD

In this section, we study the dynamic background of TLDs. We first look at some basic definitions and formulae in this connection. The critical damping ratio in general is

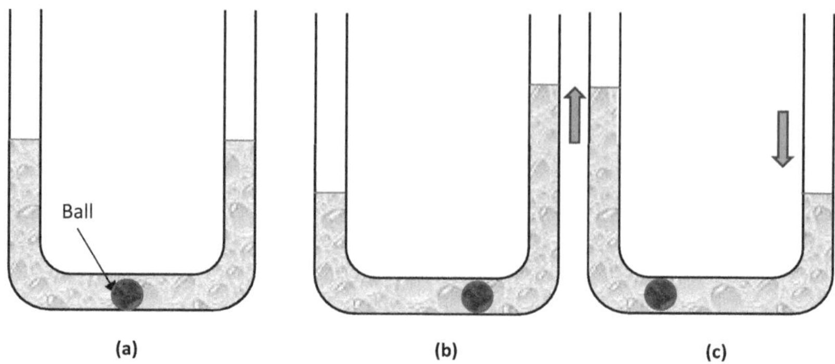

FIGURE 17.7 TLCBD: TLCD with ball.

$\xi = c/c_{critical}$. Also, we know $c_{critical} = 2m\omega$ and $\omega = 2\pi f$, Putting these for liquid of TLD, we can write Eq. (17.2.1).

$$c_{critical} = 4\pi m_L f_L \tag{17.2.1}$$

where f_L is the frequency of the liquid movement in the tank.

m_L is the mass of the liquid in TLD.

Mass ratio: The ratio of the mass of TLD to the mass of structure (m_s) is called as mass ratio (μ_m). Neglecting the mass of tank body of TLD and considering only liquid mass (m_L), the mass ratio is expressed as Eq. (17.2.2).

$$\mu_m = \frac{m_L}{m_s} \tag{17.2.2}$$

Sun et al. (1992) developed Eq. (17.2.3) for damping in TLD. In this equation, υ is the viscosity of the liquid, h is height of the liquid in the tank and L is length of the tank in the direction concerned.

$$\xi = \frac{1}{2\pi}\left(1+\frac{h}{L}\right)\sqrt{\frac{\upsilon}{\pi f_L}} \tag{17.2.3}$$

Tuning ratio: The ratio of natural frequency of the liquid in the tank to the frequency of the structure is called the tuning ratio (η) (Eq. 17.2.4).

$$\eta = \frac{\omega_L}{\omega_s} = \frac{f_L}{f_s} \tag{17.2.4}$$

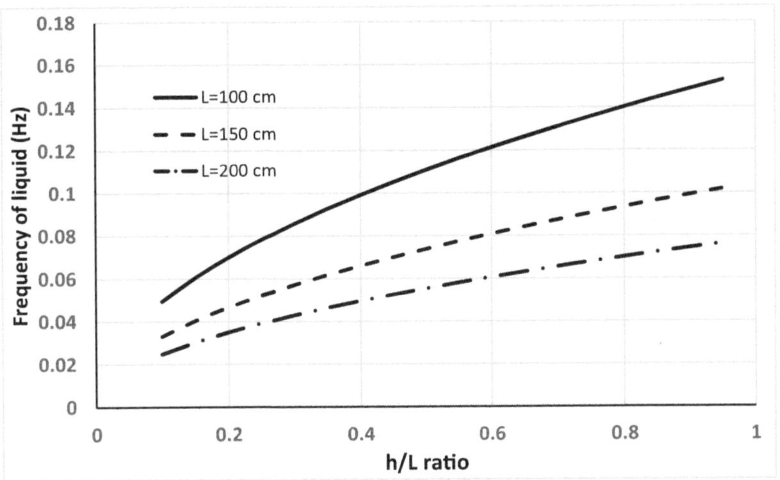

FIGURE 17.8   Variation of TLD frequency with length of the tank.

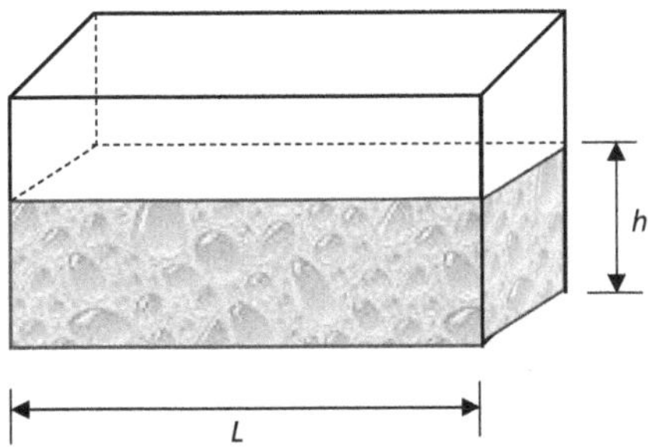

FIGURE 17.9   Liquid dimensions in TLD.

The natural frequency of liquid in the tank is given by Eq. (17.2.5) (Lamb, 1932). Here, g is acceleration due to gravity.

$$f_L = \frac{1}{2\pi}\sqrt{\frac{\pi g}{L}\tanh\frac{\pi h}{L}} \tag{17.2.5}$$

The variations of frequency as per Eq. (17.2.5) with h/L ratio for different values of L are shown in Figure 17.8. The Figure shows that the frequency increases with increase of h/L ratio.

Figure 17.9 shows a TLD with the length and height of liquid.

Depth ratio: The ratio of the depth of liquid to the length of liquid in tank is called depth ratio (δ). This is given by Eq. (17.2.6).

$$\delta = \frac{h}{L} \qquad (17.2.6)$$

17.2.1 Equation of Motion of TLD Structure System

A TLD-structure system is shown in Figure 17.10. In this figure, m_s is the mass of structure, k_s is the stiffness of the structure, m_L is mass of the liquid in the tank and k_L is the stiffness of TLD. We may combine the spring force and damping force in TLD by a variable F_L. With this, we can write the equation of motion of the TLD-structure system as Eq. (17.2.7). In this equation $\ddot{u}_g$ is ground acceleration.

$$m_s\ddot{u}_s + c\dot{u}_s + k_s u_s = -m_s\ddot{u}_g + F_L \qquad (17.2.7)$$

The sloshing force on the wall (F_{wall}) of TLD is given by Eq. (17.2.8) (Reed et al., 1998).

$$F_{wall} = \frac{\rho g B}{2}\left(h_{left}^2 - h_{right}^2\right) \qquad (17.2.8)$$

where h_{left} is the depth of water on the left (receding) side and h_{right} is the depth of water on the right (rising) side, B is the width of the tank, ρ is density of the liquid. Obviously, the force is higher when sloshing height is higher.

17.3 MODELLING OF WATER IN A TANK

Under ground motion, water in a tank on movement produces a dynamic effect on the wall of the tank. The modelling of water is a challenging concept. The dynamic force in water can be roughly modelled by dividing the total water in the tank into two categories: (i) impulsive water mass (m_i), which acts like a rigid body, and (i) convective

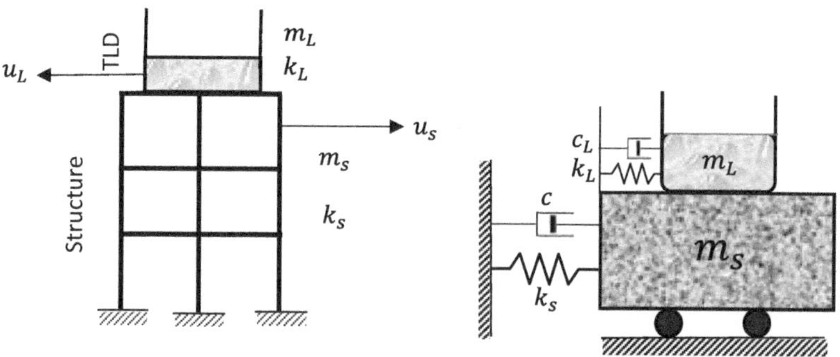

FIGURE 17.10 TLD-structure motion.

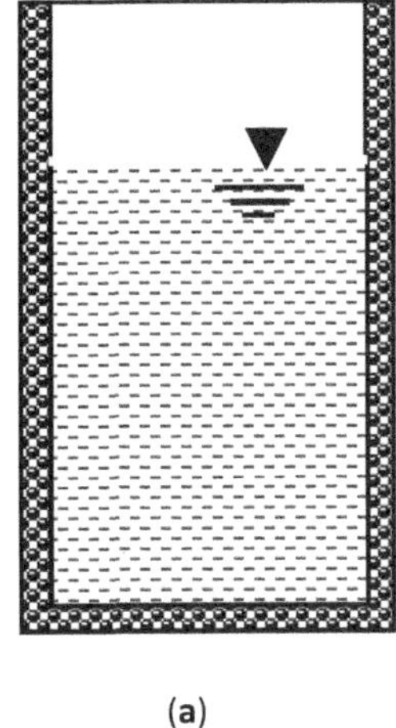

(a)

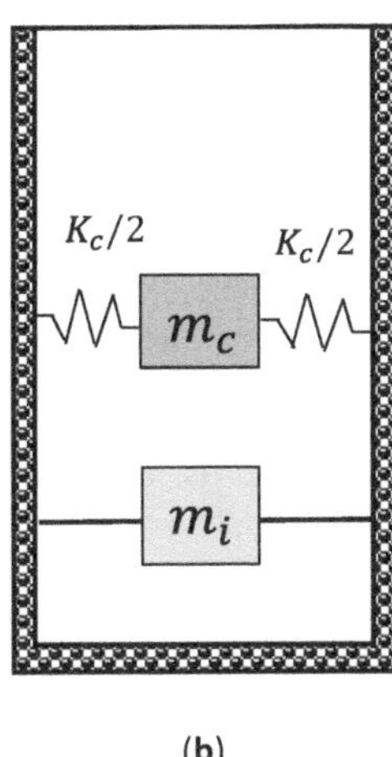

(b)

FIGURE 17.11   Modelling of water.

water mass (m_c), which is more turbulent. The impulsive mass of water interacts with the tank wall. The convective water mass works through the sloshing action. The line of action of impulsive mass from base of the tank is taken as h_i and the line of action of convective mass of water from the base of the tank is designated by h_c. Figure 17.11 illustrates the modelling.

17.4 CLOSURE

Different types of TLDs have been highlighted in this chapter. A brief idea of the working of TLD has been presented. Some modelling aspects have been discussed. If this creates interest in the mind of the reader, material available in the literature may be explored.

FURTHER READINGS

Banerji, P., Murudi, M., Shah, A.H. and Popplewell, N. (2000) Tuned Liquid Dampers for Controlling Earthquake Response of Structures, *Earthquake Engineering & Structural Dynamics,* 29, 587–602.

Banerji, P., Samanta, A. and Chavan, S.A. (2010) Earthquake Control of Vibration Using Tuned Liquid Damper. *International Journal of Advanced Structural Engineering.* 2 (2), 133–152.

Buckingham, E. (1914) On Physically Similar Systems: illustration of the Use of Dimensional Equations, *Bureau of Standards, the American Physical Society,* 4, 345–376.

Das, S. and Choudhury, S. (2017) Seismic Response Control by Tuned Liquid Dampers for Low-rise RC Frame Buildings, *Australian Journal of Structural Engineering,* DOI: 10.1080/13287982.2017.1351180, July 2017.

Das, S., Choudhury, S. and Dey A.K. (2017) Performance of Low-Rise RC Buildings with Tuned Liquid Dampers (TLDs) in the Presence of Masonry Infill, *Asian Journal of Civil Engineering,* 18(4),, 535–546.

Fujino, Y., Sun, L., Pacheco, B.M. and Chaiseri, P. (1992) Tuned Liquid Damper (TLD) for Suppressing Horizontal Motion of Structures, *Journal of Engineering Mechanics*, 118, 2017–2030.

Gardarsson, S., Yeh, H. and Reed, D. (2001) Behavior of Sloped-bottom Tuned Liquid Dampers, *Journal of Engineering Mechanics*, 127(3), 266–271.

Jin, Q., Li, X., Sun, N., Zhou, J. and Guan, J. (2007) Experimental and Numerical Study on Tuned Liquid Dampers for Controlling Earthquake Response of Jacket Offshore Platform, *Marine Structures*, 20, 238–54.

Lamb, H. (1932) *Hydrodynamics,* Cambridge University Press, London, , pp. 619–21.

Olson, D.E. and Reed, D. (2001) A Nonlinear Numerical Model for Sloped-bottom Tuned Liquid Dampers, *Earthquake Engineering & Structural Dynamics* 30, 731–743.

Reed, D., Yu, J., Yeh, H. and Gardarsson, S. (1998) Investigation of Tuned Liquid Dampers under Large Amplitude Excitation, *Journal of Engineering Mechanics* 124(4), 405–413.

Sharma, A., Reddy, G.R. and Vaze, K.K. (2012) Shake Table Tests on a Non-seismically Detailed RC Frame Structure, *Structural Engineering and Mechanics*, 41, 1–24.

Shum, K.M., Xu, Y.L. and Guo, W.H. (2008) Wind-induced Vibration Control of Long Span Cable-stayed Bridges Using Multiple Pressurized Tuned Liquid Column Dampers, *Journal of Wind Engineering and Industrial Aerodynamics* 96, 166–192.

Sun, L.M., Fujino, Y., Pacheco, B.M. and Chaiseri, P. (1992) Modelling of Tuned Liquid Damper (TLD), *Journal of Wind Engineering and Industrial Aerodynamics* 43, 1883–1894.

Tait, M.J. (2008) Modelling and Preliminary Design of a Structure-TLD System, *Engineering Structures* 30, 2644–2655.

Tait, M.J., A. Damatty, N. Isyumov, and M.R. Siddique (2005) Numerical Flow Models to Simulate Tuned Liquid Dampers (TLD) with Slat Screens. *Journal of Fluids and Structures* 20, 1007–1023.

Tamura, Y., Fujii, K., Ohtsuki T., Wakahara T. and Kohsaka R. (1995) Effectiveness of Tuned Liquid Dampers Under Wind Excitation, *Engineering Structures*, 17(9), November, 609–621.

18 Tuned Mass Dampers

18.1 INTRODUCTION

Tuned mass damper (TMD) acts on the principle of the two-degree of freedom system where the main structure is attached with an additional mass, which is tuned to some desired frequency with respect to the frequency of the structure. The mass of the damper is much smaller than the mass of the structure. The ratio of mass of the damper to the mass of the structure is called mass ratio. The characteristics of TMD largely depend on the mass ratio. TMD has its associated damping and stiffness. TMD is very efficient in reducing structural response, is easy to frame and does not generally need any external power source to activate it. However, the demerits are that it is heavy, requires large space and unless special care is taken, it is affected by frictional resistance in the assembly.

The concept TMD was supposed to be first applied to reduce the rolling motion of ships.

18.2 UNDAMPED TMD SYSTEM UNDER SINUSOIDAL EXCITATION

Figure 18.1 shows a dynamic model of structure–TMD combination. With reference to Figure 18.1 and undamped TMD system under sinusoidal excitation, let us use the following symbols:

$$m_{TMD} = \text{mass of TMD}$$
$$m_s = \text{mass of structure}$$
$$\mu_{mass} = \text{mass ratio} = m_{TMD}/m_s$$
$$k_s = \text{stiffness of structure}$$
$$k_{TMD} = \text{stiffness of TMD}$$
$$u_s = \text{displacement of structure}$$
$$u_{TMT} = \text{displacement of TMD}$$
$$\omega_s = \sqrt{\frac{k_s}{m_s}} = \text{natural frequency of the structure}$$
$$\omega_{TMD} = \sqrt{\frac{k_{TMD}}{m_{TMD}}} = \text{natural frequency of the TMD.}$$

DOI: 10.1201/9781003441090-18

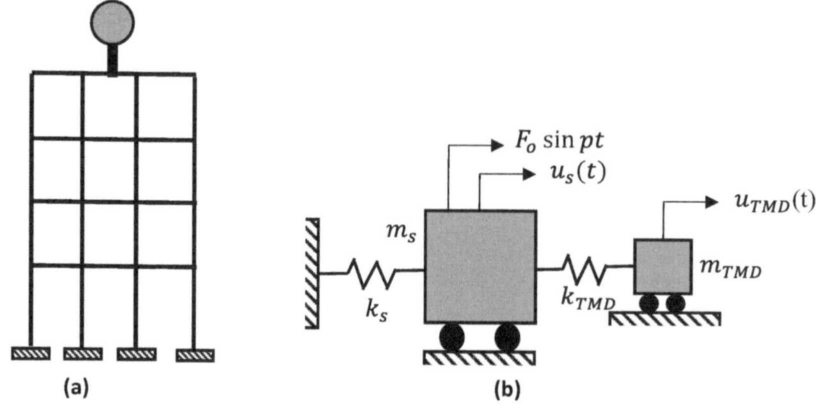

FIGURE 18.1 (a) Structure MDOF system with TMD and (b) Structure–TMD model.

The equations of motions of the structure and TMD are written as Eqs. (18.2.1) and (18.2.2).

$$m_s \ddot{u}_s + \left(k_s + k_{TMD}\right)u_s - k_{TMD}u_{TMD} = F_o \sin pt \tag{18.2.1a}$$

$$m_{TMD}\ddot{u}_{TMD} + k_{TMD}\left(u_{TMD} - u_s\right) = 0 \tag{18.2.1b}$$

Assuming trial solution as

$$u_s\left(t\right) = U_s \sin pt$$

$$u_{TMD}\left(t\right) = U_{TMD} \sin pt$$

where U_s is the peak amplitude of the structure and U_{TMD} is the peak amplitude of TMD.

Substituting these and their second derivative in Eqs. (18.2.1), we get,

$$-m_s U_s p^2 \sin pt + \left(k_s + k_{TMD}\right)U_s \sin pt - k_{TMD}U_{TMD} \sin pt \\ = F_o \sin pt \tag{18.2.2a}$$

$$-m_{TMD}U_{TMD} p^2 \sin pt + k_{TMD}\left(U_{TMD} - U_s\right)\sin pt = 0 \tag{18.2.2b}$$

Equating coefficients of sine terms in Eq. (18.2.2a),

$$-m_s U_s p^2 + \left(k_s + k_{TMD}\right)U_s - k_{TMD}U_{TMD} = F_o \tag{18.2.3a}$$

Equating coefficients of sine terms in Eq. (18.2.2b),

$$-m_{TMD}U_{TMD}p^2 + k_{TMD}\left(U_{TMD} - U_s\right) = 0 \qquad (18.2.3b)$$

Dividing Eq. (18.2.3a) by k_s,

$$-m_s U_s p^2/k_s + \left(1 + \frac{k_{TMD}}{k_s}\right)U_s - k_{TMD}U_{TMD}/k_s = F_o/k_s \qquad (18.2.4a)$$

Noting that $F_o/k_s = $ static deflection $= u_{st}$, $\quad \omega_s^2 = \dfrac{k_s}{m_s}$, $\quad \omega_{TMD}^2 = \dfrac{k_{TMD}}{m_{TMD}}$, Eq. (18.2.4a) gives

$$-U_s p^2/\omega_s^2 + \left(1 + \frac{k_{TMD}}{k_s}\right)U_s - U_{TMD}\omega_{TMD}^2 = u_{st}$$

$$\text{Or,}\quad U_s\left(1 + \frac{k_{TMD}}{k_s} - \frac{p^2}{\omega_s^2}\right) - U_{TMD}\frac{k_{TMD}}{k_s} = u_{st} \qquad (18.2.5a)$$

Dividing (18.2.3b) by k_{TMD},

$$-U_{TMD}p^2\frac{m_{TMD}}{k_{TMD}} + U_{TMD} - U_s = 0$$

$$\text{Or,}\quad -U_{TMD}p^2/\omega_{TMD}^2 + U_{TMD} - U_s = 0$$

$$\text{Or,}\quad U_s = -U_{TMD}p^2/\omega_{TMD}^2 + U_{TMD} =$$

$$\text{Or,}\quad U_s = U_{TMD}\left(1 - \frac{p^2}{\omega_{TMD}^2}\right) \qquad (18.2.5b)$$

We need to solve Eqs. (18.2.5a) and (18.2.5b) for unknowns U_s and U_{TMD}.
  Substituting Eq. (18.2.5b) in Eq. (18.2.5a),

$$U_{TMD}\left(1 - \frac{p^2}{\omega_{TMD}^2}\right)\left(1 + \frac{k_{TMD}}{k_s} - \frac{p^2}{v_s^2}\right) - U_{TMD}\frac{k_{TMD}}{k_s} = u_{st}$$

$$\Rightarrow U_{TMD}\left[\left(1-\frac{p^2}{\omega_{TMD}^2}\right)\left(1+\frac{k_{TMD}}{k_s}-\frac{p^2}{\omega_s^2}\right)-\frac{k_{TMD}}{k_s}\right]=u_{st}$$

$$\Rightarrow \frac{U_{TMD}}{u_{st}}=\frac{1}{\left(1-\frac{p^2}{\omega_{TMD}^2}\right)\left(1+\frac{k_{TMD}}{k_s}-\frac{p^2}{\omega_s^2}\right)-\frac{k_{TMD}}{k_s}} \qquad (18.2.6a)$$

Dividing both sides of Eq. (18.2.5b) by u_{st},

$$\frac{U_s}{u_{st}}=\frac{U_{TMD}}{u_{st}}\left(1-\frac{p^2}{\omega_{TMD}^2}\right) \qquad (18.2.6b)$$

Substituting the value of $\dfrac{U_{TMD}}{u_{st}}$ from Eq. (18.2.6a) in Eq. (18.2.6b),

$$\frac{U_s}{u_{st}}=\frac{\left\{1-\frac{p^2}{\omega_{TMD}^2}\right\}}{\left\{1-\frac{p^2}{\omega_{TMD}^2}\right\}\left\{1+\frac{k_{TMD}}{k_s}-\frac{p^2}{\omega_s^2}\right\}-\frac{k_{TMD}}{k_s}} \qquad (18.2.6c)$$

If the numerator of Eq. (18.2.6c) is zero, then U_s becomes zero, indicating that the structure does not vibrate. So, the condition to avoid the vibration of the structure is

$$1-\frac{p^2}{\omega_{TMD}^2}=0$$

$$\Rightarrow p=\omega_{TMD} \text{ for no motion of the structure} \qquad (18.2.7)$$

This shows that when the TMD and the structure have the same frequency of vibration, the structural vibration is absent theoretically.

18.2.1   Resonance Condition Between Structure and TMD

Considering *resonance condition* between the structure and TMD,

$$\omega_s=\omega_{TMD}$$

$$\Rightarrow \sqrt{\frac{k_s}{m_s}}=\sqrt{\frac{k_{TMD}}{m_{TMD}}}$$

$$\Rightarrow \frac{k_s}{m_s} = \frac{k_{TMD}}{m_{TMD}}$$

$$\Rightarrow \frac{k_{TMD}}{k_s} = \frac{m_{TMD}}{m_s} = \mu_{mass},$$

Here μ_{mass} is the mass ratio.

With this resonance condition, Eq. (18.2.6a) becomes

$$\frac{U_{TMD}}{\mu_{st}} = \frac{1}{\left(1 - \dfrac{p^2}{\omega_{TMD}^2}\right)\left(1 + \mu_{mass} - \dfrac{p^2}{\omega_s^2}\right) - \mu_{mass}} \tag{18.2.1.1a}$$

$$\frac{U_s}{u_{st}} = \frac{\left\{1 - \dfrac{p^2}{\omega_{TMD}^2}\right\}}{\left[1 - \dfrac{p^2}{\omega_{TMD}^2}\right]\left\{1 + \mu_{mass} - \dfrac{p^2}{\omega_s^2}\right\} - \mu_{mass}} \tag{18.2.1.1b}$$

Remember that we made the initial trial solution as

$$u_s(t) = U_s \sin pt$$

$$u_{TMD}(t) = U_{TMD} \sin pt$$

Dividing both sides by u_{st} and substituting (18.2.6c) and (18.2.6a), we get

$$\frac{u_s(t)}{u_{st}} = \frac{\left\{1 - \dfrac{p^2}{\omega_{TMD}^2}\right\}}{\left[1 - \dfrac{p^2}{\omega_{TMD}^2}\right]\left\{1 + \mu_{mass} - \dfrac{p^2}{\omega_s^2}\right\} - \mu_{mass}} \sin pt \tag{18.2.1.2}$$

$$\frac{u_{TMD}(t)}{u_{st}} = \frac{1}{\left(1 - \dfrac{p^2}{\omega_{TMD}^2}\right)\left(1 + \mu_{mass} - \dfrac{p^2}{\omega_s^2}\right) - \mu_{mass}} \sin pt \tag{18.2.1.3}$$

Eqs. (18.2.1.2) and (18.2.1.3) express the structural response and TMD response in terms of u_{st}.

The denominator of both the equations are the same. When the denominator is zero, the responses are infinite, which is the resonance condition. Equating the denominator to zero,

$$\left(1-\frac{p^2}{\omega_{TMD}^2}\right)\left(1+\mu_{mass}-\frac{p^2}{\omega_s^2}\right)-\mu_{mass}=0 \tag{18.2.1.4a}$$

As at resonance between structure and TMD, $\omega_s = \omega_{TMD}$, Eq. (18.2.1.4a) can be written as

$$\left(1-\frac{p^2}{\omega_{TMD}^2}\right)\left(1+\mu_{mass}-\frac{p^2}{\omega_{TMD}^2}\right)-\mu_{mass}=0 \tag{18.2.1.4b}$$

Let $\dfrac{p^2}{\omega_{TMD}^2} = x$; Eq. (18.2.1.4a) becomes

$$\left(1-x\right)\left(1+\mu_{mass}-x\right)-\mu_{mass}=0$$

$$\Rightarrow 1+\mu_{mass}-x-x-x\mu_{mass}+x^2-\mu_{mass}=0$$

$$\Rightarrow x^2-2x-x\mu_{mass}+1=0$$

$$\Rightarrow x^2-\left(2+\mu_{mass}\right)x+1=0$$

$$\Rightarrow x = \frac{\left(2+\mu_{mass}\right)\pm\sqrt{\left(2+\mu_{mass}\right)^2}}{2}$$

$$\Rightarrow x = \left(1+\frac{\mu_{mass}}{2}\right)\pm\sqrt{\mu_{mass}+\frac{\mu_{mass}^2}{4}}$$

$$\text{Or,}\ \frac{p^2}{\omega_{TMD}^2}=\left(1+\frac{\mu_{mass}}{2}\right)\pm\sqrt{\mu_{mass}+\frac{\mu_{mass}^2}{4}} \tag{18.2.1.5}$$

Eq. (18.2.1.5) is plotted in Figure 18.2.

Figure 18.2 shows that corresponding to any mass ratio (μ_{mass}) there are two possible frequency ratios (p/ω).

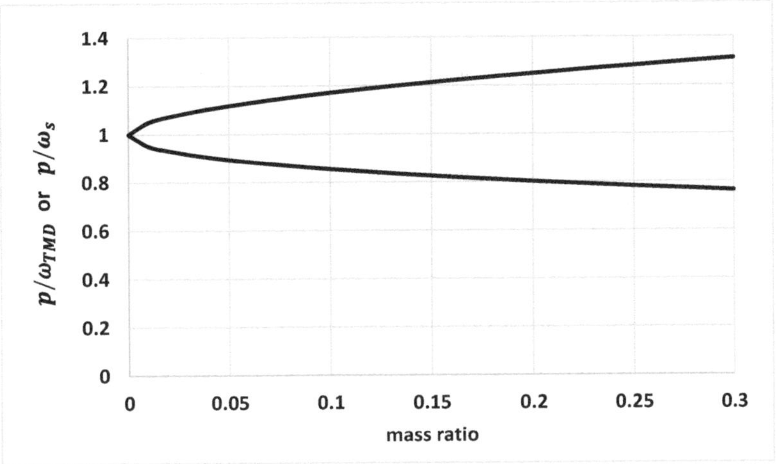

FIGURE 18.2 Variation of frequency ratio with mass ratio.

Eqs. (18.2.1.2) and (18.2.1.3) will reach the peak value when $\sin pt$ is equal to unity. So, we can write

$$\left(\frac{u_s(t)}{u_{st}}\right)_{max} = \frac{\left\{1 - \dfrac{p^2}{\omega_{TMD}^2}\right\}}{\left\{1 - \dfrac{p^2}{\omega_{TMD}^2}\right\}\left\{1 + \mu_{mass} - \dfrac{p^2}{\omega_s^2}\right\} - \mu_{mass}} \tag{18.2.1.6}$$

$$\left(\frac{u_{TMD}(t)}{u_{st}}\right)_{max} = \frac{1}{\left(1 - \dfrac{p^2}{\omega_{TMD}^2}\right)\left(1 + \mu_{mass} - \dfrac{p^2}{\omega_s^2}\right) - \mu_{mass}} \tag{18.2.1.7}$$

18.3 UNDAMPED TMD SYSTEM UNDER SINUSOIDAL BASE EXCITATION

We consider that base or ground movement $(u_g(t))$ is expressed as

$$u_g(t) = u_{go} \sin pt \tag{18.3.1}$$

where u_{go} is the peak ground displacement.

Consider Figure 18.3. The equations of motion of structure and TMD can be expressed as

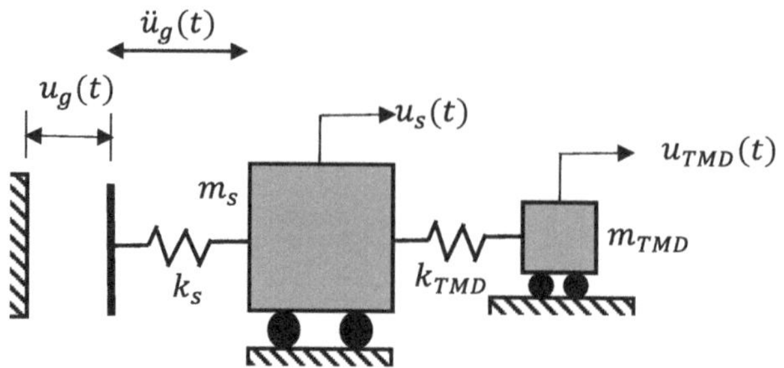

FIGURE 18.3 Structure–TMD model under base excitation.

$$m_s\ddot{u}_s + \left(k_s + k_{TMD}\right)u_s - k_{TMD}u_{TMD} = m_s p^2 u_{go} \sin pt \qquad (18.3.2)$$

$$m_{TMD}\ddot{u}_{TMD} + k_{TMD}\left(u_{TMD} - u_s\right) = m_{TMD} p^2 u_{go} \sin pt \qquad (18.3.3)$$

The right-hand sides of the above equations are obtained from the analogy of centrifugal force in circular motion.

Substitute the following trial solutions:

$$u_s\left(t\right) = U_s \sin pt$$

$$u_{TMD}\left(t\right) = U_{TMD} \sin pt$$

Eqs. (18.3.2) and (18.3.3) take the following forms:

$$-(m_s U_s p^2 + \left(k_s + k_{TMD}\right)U_s - k_{TMD}U_{TMD})\sin pt = m_s u_{go} p^2 \sin pt$$

$$\text{Or,} \quad -m_s U_s p^2 + \left(k_s + k_{TMD}\right)U_s - k_{TMD}U_{TMD} = m_s p^2 u_{go} \qquad (18.3.4)$$

And

$$(-m_{TMD}U_{TMD}p^2 + k_{TMD}\left(U_{TMD} - U_s\right))\sin pt = m_{TMD}u_{go}p^2 \sin pt$$

$$\text{Or,} \quad -m_{TMD}U_{TMD}p^2 + k_{TMD}\left(U_{TMD} - U_s\right) = m_{TMD}p^2 u_{go} \qquad (18.3.5)$$

Dividing Eq. (18.3.4) by k_s,

$$\frac{-m_s}{k_s}U_s p^2 + \left(1 + \frac{k_{TMD}}{k_s}\right)U_s - \frac{k_{TMD}}{k_s}U_{TMD} = \frac{m_s}{k_s}p^2 u_{go}$$

$$\Rightarrow \left(1 + \frac{k_{TMD}}{k_s} - \frac{m_s}{k_s} p^2 \right) U_s - \frac{k_{TMD}}{k_s} U_{TMD} = \frac{m_s}{k_s} p^2 u_{go}$$

$$\Rightarrow \left(1 + \frac{k_{TMD}}{k_s} - \frac{p^2}{\omega_s^2} \right) U_s - \frac{k_{TMD}}{k_s} U_{TMD} = \frac{p^2}{\omega_s^2} u_{go} \left[as, \frac{k_s}{m_s} = \omega_s^2 \right] \tag{18.3.6}$$

Dividing Eq. (18.3.5) by k_{TMD},

$$-\frac{m_{TMD}}{k_{TMD}} U_{TMD} p^2 + U_{TMD} - U_s = \frac{m_{TMD}}{k_{TMD}} p^2 u_{go}$$

$$\Rightarrow \left(1 - \frac{p^2}{\omega_{TMD}^2} \right) U_{TMD} - U_s = \frac{p^2}{\omega_{TMD}^2} u_{go} \tag{18.3.7}$$

Solving Eqs. (18.3.6) and (18.3.7) for U_s and U_{TMD},

$$\frac{U_s}{u_{go}} = \frac{p^2 \left[\dfrac{1}{\omega_s^2} \left(1 - \dfrac{p^2}{\omega_{TMD}^2} \right) + \dfrac{1}{\omega_{TMD}^2} \cdot \dfrac{k_{TMD}}{k_s} \right]}{\left(1 - \dfrac{p^2}{\omega_{TMD}^2} \right)\left(1 + \dfrac{k_{TMD}}{k_s} - \dfrac{p^2}{\omega_s^2} \right) - \dfrac{k_{TMD}}{k_s}} \tag{18.3.8}$$

$$\frac{U_{TMD}}{u_{go}} = \frac{\dfrac{p^2}{\omega_s^2} + \dfrac{p^2}{\omega_{TMD}^2} \left[1 + \dfrac{k_{TMD}}{k_s} - \dfrac{p^2}{\omega_s^2} \right]}{\left(1 - \dfrac{p^2}{\omega_{TMD}^2} \right)\left(1 + \dfrac{k_{TMD}}{k_s} - \dfrac{p^2}{\omega_s^2} \right) - \dfrac{k_{TMD}}{k_s}} \tag{18.3.9}$$

For a tuning condition when structural displacement amplitude (U_s) is zero, the numerator of the right-hand side of Eq. (18.3.8) is zero, which gives

$$\frac{1}{\omega_s^2} \left(1 - \frac{p^2}{\omega_{TMD}^2} \right) + \frac{1}{\omega_{TMD}^2} \cdot \frac{k_{TMD}}{k_s} = 0 \tag{18.3.10}$$

Eq. (18.3.10) leads to Eq. (18.3.11).

$$\omega_{TMD} = \frac{p}{\sqrt{1 + \mu_{mass}}} \tag{18.3.11}$$

18.4 DAMPED TMD SYSTEM UNDER SINUSOIDAL EXCITATION

Consider Figure 18.4, which shows a structure under sinusoidal excitation and TMD with damping. Using the following notations:

m_s = mass of structure

m_{TMD} = mass of TMD

μ_{mass} = mass ratio = m_{TMD}/m_s

k_s = stiffness of structure

k_{TMD} = stiffness of TMD

c_{TMD} = damping in TMD

$u_s(t)$ = displacement of structure

$u_{TMD}(t)$ = displacement of TMD

$$\omega_s = \sqrt{\frac{k_s}{m_s}} = \text{natural frequency of the structure}$$

$$\omega_{TMD} = \sqrt{\frac{k_{TMD}}{m_{TMD}}} = \text{natural frequency of the TMD.}$$

$F_0 \sin pt$ = excitation force with peak force F_0 and frequency p.

The equations of motion are given in Eqs. (18.4.1) and (18.4.2).

$$m_s \ddot{u}_s + k_s u_s + c_{TMD}\left(\dot{u}_s - \dot{u}_{TMD}\right) + k_{TMD}\left(u_s - u_{TMD}\right) = F_o \sin pt \qquad (18.4.1)$$

$$m_{TMD}\ddot{u}_{TMD} + c_{TMD}\left(\dot{u}_{TMD} - \dot{u}_s\right) + k_{TMD}\left(u_{TMD} - u_s\right) = 0 \qquad (18.4.2)$$

A trial solution strategy can be taken as follows and substitution may be done in Eqs. (18.4.1) and (18.4.2).

$$u_s(t) = C_1 e^{ipt}$$

$$u_{TMD}(t) = C_2 e^{ipt}$$

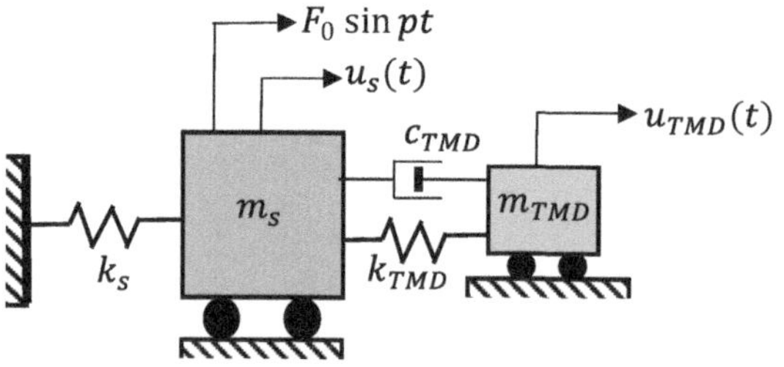

FIGURE 18.4 TMD with damping and structure under sinusoidal excitation.

Here, i stands for an imaginary parameter. The substitution leads to a response solution (not shown here).

18.5 EXERCISES

Q18.5.1 Describe the working principle of TMD.
Q18.5.2 Describe the dynamics of undamped TMD under sinusoidal excitation.
Q18.5.3 Highlight the structure–TMD resonance.

18.6 CLOSURE

TMD is a vast subject. Only an outline of TMD is given here. It may be shown that the tuned frequency for the minimum relative displacement of the primary structure for sinusoidal loading applied on the primary structure is given by Eq. (18.6.1).

$$p = \frac{1}{1 + \mu_{mass}} \tag{18.6.1}$$

FURTHER READINGS

Baz, A. (1998) Robust Control of Active Constrained Layer Damping, *Journal of Sound and Vibration, Elsevier.*

Blekherman, A.N. (1996) Mitigation of Response of High-rise Structural Systems by Means of Optimal Tuned Mass Damper, *11WCEE.*

Chang, C. C. and Yang H.T.Y. (1995) Control of Buildings Using Active Tuned Mass Dampers, *Journal of Engineering Mechanics.*

Choi, K.M., Cho, S.W., Jung, H.J. and Lee, I.W. (2004) Semi-active Fuzzy Control for Seismic Response Reduction Using Magneto-rheological Dampers, *Earthquake Engineering Structural Dynamics.*

Den Hartog, J.P. (1984) *Mechanical Vibration,* Dover

Garg Devendra, P. and Anderson, G.L (2003) Structural Vibration Suppression Via Active/ Passive Techniques, *Journal of Sound and Vibration.*

Nicholas, A.A. and Frank, S. (2009) Exploring the Performance of a Nonlinear Tuned Mass Damper, *Journal of Sound and Vibration.*

Robinson, W.H., Passive Control of Structures, the New Zealand Experience, *ISET Journal,* Dec 1998.

19 Seismic Safety of OFCs

19.1 GENERAL

Any building system consists of two major components: the structural components and the non-structural components. Non-structural components are not designed to carry loads. There can be structural–non-structural interaction under dynamic conditions. As the safety of non-structural components is to be taken care of by the structural engineer, a new name has been assigned to them, namely, operational and functional components (OFC). The OFCs necessarily involve the functional purposes of the building. As the loss of functionality of a building means downtime loss, the safety of OFCs is very important. The OFCs can be broadly classified as (a) architectural/civil, (b) mechanical, (c) electrical and (d) equipment. Very often, more than one such entity may be involved in an OFC. In some buildings like hospitals, the cost of OFCs constitutes about 85% of the total cost of the building. Thus, the protection of OFCs is important from an economic viewpoint too. The failure of OFCs may lead to inaccessibility to egress, fire hazard, chemical and gas hazard, death of patients in hospitals, content damage, economic loss, disruption of services, downtime loss and falling hazard that may lead to casualty and death, failure of the sprinkler system, spalling of outer concrete from reinforced concrete members, sliding or overturning of unanchored or wrongly anchored equipment.

The safety of OFCs is very often related to the anchoring the OFC. Any cantilever system (horizontal or vertical) needs to be properly designed at the base. IS-1893 suggests the design for five times the normal horizontal seismic action. All equipment should be properly anchored/fixed at the base. Even the hospital beds should be anchored. Some OFCs may be base isolated.

The categories of OFCs are given in Table 19.1.

The seismic safety of OFCs involve IO, LS and CP damage states.

So far as the design criteria and damageability are concerned, OFCs can be classified as:

(i) Acceleration-sensitive OFCs and
(ii) Drift-sensitive OFCs.

DOI: 10.1201/9781003441090-19

TABLE 19.1
Types of OFCs

Type of OFC	*Items*
(a) Architectural/civil	Infill walls, cornices, claddings (Veneer, glass, marbles, tiles), false ceiling, parapets, appendages, chimneys, canopies, stacks, staircase, gas conduits, air condition conduits, building joint, water tanks, oxygen supply conduits, plumbing, doors and windows, storage racks, book shelves, almirah and containers, contents, sprinkler system
(b) Mechanical	Boilers, pumps, heating–ventilation–air conditioning (HVAC) system, mechanical engines, lifts, fire-fighting system, furnaces, cranes, elevator, conveyor
(c) Electrical	Generator, water heater, elevators, computers, electrical panels, conveyors, elevator, chillers, communication equipment, transformers,
(d) Equipment	medical equipment like radiation unit, CT scan unit, X-ray unit, mammograph unit, MRI unit, OT equipment, oxygen plants, fire-fighting equipment

Acceleration-sensitive OFCs with large mass are subjected to higher inertia forces. Drift-sensitive OFCs are mounted horizontally or vertically over a large length. Some OFCs may be both drift (or deformation) sensitive and acceleration sensitive; in that case, these are called *deformation-sensitive* OFCs (FEMA-356). It may be noted that any element is simultaneously subjected to deformation and acceleration; the question is which component is more critical for the OFC.

Examples of acceleration-sensitive OFCs are boilers, generator, transformer, stairs, lift, appendages, balcony, chimneys, stacks, HVAC system, storage tanks and water heaters.

Examples of drift sensitive items are infill walls, wall-mounted vertical equipment, wall-mounted horizontal equipment, suspended ceilings, claddings and tiles.

19.2 COMPUTATION OF SEISMIC FORCE ON OFCS

FEMA-356 gives a table (Table 11.1 of FEMA-356) describing what analysis can be carried out for a particular OFC. Force analysis, deformation analysis and prescriptive analysis are the three types of analyses described in FEMA-356. Based on such a description, the force equation applicable shall vary.

19.2.1 DEFAULT FORCE WHEN FORCE ANALYSIS IS PERMITTED

The default equations are (19.2.1.1) and (19.2.1.2) for horizontal and vertical seismic forces, respectively (after FEMA-356).

$$F_h = 1.6 S_{as} IW \qquad (19.2.1.1)$$

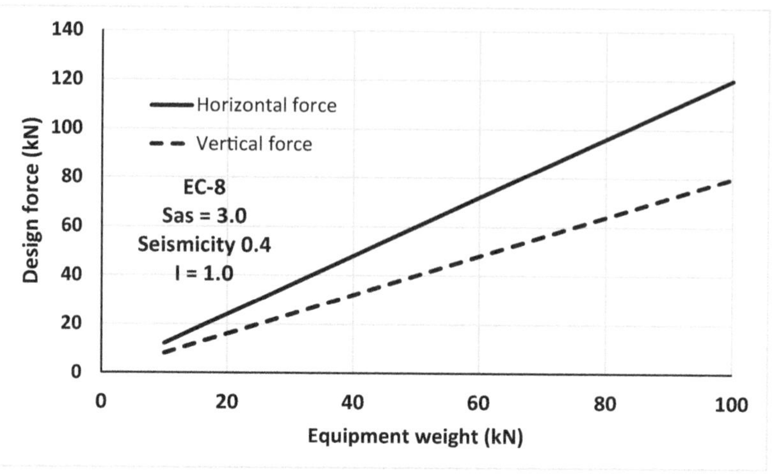

FIGURE 19.1　Representation of equipment force as per force analysis (EC-8, 0.4g).

$$F_v = \frac{2}{3} F_h \tag{19.2.1.2}$$

where　F_h = horizontal design force on OFC, acting through the centre of the mass
　　　S_{as} = spectral acceleration corresponding to a short period
　　　I = OFC performance factor, which is 1.0 for LS non-structural performance,
　　　　　and 1.5 for IO non-structural performance
　　　W = weight of OFC
　　　F_v = vertical design force on OFC acting through the centre of mass.

The typical plot of Eqs. (19.2.1.1) and (19.2.1.2) are shown in Figure 19.1, with some laid down parameters.

19.2.2　General Equation in Force Analysis: *Horizontal* Design Force

When Section 19.2.1 does not apply, the horizontal design force for OFC is given by Eq. (19.2.2.1), subject to minimum force given by Eq. (19.2.2.2):

$$F_h = \frac{0.4 a_f S_{as} IW \left(1 + \dfrac{2h}{H}\right)}{R_{OFC}} \tag{19.2.2.1}$$

$$F_{h,min} = 0.3 S_{as} IW \tag{19.2.2.2}$$

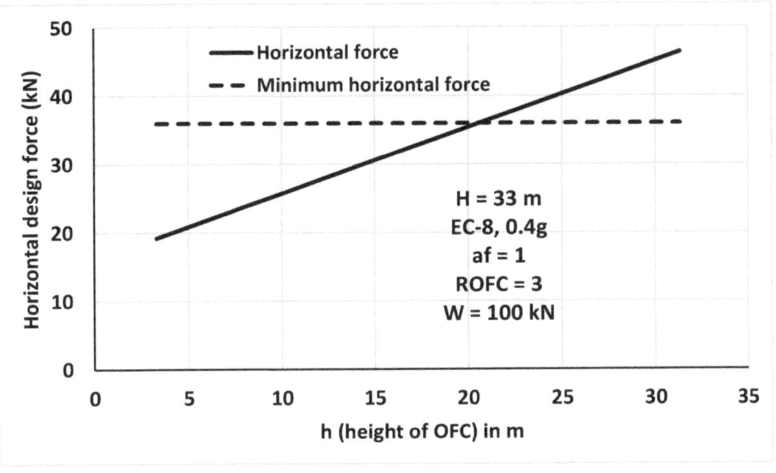

FIGURE 19.2 Horizontal force on OFC as per the general equation.

where F_h = horizontal design force on OFC, acting through centre of mass

$F_{h,\,min}$ = minimum horizontal design force on OFC, acting through centre of mass

S_{as} = spectral acceleration corresponding to a short period

I = OFC performance factor, which is 1.0 for LS non-structural performance, and 1.5 for IO non-structural performance

W = weight of OFC

h = height of OFC from grade elevation

H = average height of roof from grade elevation

a_f = OFC dynamic amplification factor

R_{OFC} = response reduction factor for OFC.

R_{OFC} varies from 1.5 to 4 depending on the type of OFC; the value of a_f varies from 1 to 2.5 (Table 11-2 of FEMA-356) (Figure 19.2).

19.2.3 GENERAL EQUATION IN FORCE ANALYSIS: *VERTICAL* DESIGN FORCE

When Section 19.2.1 does not apply, the vertical design force for OFC is given by Eq. (19.2.3.1), subject to minimum force given by Eq. (19.2.3.2).

$$F_v = \frac{0.27 a_f S_{as} IW}{R_{OFC}} \tag{19.2.3.1}$$

$$F_{v,\,min} = 0.2 S_{as} IW \tag{19.2.3.2}$$

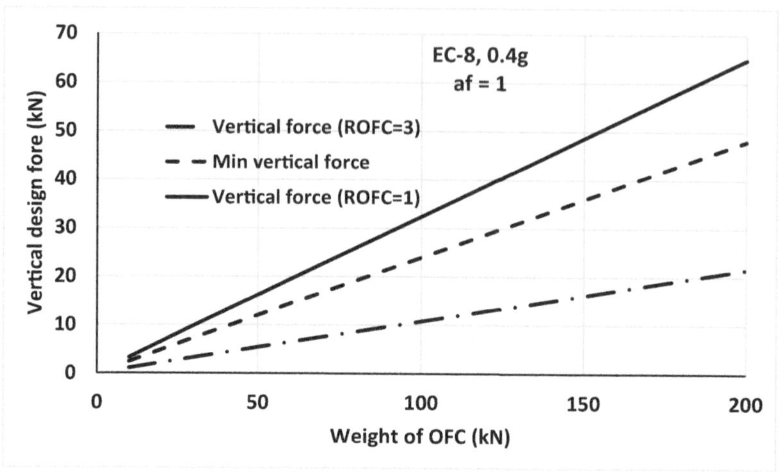

FIGURE 19.3 Variation of vertical design force with the weight of OFC.

where

F_v = vertical design force on OFC, acting through centre of mass

$F_{v,\,min}$ = minimum vertical design force on OFC, acting through centre of mass

S_{as} = spectral acceleration corresponding to short period

I = OFC performance factor, which is 1.0 for LS non-structural performance, and 1.5 for IO non-structural performance

W = weight of OFC

a_f = OFC dynamic amplification factor

R_{OFC} = response reduction factor for OFC

The graphical representation of Eqs. (19.2.3.1) and (19.2.3.2) are shown in Figure 19.3. From the figure, it is clear that the actual force will highly depend on the value of R_{OFC}. Depending on the value of RFC, the computed values may be lower or higher than the minimum vertical design force.

19.2.4 DEFORMATION ANALYSIS

When the OFC is connected to the same building, Eq. (19.2.4.1) is applicable.

$$D_{OFC} = \frac{\Delta_2 - \Delta_1}{y_2 - y_1} \tag{19.2.4.1}$$

where

D_{OFC} = drift of OFC

Δ_2 = deflection of the building at height y_2 of the higher end of OFC from grade elevation

Δ_1 = deflection of the building at height y_1 of the lower end of OFC from grade elevation

Sometimes, the OFC may be anchored in two adjacent buildings. In that case, Eq. (19.2.4.2) will apply.

$$\Delta_{OFC} = \left|\Delta_A\right| + \left|\Delta_B\right| \tag{19.2.4.2}$$

where

Δ_{OFC} = relative seismic deflection of OFC
Δ_A = deflection of building A at level of OFC
Δ_B = deflection of building B, at the same level.

19.3   FLOOR RESPONSE SPECTRA

The machine and equipment foundation often rest on the building floor at some height. For obtaining the design force for such foundation, the design spectrum of the code is not applicable. The reason is that the machine foundation, unlike the building foundation, is not ground-anchored. As the machine rests at some height from the ground, there is an amplification of acceleration response. In fact, the spectral acceleration of the floor is to be used for the design of the machine foundation on a floor. The spectral acceleration plot for floor is the plot of peak responses of the SDOF system with some damping subjected to the acceleration of the floor. As the floor may be subjected to different acceleration histories under different ground motions, a series of spectral acceleration histories of the floor is possible. It is wise to take an average of all such spectral acceleration plots and when done so, it is called the average floor spectra. Such average floor spectra can be used for designing a foundation resting on the floor. Remember that average floor spectra is floor-specific.

19.4   RESPONSE AMPLIFICATION

Floor dynamic responses like displacement response and acceleration response get magnified towards higher floors due to dynamic amplification. The following are three common amplification parameters of interest in floor response study:

(i)   Floor amplification factor (FAF)
(ii) Component amplification factor (CAF)
(iii) Total amplification factor (TAF)

FAF is the ratio of peak floor acceleration (PFA) to the peak ground acceleration (PGA):

$$FAF = \frac{PFA}{PGA} \tag{19.4.1}$$

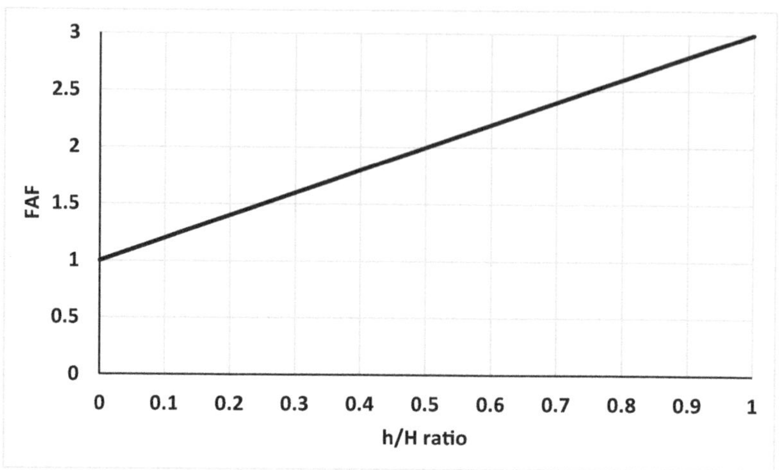

FIGURE 19.4 Plot of FAF with height ratio of OFC.

CAF is the ratio of peak floor spectral acceleration (PFSA) to the PFA:

$$CAF = \frac{PFSA}{PFA} \tag{19.4.2}$$

TAF is the PFSA to the effective peak ground acceleration (EPGA).

$$TAF = \frac{PFSA}{EPGA} \tag{19.4.3}$$

Considering ROFC as constant, the FAF can be expressed as $\left(1 + \frac{2h}{H}\right)$. Hence, the graph of FAF is drawn and shown in Figure 19.4. The graph shows that the FAF increases with height of OFC and reaches a maximum value of 3.0. FAF value of 1.0 at ground level indicates that there is no amplification at that level.

The plot for CAF and TAF needs the knowledge of the floor spectral acceleration and value of EPGA. Floor spectral acceleration is obtained from the plot of floor spectra, which are building-specific.

19.5 EXAMPLES

Example 19.5.1 Find the design force on an OFC of weight 100 kN for IO performance. Assume that force analysis is applicable. Use Indian design spectrum for Zone V.

Solution: For the Indian design spectrum, the spectral acceleration for short period $S_{as} = 2.5$.

For IO performance, $I = 1.5$. For Zone V of Indian spectrum $Z = 0.36$.

Horizontal design force $= F_h = 1.6 S_{as} IW$

$$= 1.6 \times (2.5 \times 0.36) \times 1.5 \times 100 \text{ kN}$$

$$= 216 \text{ kN.}$$

Vertical design force $= F_v = \dfrac{2}{3} F_h = (2/3) \times 216 = 144 \text{ kN.}$

Both the design forces act through the centre of mass of OFC.

Example 19.5.2 As per the general condition, find the design force on a generator of weight 150 kN located at a height 15 m of a hospital building of height 30 m. Assume EC-8 spectrum at 0.4 seismicity level.

Solution: As per Table 11-2 of FEMA-356, for Electrical equipment $a_f = 1.0$ and $R_{OFC} = 3.0$. Also, for EC-8 $S_{as} = 3.0$. Hospital building should be designed for IO performance of equipment; so, $I = 1.5$.

Horizontal seismic force on equipment $= F_h = \dfrac{0.4 a_f S_{as} IW \left(1 + \dfrac{2h}{H}\right)}{R_{OFC}}$

$$= \dfrac{0.4 \times 1 \times (3 \times 0.4) \times 1.5 \times 150 \left(1 + \dfrac{2 \times 15}{30}\right)}{3}$$

$$= 72 \text{ kN.}$$

Minimum horizontal seismic force $= F_{h,\,min} = 0.3 S_{as} IW = 0.3 \times (3 \times 0.4)\, 1.5 \times 150$

$$= 81 \text{ kN.}$$

So, horizontal design force $= 81$ kN.

Vertical design force $= F_v = \dfrac{0.27 a_f S_{as} IW}{R_{OFC}} = 0.27 \times 1 \times (3 \times 0.4) \times 1.5 \times 150 / 3 = 24.3 \text{ kN.}$

Minimum vertical seismic force $= F_{v,\,min} = 0.2 S_{as} IW = 0.2 \times (3 \times 0.4) \times 1.5 \times 150 = 54 \text{ kN.}$

So, vertical design force $= 54$ kN.

19.6 EXERCISES

Q19.6.1 Find the design force on an OFC of weight 500 kN for LS performance. Assume that force analysis is applicable. Use Indian design spectrum for Zone V.

Q19.5.2 As per the general condition, find the design force on a AC machine of weight 1500 kN located at height 5 m of a hospital building of height 30 m. Assume EC-8 spectrum at 0.4 seismicity level.

19.7 CLOSURE

The seismic safety of OFCs is equally important as that of the structure. The damage of OFCs leads to downtime loss, repairing cost and replacing cost. Adequate design is necessary for OFCs resting on building floor or mounted on walls.

FURTHER READINGS

Debnath, P.P. and Choudhury, S. (2015) Effect of Unreinforced Masonry Infill on Floor Amplification and Other Parameters in Frame-Wall Buildings, *Journal of Mechanics Based Design of Structures and Machines*, 43, 450–465, DOI: 10.1080/15397734.2015.1025961

FEMA-150 (1990) Seismic Considerations – Health Care Facilities, US Federal Emergency Management Agency.

FEMA-273 (1996) NEHRP Guidelines for the Seismic Rehabilitation of Buildings, US Federal Emergency Management Agency, Building Seismic Safety Council, Washington DC.

FEMA-349 (2000) Action Plan for Performance-Based Seismic Design, US Federal Emergency Management Agency, Earthquake Engineering Research Institute.

FEMA-356 (2000) Prestandard and Commentary for the Seismic Rehabilitation of Buildings, US Federal Emergency Management Agency.

FEMA-412 (2002) Installing Seismic Restraints for Mechanical Equipment, US Federal Emergency Management Agency, December.

FEMA-413 (2004) Installing Seismic Restraints for Electrical Equipment, US Federal Emergency Management Agency, January.

FEMA-414 (2004) Installing Seismic Restraints for Duct and Pipe, US Federal Emergency Management Agency, January.

FEMA-415 (2004) Installing Seismic Restraints for Duct and Pipe, US Federal Emergency Management Agency, January.

Foo, S. And Lau, D. (2004) Seismic Risk Reduction of Operational and Functional Components of Buildings: Research Perspective, *13th World Conference on Earthquake Engineering*, Canada, August 1–6, Paper No. 5052.

Goodwin, E., Maragakis, E. and Itani, A. (2004) "Seismic Evaluation of Hospital Piping System, 13th World Conference on Earthquake Engineering, Paper No. 1081.

Guevara, T.L. and Alvarez, Y. (2000) Functionality of the Architectural Program in the Remodeling of Existing Hospitals in Seismic Zones of Venezuela, 12th World Conference on Earthquake Engineering, Paper No. 0275.

Kelly, J.M. and Tsai, H.C. (1985) Seismic Response of Light Internal Equipment in Base-Isolated Structures, *Earthquake Engineering & Structural Dynamics,* V.13 pp. 711–732.

Kunnath, S.K., Panahshahi, N. and Reinhorn, A.M. (1991) Seismic Response of RC Buildings with Inelastic Floor Diaphragm, *Journal of Structural Engineering*, ASCE, V. 117, No. 4, Apl., pp. 1218–1237.

Lai, M.L. and Soong, T.T. (1991) Seismic Design Considerations for Secondary Structural Systems, *Journal of Structural Engineering*, 117(2), 459–472.

Lewis, J. and Wang, M. (2004) Seismic Risk Mitigation of Operational and Functional Components in Hospitals – The British Columbia Experience, 13th World Conference on Earthquake Engineering, August 1–6, Paper No. 1636.

Lin, J. and Mahin, S.A. (1985) Seismic Response of Light Subsystems on Inelastic Structures, *Journal of Structural Engineering*, 111(2), 400–416.

McGavin, G.L. (1981) *Earthquake Protection of Essential Building Equipment Design Engineering Installation*, John Wiley and Sons, Toronto.

McGavin, G.L. and Patrucco, H. (1998) Nonstructural Functional Design Considerations for Healthcare Facilities, 6th US Conference on Earthquake Engineering.

Nazer, M.H. and Elahi, F.N. (2004) Seismic Vulnerability of Nonstructural Components of Hospitals, 13th World Conference on Earthquake Engineering, Canada, 1–6 August, Paper No. 1250.

Segal, F., Rutenberg, A. and Levy, R. (1996) Earthquake Response of Structure-Elevator System, *Journal of Structural Engineering,* ASCE, 122(6), 607–616.

Shenton, H.W. (1996) Criteria for Initiation of Slide, Rock, and Slide-Rock Rigid-Body Modes, *Journal of Structural Engineering,* 122(7), 690–693.

Su, R.K.L., Chandler, A.M., Sheikh, M.N. and Lam, N.T.K. (2005) Influence of Non-Structural Components on Lateral Stiffness of Tall Buildings, *The Structural Design of Tall and Special Buildings*, 14, 143–164.

Tsai, H.C. and Kelly, J.M. (1989) Seismic Response of the Superstructure and Attached Equipment in a Base-Isolated Building, *Earthquake Engineering and Structural Dynamics*, 18, 551–564.

Ventura, C.E. and Kharrazi, M.H.K. (2004) Performance of OFC's in Earthquakes by Shake Table Tests, 13th World Conference on Earthquake Engineering, 1–6 August, Canada, Paper No. 80.

Villevarde, R. (1997a) Method to Improve Seismic Provisions for Nonstructural Components in Buildings, *Journal of Structural Engineering, ASCE*, 123(4), pp. 432–439.

Villevarde, R. (1997b) Seismic Design of Secondary Structures: State of the Art, *Journal of Structural Engineering, ASCE*, 123(8), 1011–1019.

Villevarde, R. (2000) Design Oriented Approach for Seismic Nonlinear Analysis of Nonstructural Components, 13th World Conference on Earthquake Engineering, 1–6 August, Canada, Paper No. 1979.

Yao, G.C. (1996) Seismic Capacity Evaluation of Equipment in a Modern Hospital at Southern Taiwan, *11th World Conference on Earthquake Engineering*, Paper No. 283.

20 Miscellaneous Topics

20.1 GENERAL

In this chapter, some additional items related to seismic safety and design are discussed. The topics include basics of probability in seismic design and seismic fragility analysis of structures, damage index and seismic base isolation.

20.2 PRELIMINARY OF PROBABILITY

20.2.1 STATISTICAL TERMS

Probability is a mathematical concept which finds application in diverse fields. In fact, in most cases, our knowledge about the world is more of a probabilistic nature rather than deterministic. Recent advancements in performance-based seismic design are highly probability oriented. When we design a building for IO, LS or CP, there arises the question as to how much is the probability that the building will show the target PL. This is another way of determining the reliability of the structure. Set theory is intimately related to the probability study. We explain some of the basic terms in probability.

Probability: Probability means the chance of an event occurring in some chosen way.

It may be noted that the number of ways an event can occur in a chosen way depends on the total number of ways the event can occur. In the famous head–tail example of a coin, there are only two ways the face can show up; so, the probability of a head (or tail) showing up is ½ or 50%. A six-faced die can show one face at a time. So, the probability of showing any face (say, pip 4) is 1/6.

Definition: If a particular event takes place n times out of total N times of total occurrence, the probability of occurrence of the event is n/N.

Sample space: The set of all possible occurrences of some phenomenon is called sample space (S). An element (a particular outcome) of S is called sample point.

The parameters that can describe the trend and characteristics of a sample are mean, median, mode, standard deviation and variance.

DOI: 10.1201/9781003441090-20

Mean (also called the expected value) is the average of the algebraic sum of all elements in a sample space. If there are N elements in sample space and any i-th element is denoted by x_i, then the mean is given by Eq. (20.2.1.1).

$$\text{Mean}(\mu) = \frac{\sum_{i=1}^{n} x_i}{N} \tag{20.2.1.1}$$

Mode is the element with the highest number of occurrences.

For example, in a shoe store, the maximum sale may be for shoe with 'number 7' designation.

Median is the mid-value of a sample space. All the data are arranged in an increasing or decreasing order initially. Then, if the number of samples is N.

(i) If N is odd then, $\text{median} = \left(\dfrac{N+1}{2}\right) th$ observation value.

(ii) If N is even then, $\text{median} = \dfrac{\left(\dfrac{N}{2}\right) th + \left(\dfrac{N+1}{2}\right) th}{2}$ observation value.

Standard deviation is a measure of dispersion of the data. It indicates how the data is distributed about the mean. The expression for standard deviation of (large) *population* is given by Eq. (20.2.1.2).

$$\sigma_{population} = \sqrt{\frac{\sum_{i=1}^{n} (\mu - x_i)^2}{N}} \tag{20.2.1.2}$$

For (smaller) *sample* out of (large) population, the expression for standard deviation (σ) is

$$\sigma_{sample} = \sqrt{\frac{\sum_{i=1}^{n} (\mu - x_i)^2}{N-1}}$$

Variance is the square of the standard deviation.

$$\text{Var} = \sigma^2 \tag{20.2.1.3}$$

Note that the above definitions will take slightly different shape when grouped data are considered.

The data in a sample space may be random or deterministic. Most of the real data are random in nature.

A real-valued function defined on the sample space is called a *random variable*. Mathematically, we can write a random variable (X) as:

$$X: S \rightarrow \mathbb{R}, \text{ where S is sample space and } \mathbb{R} \text{ is set of real numbers.}$$

Random variable may be *discrete random variable* or *continuous random variable*.

20.2.1.1 Discrete Random Variable

A random variable that takes on finite or countably infinite number of values is called a discrete random variable.

20.2.1.2 Probability Mass Function (PMF)

The function $f(x)$ is called a probability mass function when

$$P(X = x_k) = f(x_k), \text{ when } k = 1, 2, 3 \ldots N$$

$$\text{and } P(X = x) = f(x) = 0, \text{ when } x \neq x_k$$

where X is a discrete random variable that assumes the values $x_1, x_2, x_3, \ldots x_N$.

$$P(X = x) \text{ stands for probability of } X \text{ taking the value } x.$$

In general, $f(x)$ is a PMF if,

(a) $f(x) \geq 0$ for all x.
(b) $\sum_x f(x) = 1$.

20.2.1.3 Cumulative Distribution Function of PMF

The cumulative distribution function (CDF) of PMF is defined as

$$F(x) = P(X \leq x) = \sum_{u \leq x} f(u), \text{where}, f \text{ is PMF of } X. \qquad (20.2.1.4)$$

20.2.1.4 Continuous Random Variable

A random variable that takes on non-countably infinite number of values is called continuous (or non-discrete) random variable.

If X is a random variable, X is said to be continuous if its distribution function is given by

$$F(x) = P(X \leq x) = \int_{-\infty}^{x} f(u)\, du, \text{where} -\infty < x < \infty. \qquad (20.2.1.5)$$

20.2.1.5   Probability Density Function (PDF)

Referring to Eq. (20.2.1.5), $f(x)$ is a PDF if,

$$\left.\begin{array}{l} f(x) \geq 0 \text{ for all } x \\[2mm] \displaystyle\int_{-\infty}^{\infty} f(x)\,dx = 1 \end{array}\right\} \tag{20.2.1.6}$$

20.2.1.6   Cumulative Distribution Function of PDF

The cumulative distribution function (CDF) of PDF is defined as

$$F(x) = \int_{-\infty}^{x} f(u)\,du \tag{20.2.1.7}$$

20.3   STATISTICAL DISTRIBUTIONS

Statistical distribution of data means the way the data are scattered in the sample space. Data may be distributed evenly on both sides of the mean; or the data may be skewed in one side of the mean value.

20.3.1   BERNOULLI DISTRIBUTION

Bernoulli distribution is a discrete probability distribution. Historically, Bernoulli distribution was the earliest discussed distribution. It deals with the case when there are only two possible outcomes of any event (success or failure).

If X is a discrete random variable having only two possible outcomes, such that

$X = 1$, if result of event is a success
$X = 0$, if result of event is a failure.

If p denotes the probability of success of an event, then the probability of failure is $(1 - p)$.

The PMF of Bernoulli distribution is given as

$$P(X = k) = p^k (1 - p)^{1-k} \tag{20.3.1.1}$$

where $k = 0$ or 1.

The Bernoulli distribution is shown in Figure 20.1.

For Bernoulli distribution, the statistical parameters are

$$\left.\begin{array}{l} \text{Mean } E(X) = p \\[2mm] \text{Standard deviation } \sigma = \sqrt{p(1-p)} \\[2mm] \text{Variance} = V(X) = p(1-p) \end{array}\right\} \tag{20.3.1.2}$$

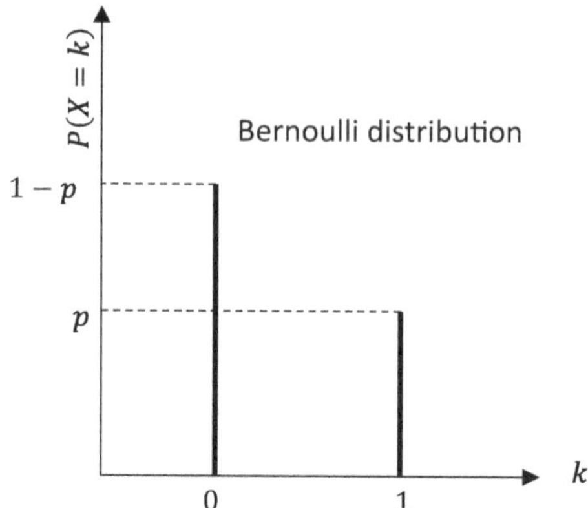

FIGURE 20.1 Bernoulli distribution.

20.3.2 BINOMIAL DISTRIBUTION

Binomial distribution is a discrete probability distribution involving trials of repetitive nature, in which the occurrence (success) or non-occurrence (failure) is of interest. Binomial distribution represents the generalized form of Bernoulli trials.

If X is a discrete random variable having only two possible outcomes in 'N' number of trials, where X can take values 0, 1, 2, 3, ... N.

If p denotes the probability of success of an event, then the probability of failure is $(1-p)$.

The PMF of binomial distribution is given as

$$P(X = k) = \binom{N}{k} p^k (1-p)^{N-k} \quad \left[\binom{N}{k} \text{ is binomial coefficient} \right] \tag{20.3.2.1}$$

where $k = 0, 1, 2, 3, ... N$.

For binomial distribution, the statistical parameters are

$$\left. \begin{aligned} Mean \ E(X) &= Np \\ Standard \ deviation \ \sigma &= \sqrt{Np(1-p)} \\ Variance = V(X) &= Np(1-p) \end{aligned} \right\} \tag{20.3.2.2}$$

Bernoulli distribution is shown graphically in Figure 20.2.

The figure shows the Binomial distribution for different values of p.

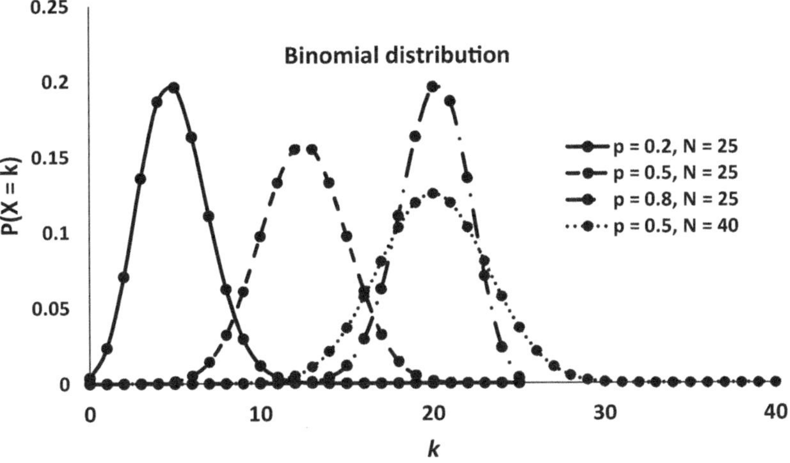

FIGURE 20.2 Binomial distribution.

Example 20.3.1 *In a toss of a biased coin (defective coin) tails come up with probability 0.25. Find the probability of (a) seeing exactly 3 tails in 15 tosses, (b) seeing exactly 9 tails in 15 tosses and (c) at least 5 tails in 15 tosses.*

Solution

Let X be the random variable of the occurrence of a tail in the tossing of the given biased coin. Given that the probability of tail showing up, $p=0.25$.

Number of trials, $N=15$.

Therefore, the probability of seeing exactly k tails in N tosses is given by

$$P(X = k) = \binom{N}{k} p^k (1-p)^{N-k}$$

(a) The probability of seeing exactly 3 tails in 15 tosses is

$$P(X = 3) = \binom{15}{3} 0.25^3 (1-0.25)^{15-3} = 0.2252.$$

(b) The probability of seeing exactly 9 tails in 15 tosses is

$$P(X = 9) = \binom{15}{9} 0.25^9 (1-0.25)^{15-9} = 0.0034.$$

(c) The probability of seeing at least 5 tails in 15 tosses is given by

$$P(X \geq 5) = 1 - P(X < 5)$$

$$= 1 - \left[P(X=0) + P(X=1) + P(X=2) + P(X=3) + P(X=4) \right]$$

$$= 1 - \left[\binom{15}{0} 0.25^0 (1-0.25)^{15-0} + \binom{15}{1} 0.25^1 (1-0.25)^{15-1} \right.$$

$$+ \binom{15}{2} 0.25^2 (1-0.25)^{15-2} + \binom{15}{3} 0.25^3 (1-0.25)^{15-3}$$

$$\left. + \binom{15}{4} 0.25^4 (1-0.25)^{15-4} \right]$$

$$= 0.3135$$

20.3.3 Poisson Distribution

Poisson distribution is a discrete probability distribution where the event is associated with a fixed time period and a constant mean value, in which the events occur independently without influencing other events.

Let X be a discrete random variable, which can assume the values 0, 1, 2, 3, 4 … such that the PMF of X is given by

$$P(X=k) = \frac{\lambda^k e^{-\lambda}}{k!} \tag{20.3.3.1}$$

where k can take values 0, 1, 2, 3, 4 … and λ is a given positive constant. Here X is said to be Poisson distributed.

The CDF of PMF for Poisson distribution is given as

$$F(k) = P(X \leq k) = \sum_{i=0}^{k} \frac{\lambda^i e^{-\lambda}}{i!} \tag{20.3.3.2}$$

The sample curve for Poisson distribution is shown in Figure 20.3. The figure shows that with the increase of the value of λ, the distribution becomes flatter.

For Poisson distribution the statistical parameters are

$$\left. \begin{array}{c} \textit{Mean } E(X) = \lambda \\ \textit{Standard deviation } \sigma = \sqrt{\lambda} \\ \textit{Variance} = V(X) = \lambda \end{array} \right\} \tag{20.3.3.3}$$

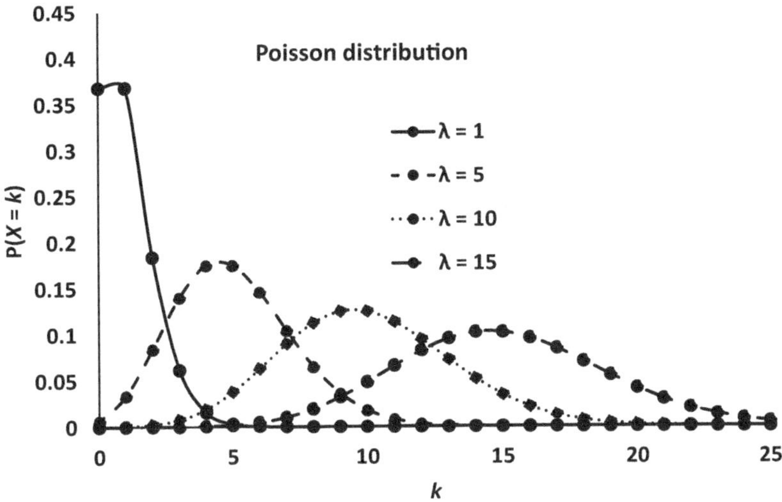

FIGURE 20.3   Poisson distribution.

20.3.4   NORMAL DISTRIBUTION (GAUSSIAN DISTRIBUTION)

Normal, or Gaussian, distribution is a continuous probability distribution where the event is distributed symmetrically about the mean. Many natural phenomena follow normal distribution. For example, if you test a large number of specimens in the laboratory for its tensile strength or compressive strength, the results will be normally distributed.

The functional expression of normal distribution is given by Eq. (20.3.4.1).

$$f(x) = \frac{1}{\sigma\sqrt{2\pi}} e^{-0.5\left(\frac{x-\mu}{\sigma}\right)^2} \tag{20.3.4.1}$$

Eq. (20.3.4.1) can be put in standard normal distribution form by substituting $z = \dfrac{x-\mu}{\sigma}$, when it takes the form as in Eq. (20.3.4.2).

$$\phi(z) = \frac{1}{\sigma\sqrt{2\pi}} e^{-0.5z^2} \tag{20.3.4.2}$$

Eq. (20.4.38) has the genesis that as the data samples are normally distributed, the positive data becomes equal to the negative data in number and magnitude, thus making the mean μ as zero. This also leads to $\sigma = 1.0$.

For normal distribution, the statistical parameters are

$$\left.\begin{array}{l} \textit{Mean } E(X) = \mu \\ \textit{Standard deviation } = \sigma \\ \textit{Variance} = V(X) = \sigma^2 \end{array}\right\} \tag{20.3.4.3}$$

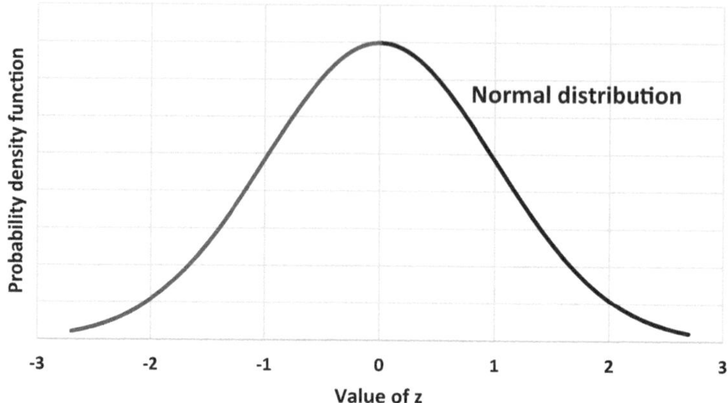

FIGURE 20.4   Normal distribution curve.

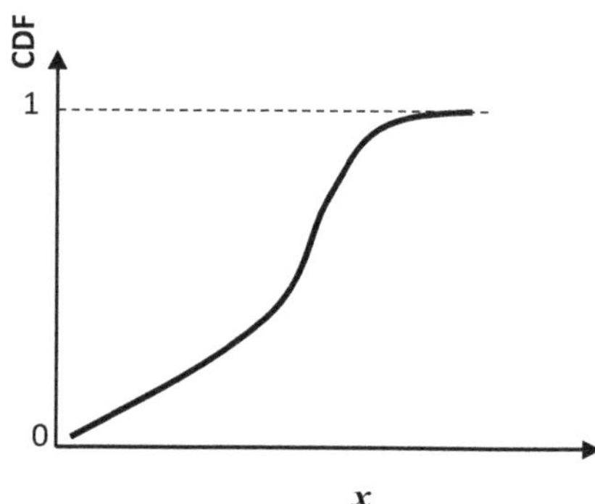

FIGURE 20.5   General shape of a CDF of a normal distribution.

A normal distribution curve is shown in Figure 20.4.

The CDF of normal distribution is given as

$$\text{CDF} = \frac{1}{\sqrt{2\pi}} \int_{-\infty}^{x} e^{t^2/2} \, dt \tag{20.3.4.4}$$

The general shape of such CDF is shown in Figure 20.5.

20.3.5   EXPONENTIAL DISTRIBUTION

Exponential distribution is a continuous probability distribution where event happens continuously at an average fixed rate. Practical examples are average rainfall (daily, monthly or annual) and radioactive decay.

The PDF of exponential distribution is given as

$$f(k) = \begin{cases} \lambda e^{-\lambda k} \; for \, k \geq 0 \\ \quad 0 \; for \, k < 0 \end{cases} \qquad (20.3.5.1)$$

where λ is a positive constant.

For exponential distribution, the statistical parameters are

$$\left. \begin{array}{l} Mean \; E(X) = 1/\lambda \\ Standard \; deviation \; \sigma = 1/\lambda \\ Variance = V(X) = 1/\lambda^2 \end{array} \right\} \qquad (20.3.5.2)$$

A typical exponential distribution curve is shown in Figure 20.6.

20.3.6 Uniform Distribution

Uniform distribution is a continuous probability distribution where all events have same likelihood of occurrence.

If X is a random variable with uniform distribution in interval $-\infty < a < k < b < \infty$, it has probabilities of equal values. An example is random numbers in a large interval.

The PDF of exponential distribution is given as

$$f(k) = \begin{cases} \dfrac{1}{b-a} \; for \, a < k < b \\ \quad 0 \quad otherwise \end{cases} \qquad (20.3.6.1)$$

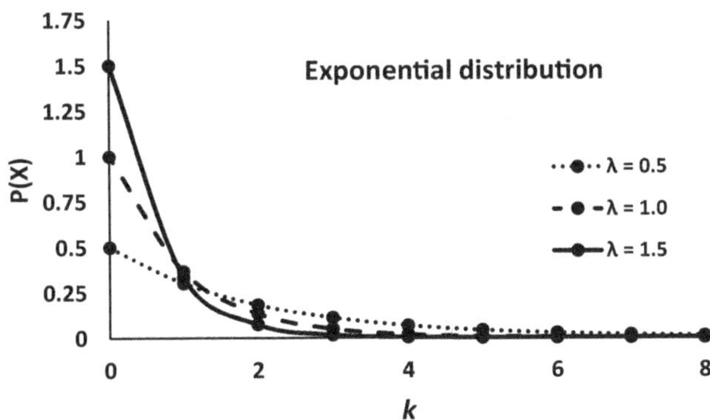

FIGURE 20.6 Exponential distribution.

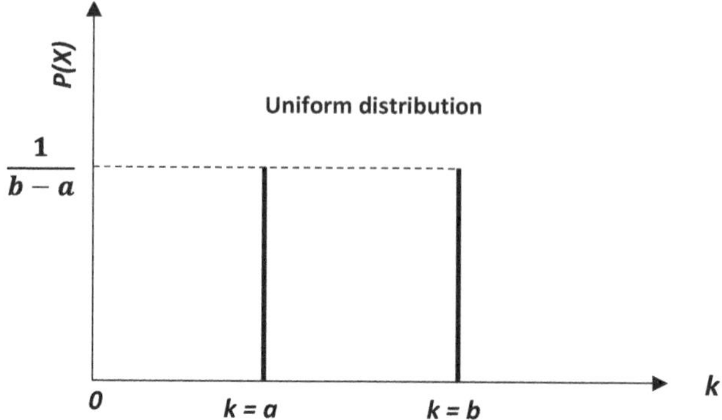

FIGURE 20.7   Uniform distribution curve.

For uniform distribution, the statistical parameters are

$$\left.\begin{aligned}
\textit{Mean } E(X) &= \frac{a+b}{2}\\[4pt]
\textit{Standard deviation } \sigma &= \frac{b-a}{\sqrt{12}}\\[4pt]
\textit{Variance} = V(X) &= \frac{(b-a)^2}{12}
\end{aligned}\right\}
\qquad (20.3.6.2)$$

The uniform distribution is shown in Figure 20.7.

20.3.7   LOGNORMAL DISTRIBUTION

A lognormal distribution is a continuous probability distribution of a random variable in which the logarithm of the variable is normally distributed. Lognormal distribution is very much like the normal distribution. It has wide applications in engineering, medical sciences and economics.

If Z is a random continuous variable and μ and σ and are two real positive numbers, then the distribution of variable X as given by Eq. (20.3.7.1) is called a lognormal distribution.

$$X = e^{\mu + \sigma Z} \qquad (20.3.7.1)$$

The PDF is given by Eq. (20.3.17).

$$f(X) = \frac{1}{x\sigma\sqrt{2\pi}}\, e^{\left(-\frac{(\ln x - \mu)^2}{2\sigma^2}\right)} \qquad (20.3.7.2)$$

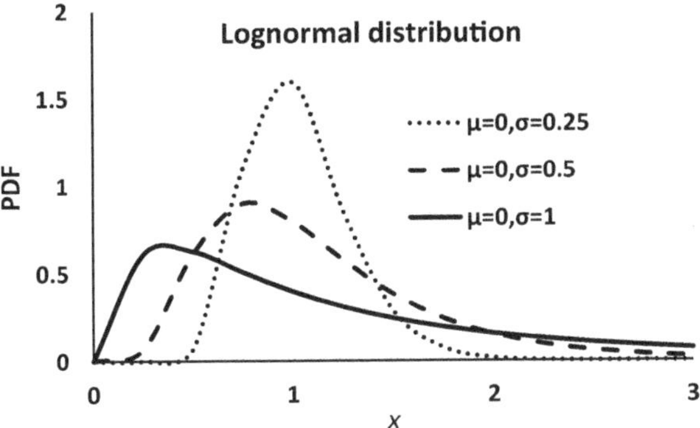

FIGURE 20.8 Lognormal distribution curve.

A distribution of pre-decided mean μ_p and predefined variance σ_p^2 can be obtained using Eq. (20.3.7.3).

$$\left.\begin{aligned}
\mu &= \ln\left(\frac{\mu_p^2}{\sqrt{\mu_p^2 + \sigma_p^2}}\right) \\
\sigma^2 &= \ln\left(1 + \frac{\sigma_p^2}{\mu_p^2}\right)
\end{aligned}\right\} \tag{20.3.7.3}$$

The graph for lognormal distribution is shown in Figure 20.8.

20.4 DAMAGE INDEX

Damage Index (DI) is a quantitative measure of damage in a structure. Many damage indices have been suggested by researchers of which Park and Ang's (1985) DI (Eq. (20.4.1)) has gained popularity. This DI was modified by Kunnath et al. (1992) and this is known as modified DI (Eq. (20.4.4)). The damage indicators are: lateral plastic sway, plastic rotation, energy dissipation through hysteresis and strain in material. Initial developments considered drift and ductility in the damage and the corresponding DI is called non-cumulative DI. When the importance of dissipated energy in DI was appreciated and incorporated, it led to what is known as cumulative DI.

Park and Ang (1985) DI is given by Eq. (20.4.1).

$$DI = \frac{\theta_m}{\theta_u} + \frac{\beta}{M_y \theta_u} \int dE \tag{20.4.1}$$

where θ_m is the maximum rotation of element under seismic condition, θ_u is the maximum rotation of element under monotonic loading, β is a constant, generally taken as 0.4, is yield moment, $\int dE$ is hysteretic energy per loading cycle.

A similar DI was proposed by Niu and Ren (1966) as shown in Eq. (20.4.2).

$$DI = \frac{\theta_m}{\theta_u} + \alpha \left(\frac{E}{E_u} \right)^{\beta} \tag{20.4.2}$$

In Eq. (20.5.2), generally $\alpha = 0.1387$ and $\beta = 0.0814$. E and E_u are energy parameters.

A non-cumulative DI based on ductility was given by Powell and Allahabadi (1988), as given in Eq. (20.4.3).

$$DI = \frac{\mu_m - 1}{\mu_u - 1} \tag{20.4.3}$$

where μ_m and μ_u are ductility pameters.

Kunnath et al. (1992) modified the Park and Ang equation and proposed Eq. (20.4.4), which is commonly known as the modified Park and Ang equation.

$$DI = \frac{\theta_m - \theta_y}{\theta_u - \theta_y} + \frac{\beta}{M_y \theta_u} \int dE \tag{20.4.4}$$

where θ_m is the maximum rotation of element under seismic condition, θ_u is the maximum rotation of the element under monotonic loading, θ_y is the yield rotation of the element under monotonic loading, β is a constant, generally taken as 0.4, M_y is the yield moment and $\int dE$ is the hysteretic energy per loading cycle.

Each of the DIs given in Eq. (20.4.1) to (20.4.4) is for members only and hence are called local damage index. We also have storey damage index (SDI). On the other hand, we also have global damage index (GDI). SDI is given by Eq. (20.5.5) and GDI is given by Eq. (20.4.6).

$$SDI = \frac{\sum_{i=1}^{N} D_i E_i}{\sum_{i=1}^{N} E_i} \tag{20.4.5}$$

where D_i is DI of i-th member, E_i is energy dissipated in hysteresis by i-th member and N is the total number of yielding members in the structure.

$$GDI = \frac{\sum_{storey,i=1}^{N} D_{storey,i} E_{storey,i}}{\sum_{i=1}^{N} E_{storey,i}} \tag{20.4.6}$$

where $D_{storey,i}$ is SDI of i-th storey, $E_{storey,i}$ is the energy dissipated in hysteresis by i-th storey and N is total number of stories in the building.

Example 20.5.1 An RC beam has yield moment capacity as 100 kN-m and yield rotation of 0.01 radian. In the pushover analysis, the beam showed a maximum rotation of 0.03 radian. In time history analysis, the same beam showed a maximum rotation of 0.021 radian and maximum hysteretic energy dissipation as 8 kN-m. Find the damage index (i) as per Park and Ang method (ii) as per modified Park and Ang method.

Solution:

Here, $M_y = 200$ kN-m, $\theta_y = 0.001$ radian, $\theta_m = 0.006$ radian, $\theta_u = 0.03$ radian, $\int dE = 4$ kN-m. β is taken as 0.4.

(i) As per Park and Ang, $DI = \dfrac{\theta_m}{\theta_u} + \dfrac{\beta}{M_y \theta_u} \int dE$

$$= \frac{0.006}{0.03} + \frac{0.4}{200 \times 0.03} \times 4 = 0.467.$$

Within a scale of 0 (no damage) and 1.0 (full damage), the present case is a median type of damage.

(ii) As per modified Pak and Ang method,

$$\frac{\theta_m - \theta_y}{\theta_u - \theta_y} + \frac{\beta}{M_y \theta_u} \int dE$$

$$= \frac{0.006 - 0.001}{0.03 - 0.001} + \frac{0.4}{200 \times 0.03} \times 4$$

$$= 0.439.$$

The results of Park and Ang and modified Park and Ang methods are close to each other.

Some classification of structural damage states are available (Sinha and Shiradhonkar, 2012) as shown in Table 20.1.

20.5 SEISMIC FRAGILITY ANALYSIS

Seismic fragility analysis involves finding the probability contour of some damage state corresponding to the variation of some intensity measure. Some of the terms associated with the topic are explained below.

TABLE 20.1

Damage state categories for building elements

Damage state	Beam	Column
S0	No damage seen	No damage seen
S1	Fine crack	Fine crack
S2	Shear cracks visible	Cracks visible
S3	Cover concrete spalls	Cover concrete spalls
S4	Bond failure, core concrete damaged	Tie bars yield and open longitudinal bar bulge
S5	Crushing of concrete, plastic deflection	Crushing of inner concrete, plastic deformation

20.5.1 INTENSITY MEASURE

In earthquake engineering, intensity measure (IM) stands as a link between probabilistic hazard outcome and probabilistic structural response parameters. The structure responds differently under different IMs. Common IMs are:

(i)   Peak ground acceleration (PGA) of the earthquake
(ii)   Peak ground velocity (PGV) of the earthquake
(iii) Peak ground displacement (PGD) of the earthquake
(iv) Spectral acceleration (S_a) response of the structure
(v)   Spectral velocity (S_v) response of the structure
(vi) Spectral displacement (S_d) response of the structure
(vii) Predominant period.

20.5.2 ENGINEERING DEMAND PARAMETER

Another important term in fragility analysis is the engineering demand parameter (EDP). An EDP is a structural response parameter. The normal EDPs are interstorey drift ratio (IDR), rotation and plastic rotation of members, deflection, floor acceleration, strain in members and crack width.

While discussing IMs, we need to discuss the engineering demand parameters (EDP). ATC 58 defines EDP as "EDPs are structural response quantities that can be used to estimate damage to structural and non-structural components and systems".

Choosing an IM is the choice of the user. An IM may be good or bad depending on how it correlates with any structural response. It is said to be good when it can correlate with an EDP. The correlation is called a fit. The level of fit (good or bad) can be mathematically obtained through regression analysis.

20.5.3 DAMAGE STATE

Damage state (DS) represents the limit state or damage level of the response parameter (EDP) of the structure. For example, damage state of the EDP (IDR) is reported in terms of Immediate Occupancy (IO), Life Safety (LS) and Collapse Prevention (CP) performance levels.

20.5.4 Incremental Dynamic Analysis

Incremental dynamic analysis (IDA) is a process of getting response of structure over increasing levels of hazards, as a plot between some EDP (like IDR, plastic rotation, displacement etc.) and IM (like PGA, spectral acceleration etc.). A suite of hazard is considered typically 22 pairs of ground motions as prescribed by FEMA P-695. For each ground motion, the level is scaled up gradually (say, 0.1g, 0.2g ... 1.0g scaling) and response of the structure is evaluated. The system goes from linear condition to nonlinear condition and finally the collapse occurs. The series of responses are plotted in what is called IDA curves. A typical IDA curve between IDR and PGA is shown in Figure 20.9.

20.5.5 Seismic Fragility

Seismic fragility is a function of ground motion intensity (more precisely intensity measure), which provides probability of failure (or non-failure) for any level of the intensity.

Mathematically, fragility is given by Eq. (20.5.5.1):

$$\text{Fragility} = P(D \geq C | IM = x) = \Phi\left(\frac{\ln\left(\frac{x}{\theta}\right)}{\beta}\right) \tag{20.5.5.1}$$

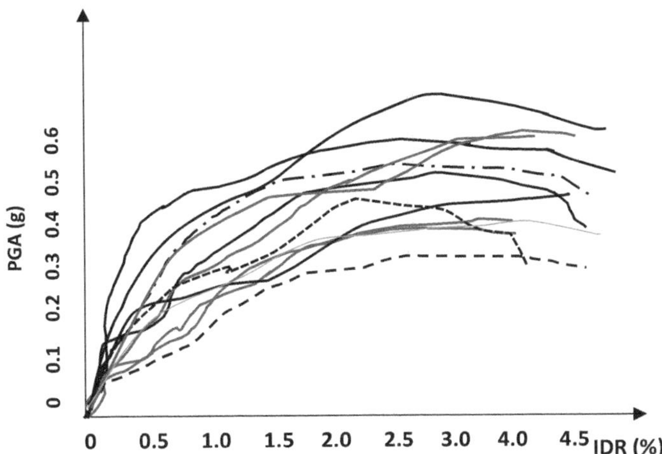

FIGURE 20.9 Typical IDA curves.

where P stands for probability, D stands for a damage state, C is a quantity describing damage state, x is the level of IM and Φ is the standard normal CDF of the distribution.

The right-hand side of Eq. (20.5.5.1) is difficult to solve. As an alternative, the following double integral is used to find fragility.

Fragility =

$$P(DV > x|IM) =$$

$$\int_{DS}\int_{EDP} P(DV > x|DS) f(DS|EDP) f(EDP|IM) dEDP dS \qquad (20.5.5.2)$$

$$\text{And } \lambda(DV > x) = \int_{IM} P(DV > x|IM)|d\lambda(IM)|. \qquad (20.5.5.3)$$

where DV is a decision variable.

20.5.6 STEPS IN FRAGILITY ANALYSIS

1. Model and design the building in some software.
2. Select the desired EDP (e.g., IDR, plastic rotation), DS of the EDP (e.g., slight damage, moderate damage etc.) and intensity measure (e.g., PGA, spectral acceleration).
3. Perform incremental dynamic analysis (IDA) on the building using selected ground motions.
4. Note down the level of EDP at each level of scaling and draw the IDA curves until it reaches the target DS values for each ground motion.
5. Get the intensity measure IM_i values at which the building reaches the target DS values for each ground motion.
6. Calculate the mean (θ) and standard deviation (β) of the IM_i values, when using Eqs. (20.5.6.1) and (20.5.6.2).

$$ln\theta = \frac{1}{n}\sum_{i=1}^{n} lnIM_i \qquad (20.5.6.1)$$

$$\beta = \sqrt{\frac{1}{n-1}\sum_{i=1}^{n}\left(ln\left(IM_i/\theta\right)\right)^2} \qquad (20.5.6.2)$$

7. Calculate fragility as a lognormal CDF:

$$P(C|IM = x) = \Phi\left(\frac{ln(x/\theta)}{\beta}\right) \qquad (20.5.6.3)$$

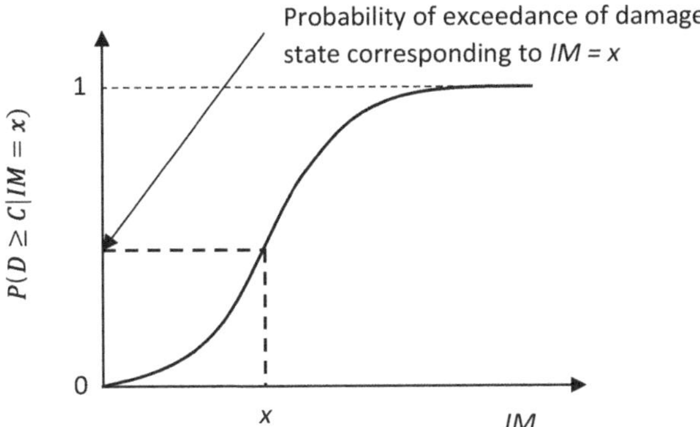

FIGURE 20.10 Typical fragility curve.

where

$P(C \mid IM = x)$ is the probability of exceeding a limit state for ground motion intensity $IM = x$; θ is the median of the fragility function and β is the standard deviation of $\ln IM$.

A typical fragility function is shown in Figure 20.10.

20.6 BASE ISOLATION SYSTEM

20.6.1 GENERAL

Base isolation is the technique to decouple the structure from seismic excitation through the application of an isolator at the base of the structure. As a result of base isolation, the superstructure does not undergo the same acceleration as that of ground, but undergoes much reduced acceleration. Thus, the damage causing vibration in the structure is reduced. Base isolation is widely used in bridges and buildings. Base isolation is a costly solution for seismic mitigation. As such, it is used for protecting historical buildings and other structures, whose importance is invaluable. Base isolation is also a solution for those buildings which have high post-earthquake importance. A hospital is such a building. After an earthquake, injured people rush to hospitals. So, a hospital must survive a big earthquake whilst maintaining operational level performance. Similarly, other emergency structures like firefighting stations, buildings containing hazardous materials, etc, also may be base isolated. Base isolation has been successfully applied in New Zealand, Japan and the United States. Initially steel was tried as a material for an isolator. Later on, the advantages of lead in isolators was discovered. Lead is a material with high damping characteristics. Thus the lead extrusion damper (LED) was developed. In bridges, lead rubber bearing (LRB) was used, where the bearing or isolator is made up of lead and rubber put in layers. Natural rubber is also used as an isolator. High damping rubber bearing (HRB) is also in use. Sliding bearing systems and friction pendulum systems are also isolators.

Base isolation makes a stiff building flexible. So, if the building is already flexible, base isolation will do no good. Base isolation increases the building period.

For seismic evaluation, at least seven ground motions suggested (ASCE-SEI-41-17).

The isolator carries the vertical load arising out of dead load, live load along with any other superimposed dead load. So, the vertical load-carrying capacity of the isolator is to be ascertained. The isolation system is also to take the seismic load in addition to gravity load. The seismic load is reversible in nature, so the system is to be checked for critical load combination. The isolation should never get disengaged from the main structure. Overturning is also a major source of concern in isolated system.

20.6.2 Types of Isolation Systems

Base isolation systems can be broadly divided into two categories:

1. Elastomeric bearings
2. Sliding isolators

Elastomeric bearings consist of alternative layers of steel (or lead) and synthetic rubbers. Such bearings have long been used in bridges and machines. In the 1960s, these were introduced to buildings. It may be noted that in comparison with machines, cars and even bridges, the gravity load from building is much higher. So, it was challenging to introduce bearings in buildings. In elastomeric bearing, the rubber layers impart energy dissipation through shear deformation, while the lead layers impart rigidity, compression-carrying capacity and prevent bulging of bearing under a strong seismic action.

The elastomeric bearings are of three types:

(i)   Natural rubber bearing (NRB)
(ii) Synthetic rubber bearings (SRB)
(iii) Lead rubber bearing (LRB)

Sliding plate works on the principle of friction.

An articulated friction isolator is shown in Figure 20.11. Such isolators are widely used.

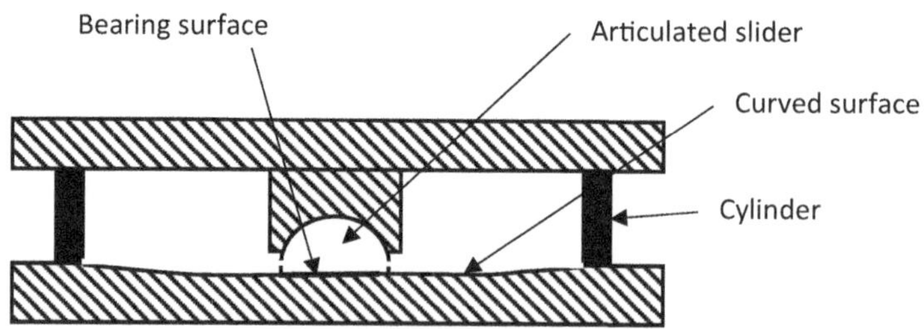

FIGURE 20.11   Friction pendulum isolator.

20.7   CLOSURE

Probability is a core idea in seismic analysis and design. Some preliminary ideas on probability have been explained in this chapter. Seismic fragility is an important study for predicting the damageability of structures under some projected hazard level. Fragility analysis requires incremental dynamic analysis. Some discussion has been made on base isolation.

FURTHER READINGS

Baker, J.W., Bradley, B.A. and Stafford P.J. (2021) *Hazard and Risk Analysis*, Cambridge University Press.

Markis, N. and Chang, S.P. (December 1998) Effect of Viscous, Viscoelastic and Friction Damping on the Response of Seismic Isolated Structures, *ISET Journal*, 35 (4), 113–141.

Qamaruddin, M. (December 1998) A State of the-Art Review of Seismic Isolation scheme for Masonry Buildings, *ISET Journal*, 35 (4), 77–93.

Sinha, R. and Shiradhonkar, S.R. (2012) Seismic Damage Index for Classification of Structural Damage – Closing the Loop, *WCEE-15*, Lisbon.

Spigel, M.R., Schiller, J. and Srinivasan, R.A. (2000) *Probabilty and Statistics*, McGraw Hill.

Index